KING COTTON
IN MODERN AMERICA

KING COTTON
IN MODERN AMERICA

A Cultural, Political, and Economic History since 1945

D. Clayton Brown

University Press of Mississippi/Jackson

www.upress.state.ms.us

The University Press of Mississippi is a member of the Association of American University Presses.

Copyright © 2011 by University Press of Mississippi
All rights reserved
Manufactured in the United States of America

First printing 2011

∞

Library of Congress Cataloging-in-Publication Data

Brown, D. Clayton (Deward Clayton), 1941–
King cotton in modern America : a cultural, political, and economic history since 1945 / D. Clayton Brown.
  p. cm.
Includes bibliographical references and index.
ISBN 978-1-60473-798-1 (cloth : alk. paper) —
ISBN 978-1-60473-799-8 (ebook)
1. Cotton trade—United States—History. 2. Cotton manufacture—United States—History. 3. Cotton growing—United States—History. 4. Cotton textile industry—History. I. Title.
  HD9075.B82 2011
338.1′73510973—dc22   2010022781

British Library Cataloging-in-Publication Data available

To Kay, Carolyn, Richard, and my parents

CONTENTS

PREFACE IX

PROLOGUE 3
The Power of Cotton

1. **THE CULTURAL IMAGE OF COTTON, 1945** 8

2. **THE NEW POLITICS OF COTTON** 31

3. **THE COTTON CONFERENCE** 47

4. **A NEW ERA BEGINS** 62

5. **AMBASSADORS OF FOREIGN POLICY, 1945–1950** 78

6. **THE DINNER TABLE WAR** 105
   Postwar Struggles

7. **THE SOUTH TRANSFORMED** 125
   Cotton's Mechanization, 1945–1970

8. **THE WHITE GOLD RUSH** 147
   Cotton Moves West

9. **BOLL WEEVILS, WORMS, AND MOTHS** 169
   A Hundred-Year War

10. **MEMPHIS** 190
    The Epicenter of the Cotton Belt

11. **"THE FABRIC OF OUR LIVES"** 213
    Cotton Incorporated

12. **THE TEXAS PLAINS** 233
    America's Cotton Patch

13. **THE QUESTION OF SUBSIDIES** 251

14. **CROP LIEN TO FUTURES** 270
    Financing Cotton

15. **THE ROLE OF TEXTILES** 290

16. **RESEARCH** 311
    The Key to Viability

17. **CHALLENGES ANEW** 326

18. **THE GLOBALIZATION OF COTTON** 350

19. **THE NEW COTTON CULTURE** 372

   NOTES 387

   GLOSSARY 423

   BIBLIOGRAPHIC ESSAY 425

   INDEX 429

PREFACE

Cotton has always been a complex and intriguing subject, and my efforts to explain its history since 1945 proved challenging. To cover the subject with a broad perspective, it was crucial to incorporate the cultural, political, and economic dimensions of the subject, because cotton always meant more than crop production—it was the southern way of life. It was necessary to understand the cotton culture associated with the southern United States, but it was also important to realize that cotton holds a place in the greater American culture. Since World War II, King Cotton and small-scale farming lost their overpowering importance in the economy of the South, a result of the increasing use of mechanization and high-tech farming. This development owed much to the movement of cotton farming into the West, where large-scale agriculture predominated. After World War II, tenant farming and sharecropping, the socioeconomic practices that accounted for much of the negative image of the cotton culture, ended. With these far-reaching developments, the cultivation and marketing of cotton became a feature of modern agriculture in which growers adopted the ethic of businessmen. But new challenges arose in the cotton kingdom, the rise of the globalized economy that caused American growers to lose their long-held monopoly as the world's supplier of the natural fiber, and the rise of synthetic fibers such as rayon and polyester that threatened to relegate cotton to a minor and inconsequential part of the economy. My purpose in this study is to show that a new determination arose within the cotton industry to adapt to modernity, face the new threats, and remain a viable part of the world economy.

Extraordinary social and cultural overtones in cotton give the subject a humanist dimension. Throughout the Cotton Belt, there is an admiration for farming, a sense of satisfaction and contentment induced by working the earth to yield the fluffy white fiber. Life on a small plot of cotton gave identity, a sense of living among the like-minded, and a feeling of sharing accomplishments and failures. A bitter taste for cotton farming also prevailed, particularly among those who toiled in the fields when hand labor and animal power were in use. Music and literature expressed both the fondness and distaste for life on a cotton farm, and these expressions came from each identifiable culture in the cotton economy, Anglo, African American, and Latino. The history of cotton is more than agricultural progress; it is a rich mixture of political activism, government policy, and socioeconomic life with multiple perspectives, all in the context of an increasingly global economy.

In undertaking this study, I had the cooperation of many individuals who spent their life's work in the growing and marketing of cotton, and I benefited from the support of organizations founded for the purpose of promoting the general welfare of American cotton growers. Without their assistance, this book would have been impossible. Particular attention should go to the National Cotton Council, which opened its archives to me and whose staff provided assistance at my request. At no point did the council restrict my access to its archives or attempt to influence my work. I wish to express my appreciation to Philip Burnet, Gaylon Booker, Mark Lange, Earl Sears, Fred Johnson, Gary Adams, William Gillon, Drayton Mayers, Andrew Jordon, Frank Carter, Dabney Wellford, Carolyn Willingham, Lance Murchison, Charles Yarwood, Buzz Shellabarger, Nancy Cox, Harrison Ashley, and Tammy Pegram. Cotton Nelson was my liaison, and he furnished every possible means of support. Several pioneers of the National Cotton Council granted interviews and furnished portions of their personal archives, and I am indebted to their kindness: Read Dunn, Robert (Bob) Jackson, and MacDonald Horne.

To John Barringer, I am indebted for use of the personal files of his father, Lewis Barringer. Outstanding support came from Bill Houston, Frank Mitchener, and Bill Pearson. To Roger Milliken, I owe thanks for allowing me to tour spinning, weaving, and dyeing mills in Millikan and Company.

Special thanks must go to Morgan Nelson, Jack Stone, Don Anderson, Barry Aycock, and Paul Harvey Jr. To Keith and Alison Deputy, I owe a special debt for allowing me to visit their farm. H. G. Dulaney provided insights about Congress. A firsthand explanation of "factoring" came from Tom Miller. To Calvin Turley and Daniel Lyons, I am particularly indebted for their help in numerous instances. Perre Magness gave insights on the Memphis Cotton Carnival. To Charles Gibson and Dallas and Hortense Russom, I owe my gratitude for their explanations of cultivation practices. My friend in the Texas Christian University English Department, David Vanderwerken, shared his knowledge of southern literature.

Individuals from organizations who gave valuable assistance included Rick Lavis of the Arizona Cotton Growers Association, Jesse Curlee of Supmia Association, Chip Morgan of the Delta Council, Tom Smith of CALCOT, Woods Eastland of Staplcotn, Roger Haldenby of Plains Cotton Growers, Emerson Tucker of Plains Cotton Cooperative Association, Billy Dunavant of Dunavant Enterprises, Janice Person of Delta Pine and Land, and Paul

Dixon and Kenny Day of the USDA Agricultural Marketing Service. Valuable assistance on the topic of cotton subsidies and fibers came from individuals at Texas Tech University: Don Ethridge of the Cotton Economics Research Institute and Michael Stephens of the Fiber and Biopolymer Research Institute. At Texas A&M University, I received explanations about integrated pest management from Perry Adkisson, Ray Frisbee, and Knox Walker; Carl Anderson gave insights into cotton marketing. Charles Glover of New Mexico State University provided an explanation of cottonseed research and development.

I used libraries across the Cotton Belt in my research, particularly their special collections. The staff of Texas Christian University Library, especially Brenda Barnes, Cheryl Sassman, Alyce Stringer, and Delores McGhee, were patient and helpful throughout my endeavors. The following university libraries furnished both general publications and archival materials for this project: Baylor University, Mississippi State University, Arizona State University, Memphis University, New Mexico State University, Rice University, Texas Tech University, Texas A&M University, and the University of Southern Mississippi. Randy Sowell helped me obtain archival materials from the Harry Truman Presidential Library in Independence, Missouri, and Wayne Dowdy of the Memphis Public Library furnished assistance with records and photographs of Memphis. The National Archives in College Park, Maryland, and the National Agricultural Library in Beltsville, Maryland, supplied critical records. The following private and public libraries contained archival material essential to this work: Sam Rayburn Library, Memphis Cotton Museum, and Houston Public Library. Randy Boman of the Texas AgriLife Extension Service furnished assistance with photographs.

Critical financial support came from Texas Christian University: the AddRan College of Liberal Arts, the Office of the Associate Provost, the Office of Sponsored Research, and the Department of History. Their generous backing was essential in this endeavor.

The University Press of Mississippi provided guidance in bringing this work to fruition, and I am indebted to Craig Gill, Sophia Halkias, and Anne Stascavage. To the late G. C. Hoskins of Southern Methodist University and to my mentor at UCLA, the late Theodore Saloutos, I am forever indebted.

My children, Richard and Carolyn, patiently waited for this undertaking to end. My wife, Kay, helped in many instances with copying, taking notes, reading sources, clipping articles, printing photographs, and being a constant companion on trips across the Cotton Belt. Without her love and devotion, this work would have been impossible and no fun.

I alone bear the responsibility for any errors, misjudgments, faulty analyses, and the interpretations in this volume.

PROLOGUE
The Power of Cotton

During Thanksgiving weekend in 1989, television viewers saw the release of a new advertising campaign, "The Fabric of Our Lives," which promoted cotton as a natural fabric. With music by the rock singer Richie Havens, the campaign with its catchy slogan appeared on NBC's *Today Show*, ABC's *Good Morning America*, and *Macy's Thanksgiving Day Parade*.[1] These soft-sell ads appealed to the preference for a naturally grown material rather than an artificial fiber like polyester. With its ease of wear, durability, and versatility for all seasons, cotton has long been the most popular fabric for clothing and household furnishings. Consumers admire the soothing feel of its soft lint, and just as bread is the staff of life, so cotton is the fabric of life.

Cotton became popular with the beginning of the Industrial Revolution. New inventions and advances in spinning and weaving in the eighteenth century created a demand for vast supplies of inexpensive fiber, and Eli Whitney's invention of the cotton gin in 1793 made it possible to meet the demand. Cotton fabric was not expensive like silk and more abundant and comfortable than wool. It democratized clothing and eased the distinction based on the raiment of spun cloth. Britain and Europe, as well as the fast-growing textile mills in New England, depended on cotton grown in the United States for their spinning machines, so cotton farming quickly became a means of livelihood or even a path to wealth. Opportunity drove large-scale planters and small farmers alike into the southern territories and states where fertile soil and semitropical weather made it possible to reap fortunes. By the mid-nineteenth century, cotton became the chief commercial crop of the United States. It supplied half the exports and was the basis for the inter-regional economy. It accounted for the growth and wealth of the antebellum South, and an empire based on the white lint stretched across the South, which came to be known as the cotton kingdom.

To nourish the plant from its soil required hand labor, and planters turned to the use of slaves, so that as the cotton economy grew, slavery expanded. Wherever southerners went, they took slaves with them; black labor and white cotton went hand in hand. A planter accompanied by his black attendant or the sight of toiling field hands was a common occurrence, and nearly all mannerisms of southern life reflected the region's dependence on the tropical plant and captive workers. As the issue of slavery divided the United States, however, cotton became associated with the brutality of forced labor and the suppression of individual liberty. Plantations were seen as work camps where black labor was exploited for the profits of growing cotton, the "white gold." The connection with forced labor gave cotton an identity peculiar to itself in our culture. America's clash over slavery was an epochal struggle that, like the American Revolution, further defined the meaning of freedom but made the cotton-based southern way of life synonymous with racism and exploitation.

Although the Civil War turned the South upside down, the power of cotton to energize and motivate did not diminish. People continued to migrate, and new lands fell to the plow, especially Texas, where virgin soil yielded fiber with long staple. The number of farms, whether large or small, increased throughout the Cotton Belt, and cotton production surpassed the slave era. A new generation of planters appeared, and the export trade resumed. A New South was born that had factories with smokestacks, but fields white with cotton continued to drive the southern economy and dominate the culture. In town squares, churches, and bottomlands, men spoke of weather and land, of plowing and picking. They kept their fields and themselves prepared for the rush to plant each spring and give the cotton plant the two hundred days free of frost that it needed to yield its fluffy lint. Light shone again in Dixie, made bright by white gold. By World War I, King Cotton sat comfortably on its throne.[2]

An ominous figure however, lurked in the kingdom. The former slaves had no alternative employment, and planters needed labor, which led to a type of tenant farming known as sharecropping. The cropper, including his wife and children, provided the labor to grow a twenty- to thirty-acre crop, while planters furnished the land and all the equipment, including a small "shotgun house" or shack. They divided the proceeds of their crop generally on a fifty-fifty basis. It was contract farming carried out loosely and informally. Croppers drew advances on their portion of the cotton to obtain food, clothing, and the necessities of living, and they regularly stayed in debt. Both blacks and whites lived as sharecroppers, and though none were bound to the land by law, they were chained nonetheless by the lack of alternative work and crushing poverty. Nothing offered escape in the undeveloped economy and strict environment of racial and social order. As they trod the rows, hoeing and picking, they sang of sadness and a triumphant day when their suffering would end. Their misery cast a shadow over the Cotton Belt.

For good reason, cotton farming was called the "southern way of life." Nothing else so effectively drove behavior, for nearly all manner of things centered on or revolved around the seasonal life of the cotton plant. Schools opened and closed so that children could help pick the crop; gins and compresses opened and shut down with the fall picking; waves of migrant workers swept through the fields and filled towns; and railroads sent extra rolling stock into the South each year to transport bales by the millions. Bankers and retailers followed the same ebb and flow. Textile mills spun and wove day and night. Work determined place in the social strata, with planters and agricultural businessmen at the top and all others somewhere below. White middle-class farmers and some black landowners flourished between the extremes. The cotton kingdom was a complex web of cultural, economic, and social behavior that could be experienced and known but not understood. Outsiders at best peered through a windowpane as they tried to grasp the whole panorama. They saw love and admiration for this way of life just as they noticed the resentment and bitterness for being caught in it. So powerfully did the region lie in the grip of the plant with fluffy bolls that it lived in a "cotton culture." In his classic work of 1929, *Human Factors in Cotton Culture*, Rupert Vance quoted a journalist: "Cotton is Religion, Politics, Law, Economics and Art."[3] From the anguish of blacks toiling in the fields there emerged a distinct form of music, the blues.

The blues, so Americana, began with black laborers keeping themselves company as they worked together in groups picking cotton or walking alone all day behind a plow mule. Originating as "hollers," singing to oneself or in unison among a group, the pulse of these chants followed the swing of a hoe. Hollers grew into stanzas marked by strong and repetitive rhythms, and the swing of the hoe became the beat of the blues. Lyrics reflected hardship and despair, disappointments and broken hearts, the words of poverty and anger. The legendary bluesman B. B. King believes that "the songs were made up by the heart, nothing written or rehearsed, music meant to take the ache out of our backs and burden off our brains."[4] In the twenty-first century, a hundred years after the birth of the blues, when sharecroppers have passed into history, the blues live in juke joints and on radio, in concert and onstage.

Whites expressed their woes through country, western, and folk music. They sang of bad luck rather than entrapment. It was, however, the black vocalist Huddie "Lead Belly" Ledbetter, born on a Louisiana plantation, who wrote "Cotton Fields," the ballad so popular with white musicians. In 1970 the Beach Boys, one of the foremost quintets in their era, made the song a hit, with Lead Belly's lyrics evoking the fondness for a childhood spent in Louisiana, "In them old cotton fields back home." Another ballad popularized by Carl Sanburg, Woody Guthrie, and Burl Ives was "Boll Weevil Blues," a song about the devastating pest that wrought havoc throughout the South. More recently the contemporary artist Billy Joe Shaver wrote and sang about his life near Corsicana, Texas, "When Jesus was our Savior and cotton was our King."[5]

This music, whether associated with whites or blacks, had roots in the fields of the South. It was the voice of people trapped in a quagmire. From the beginning, the music was modified and moved into the mainstream of American culture until the connection with cotton faded from the popular mind. However remote the link may now appear, it demonstrates the power of cotton in our background.

In the latter twentieth century, when agriculture had declined as a way of life and the status of farming had slipped precipitously in the social strata, cotton continued to hold power in drama and literature. In 1984 the movie *Places in the Heart*, starring Sally Fields and Danny Glover, related the universal theme of single mothers struggling to raise families and the reward for people of different races cooperating. The drama takes place on a family cotton farm where the protagonists face the risks of nature as they race to save the farm from foreclosure. Subthemes abound, such as the prejudice toward blacks and the handicapped, but the primary message is the widowed farm wife's growth of self-confidence. The screenwriter Robert Benton placed the story in the cotton South, where audiences might commonly recognize that the likelihood of a young widow overcoming the odds against her were slim.

In 2001 John Grisham used cotton as the backdrop for his novel *The Painted House*. Another universal theme, a youth's passage from innocence to experience, plays out among a family of cash tenants as they try to grow a crop. Racist attitudes again surface, in this case between whites and Mexicans. Hope and fear are mixed as the boy's family, with the help of migrant workers, attempts to save eighty acres of cotton from rain and flood. In both stories, a cotton farm remained the setting for the pathos of human suffering and triumph.

Not all servants of King Cotton longed to escape its domain. Across the realm lived many who found prosperity or at least contentment in the yearly routine of picking and plowing. In the world of dollars and cents, a meager existence on a small plot of land might seem foolish, but for many, farming brought pride and satisfaction even when the returns were slim. Country folk submitted to the pull of rural life by choice rather than the compulsion of necessity. A cotton farm was home, a place of belonging, an identity in a community where all prospered or suffered alike. Such people lived a life of sacrifice, especially in the generation after World War I, but they were compensated by the solitude of farming, drawing satisfaction from dwelling close to the earth and knowing their labors bore the earth's fruit. The ring of the cash register meant less than their independence as yeomen of the land.

For those who see cotton as the opportunity and means for the betterment of their lives, pride and celebration are in order. Whether through festivals and charity balls, or outdoor barbeques and country hoedowns, they pay homage to King Cotton. These loyal servants renew their allegiance through events as varied as the Grand Krewes balls of the Memphis Cotton Carnival, the Annual Redneck Barbeque of the Cotton Festival in Sikeston, Missouri, or the presentations to schoolchildren by the Casa Grande Valley Cotton Women of Pinal County, Arizona. Through these mediums, they assert the legitimacy and worthiness of cotton farming, and however small and unrecognized the celebration, it sends a positive and enduring message.

Cotton has the power to provoke and incite controversy, a feature not found in other commodities like wheat or corn. Only tobacco has managed to rouse emotions so fervently. When the mechanical cotton picker at last seemed available, it was hailed as a revolution in agriculture, but it also provoked anxiety that sharecroppers would flood cities. A fight broke out after World War II between cotton and dairy interests over the use of oleomargarine, and public impatience with agricultural subsidies led to the first limitations imposed on federal crop supports in 1970, sparked by the public realization that cotton growers received the largest share. In 2003 cotton captured international headlines when the negotiations of the World Trade Organization broke down over the impasse of cotton subsidies. Within the United States charges were thrown at cotton interests for allegedly prolonging the poverty of African farmers.[6]

Because cotton is often the subject of public discussion and occasionally becomes a target for cultural bashing, it holds a place in American agriculture as an object of pride or scorn. It appears and reappears in the maelstrom of shifting values and priorities, be it slavery, the family farm, or trade policy. Much attention has focused on the negative features of cotton that cannot be denied, but the force of this common plant in the welfare of mankind keeps it relevant in our national culture. Its recurring and dynamic role, woven into the web of our history, reflects its lasting power.

CHAPTER 1

THE CULTURAL IMAGE OF COTTON, 1945

In the closing months of World War II, the U.S. cotton industry was nervous about the future. It anticipated technological and scientific breakthroughs that would lead to higher production and more efficient farming, but it feared the loss of world markets and the rising popularity of synthetic fibers in the consumer market. These developments had national implications because not only did the 1.5 million growers across the Cotton Belt produce the country's most important commercial crop, but cotton farming was the southern way of life.

In 1945 the industry could not know the future, but it understood the past. It was rich in history, having its own identifiable culture and embodying much that defined America: the quest for wealth and success exemplified by the planters, the self-reliance of the yeoman farmer, and a way of life based on the agrarian ideal of land and soil. The white sea of cotton that engulfed the fields each fall had provided the foundation for the South's recovery after the Civil War, making the region flourish with commercial centers, literary achievements, and political power beyond its numerical justification. For generations the cycles of plowing the ground, planting the seeds, tending the plants, and picking the snowy fiber from the shrublike stalks dictated life in the Cotton Belt. The production of cotton set the seasonal routine of farmers, ruling the patterns of commerce in small towns and cities, and even closing the schools to allow children to pick cotton during the annual shutdown known as "cotton vacation." This commitment to nurturing such a common plant injected pride into southerners, who felt a sense of accomplishment as they stirred the soil to bring forth from the earth the fiber prized around the world. They celebrated their toil with carnivals and jubilees and through song and poetry. "King Cotton" was the term that expressed this admiration for, and subservience to, the plant that yielded a livelihood for millions of Americans in the coastal plains and deltas of the South.

Yet the past was scarred by the practice of slavery in the Old South, hardened by the impoverishment of a large mass of farmers since the Civil War, and stigmatized after World War I as socially and culturally backward. Cotton and slavery had split the Union, and whether commercial planters or small farmers, nearly all were destitute at the war's end. Small landowners had extracted a meager living by producing foodstuffs, but cotton provided a means for obtaining cash, so they commonly raised a few bales each year. Owners of midsized farms fared better but customarily produced portions of their own food. For the classes at the bottom of cotton's social ladder, such as small landowners, cash tenants, and sharecroppers, life was harsh, and their livelihood precarious, and only rarely did they experience a plentiful and prosperous year.[1]

Despite the hardship for many, cotton brought wealth to the United States throughout the half century after the end of Reconstruction, until the mid-1920s. In this sense, the past was not wholly burdensome. Cotton had been the principal source of clothing around the world and the first global commodity. As the science of chemistry advanced, scientists by the eve of World War I had developed a variety of household and industrial uses of lint and seed, from oleomargarine to gunpowder, from paint to automobile tires. Food products made from cottonseed oil added to the value of the crop. Because cotton grows in semitropical climates, the leading industrial nations of Europe and Japan were forced to import their cotton supplies, and trade flourished in cities such as Liverpool and Bremen, where cotton exchanges handled millions of bales each year. In 1903, when the Wright brothers successfully achieved flight at Kitty Hawk, raw cotton was the most internationally traded item, and the United States supplied 75 percent of the world market. Cotton was America's chief export and cash crop; the famed orator Henry Grady called it the "royal plant." It did much to make the country a viable trading nation and brought wealth to planters and the various industries involved in ginning, shipping, and spinning yarn. So extensively did it cross oceans and borders "that cotton came to serve," wrote two scholars of the Manchester Metropolitan University Business School, "as a barometer of world trade," and "the marketing of cotton became the paradigm of a new economic order."[2] The plant, *Gossypium hirsutum*, which originated in Mexico, grew almost exclusively in the southern states and became known as upland cotton. After the Civil War, acreage increased as evidenced by the development of the Yazoo-Mississippi Delta, the Missouri Bootheel, and the blacklands of Texas. Acreage reached a pre–World War I peak in 1914 of

35,615,000 acres, and production tripled from five million bales in 1880 to over fifteen million in 1911.³

During the same period, the United States industrialized; built cities, factories, and ports; and expanded its railroads and warehouses. It developed the modern corporation. In the South, an urban business class emerged that managed new factories and developed a commercial atmosphere geared toward money, smokestacks, and steel. Lumber, textiles, railroads, and later oil joined the list of southern industries employing thousands of workers and generating new wealth, so that by World War I material success had begun to manifest itself throughout the region with office buildings, grand railroad depots, museums, colleges and universities, palatial homes, and numerous other indications of progress.

But the constructive zeal of industrialization achieved much less for agriculture throughout the nation, which had steadily lost wealth and influence to the rising forces of business and commerce. Cotton prices fluctuated, and the natural variations of weather, which in combination with the ravages of the boll weevil that invaded the South in the early 1890s, made prosperity uncertain for growers. Worst of all, more and more southerners fell into sharecropping, creating a large class of landless farmers, both black and white, beholden to the land. Sharecroppers had the most deplorable living conditions in the United States, a severe ramification for the country because of the sheer number living in rural squalor. "Thus the risks of the cotton market," wrote a noted historian of southern culture, "combine with the risks of the weather, and the weevil to make the climb to ownership all the more difficult."⁴ Hardship existed in all the states, but it was worst where cotton grew in the South.

Mining the soil for the white gold was a risky venture. For the planters, cotton doggedly remained an investment riddled with risks, just as it gave the vulnerable middle-class farmers no assurance from year to year. For the tenants and sharecroppers, it offered only a hand-to-mouth existence. Cotton prices were volatile, depending wholly on supply and demand. Production costs steadily rose, however, because farmers began using fertilizer to cope with the soil worn out by repeated plantings of cotton, and the arrival of the boll weevil added the new expense of pesticides as farmers tried to save their crops, or at least minimize their losses, from the long-snouted weevil, which hit the South like an apocalyptic scourge. Other costs, some hidden from cotton statistics, drove up expenses: the cost of mules went up, a real factor because cotton was not mechanized.⁵ Land prices inched upward. And the infamous crop-lien system developed because of the shortage of capital in the South, resulting in unusually high credit costs for all classes of farmers, from sharecroppers to planters. Household costs went up, too. In other words, the ratio of cotton prices to industrial goods was not favorable, a condition that came to be known as parity.

Despite the hardship, the South remained devoted to cotton farming. It was the principal source of money for much of the rural upper crust, and small landowners depended on cotton for the only hard money they saw. Tenants who farmed small plots on a cash basis and the lowly sharecroppers fared less well, but even among the downtrodden was a belief in the superiority of cotton as a means for advancement. The South's agricultural ladder always had a strict order, but each class "never quite lost its capacity to believe the thing it was most anxious to believe: that come next year, cotton at last would begin to fetch always more handsome prices. And once prices had actually begun to improve a little, that romantic faith took fire anew."[6] These words by the writer W. J. Cash capture the single-minded devotion to cotton farming. But compared with advances in the general economy for the half century before World War I, cotton remained stagnant: farmers continued to rely on hand labor and animal power, prices fluctuated and profits were unpredictable, and many lived at the subsistence level. Except for large landowners, a livelihood based on cotton offered only the status quo with little chance for upward mobility.

THE CRISIS

During the 1920s, particularly for urban wage earners, the quality of life improved with increases in per capita income, a shorter workweek, a longer life expectancy, and a lower infant mortality rate. Life in the cities became more sophisticated with the increased use of automobiles, the greater availability of movies, and the spread of suburban neighborhoods. New household conveniences—radios, refrigerators, and telephones—became standard features of urban life. The mass consumerism of modern America began to take hold, and the traditional ethic of thrift and sacrifice gave way to a new standard of prestige based on the acquisition of consumer goods. "The prosperity bandwagon rolled down Main Street," wrote Frederick Lewis Allen, and the new mass-production consumerism reigned triumphant.[7]

Agriculture missed the bandwagon. "Mighty few farmers could get so much as a fingerhold upon it," continued Allen.[8] Foreign demand for agricultural products declined soon after World War I and created surpluses in the staple crops of corn, wheat, and cotton. Prices fell, and growers tried to offset them with greater production, causing commodity prices to spiral downward even more. Land values fell while the total of mortgaged farm property went up. "The farmers lost all that they had gained," wrote an economist, "and were at a level below that which they had occupied before the war started."[9]

Throughout America, rural youth clamored for the more exciting and glamorous atmosphere of the cities—and understandably so. Opportunities for social and economic mobility were better there, and the comforts of daily living were superior. For lack of electrical service, rural inhabitants continued to follow preindustrial routines akin to those of the nineteenth century. As late as 1935, only 10 percent of rural homes and farms in the United States had electricity, but the figure reached only 3 percent across the South. This condition forced families to live without electric lights, running water, refrigerators, and indoor bathrooms. And like the term "Bible Belt," the term "Pellagra Belt" applied to the South because of the high frequency of pellagra, the nutritional disease concentrated in the cotton-growing areas. A similar condition existed with hookworm, an intestinal parasite found commonly in the rural South, where the combination of warm moist soil and the outdoor privy created a habitat for the blood-sucking parasites. Travelers in the remote areas of the region noted the sight of children with stunted growth caused by these diseases. The undeveloped economy of the cotton South accounted for this wretched situation, demonstrating how single-crop agriculture impacted the way of life. No wonder Americans began abandoning farms, justifying Allen's statement that "by the hundreds of thousands they left the farm for the cities."[10]

Substandard conditions on farms were common throughout the United States, but the extreme hardship in the Cotton Belt began to draw attention. Commercial and industrial development advanced through the South, but southerners continued to be seen and to see themselves as distinct and their homeland different from the rest of the United States. This distinction and status as a "cross-eyed stepchild," which rested greatly on the region's burdensome past, attracted the attention of writers and analysts who began to explore the South with a reinvigorated sense of interest. They found cotton to be a particularly intriguing subject, and their reporting contributed to shaping the image of cotton in 1945.

COTTON IN LITERATURE

In 1924 a young professor at Columbia University, Frank Tannenbaum, published *Darker Phases of the South*, a stinging indictment of the region's social, economic, and cultural life. He saw southerners as a people beset with poverty, racism, ignorance, prejudice, and spiritual stagnation, the "darker side" of what W. J. Cash later called the savage South. For Tannenbaum the South was handicapped by the Civil War, the Lost Cause, and hardship, but he felt it exhibited a spirit of defiance and maintained a general unprogressive environment. He focused on the Ku Klux Klan, mill villages, prisons, racism, and cotton.

Tannenbaum referred to cotton as a "white plague" that "instead of being a blessing, a beneficent, fortunate thing, is almost a curse."[11] He laid most of the South's ills at the feet of cotton, making it responsible for farm tenancy, soil erosion, lower rural living standards, the migration of sharecroppers from farm to farm, poor diets and ill health, and "the eternal friction between debtor and creditor," a reference to the crop-lien system. Tannenbaum expressed the common idea that single-crop farming held back economic advancement and encouraged poverty. For several generations, reformers and innovators had urged crop diversification on the grounds that it would raise farm incomes, but Tannenbaum went further and linked cotton with social ills: "the neglect of politics," "the lack of civic interest and of civic pride," and "spiritual stagnation." For him, single-crop farming had brought isolation and monotony; it led to idleness and "made the rural community in the South much more a burden spiritually and has meant much greater need for external excitement, partly expressed in intense religious emotions and protracted meetings."[12]

Behind these "darker phases," Tannenbaum continued, lay the rise of industrial farming, factory farms that relegated the farmer to the life of a city laborer. Industrialization and single-crop farming went together and destroyed the "cherished, self-sufficient and independent farm community upon the traditions of which we have been reared." Tannenbaum felt that lack of crop diversification had the same detrimental effect whether "tobacco, wheat, sugar, or corn." Since cotton dominated southern rural culture, however, he coupled it with the "darker phases," the South's worst ills. For him cotton made the region backward and deprived southerners of cultural, civic, economic, and spiritual viability. It represented "a burden and drag upon the life and spirit of the people in the South." Tannenbaum wrote with an imputation of guilt or shame, but his incriminating style was already becoming common.[13]

Certainly the best-known critic of the South, one who proudly mocked it, was H. L. Mencken, author of the famous essay "Sahara of the Bozart" (1917), in which he described the region as devoid of any intellectual creativity. In 1924 he coined the phrase "Bible Belt" to demonstrate his view of the South as "the bunghole of the United States, a cesspool of Baptists, a miasma of Methodism, snake-charmers, phony real-estate operators, and syphilitic evangelists."[14] Unlike Tannenbaum, Mencken did not focus specifically on cotton but portrayed the South as a barren land lacking modern culture and economic development. In his writings, however, Mencken did address the substandard living and working conditions among cotton farmers, which were becoming increasingly known in popular literature.

Similarly, in 1924 William H. Skaggs published *The Southern Oligarchy*, "one of the most thoroughgoing indictments of Southern villainy ever published," according to a noted historian.[15] Skaggs did not focus exclusively on cotton but nonetheless devoted considerable attention to it, reiterating the growing theme that sharecroppers lived in peonage and serfdom. He castigated landlords: "The tenant is in fact neither a citizen or a freeman; he is a serf, bound to the soil at the will of the landlord."[16] Skaggs alleged that the "southern oligarchy," the privileged class in which he included planters as part of the upper crust, maintained a culture of corruption and lawlessness and permitted little change or challenge to the established order.

Cotton was a key part of Skaggs's larger assessment, since it was the principal cash crop of the United States and fundamental to the economy of southerners. He considered it a disgraceful industry that had produced no innovations or "achievement in statecraft, or other product of the brain," and kept labor in "economic serfdom and political subjection." He regarded exploitation of sharecroppers as a major characteristic of the cotton culture, which ensured the supremacy of the oligarchy while producing no solutions for the ills of southern agriculture. For Skaggs, the terrible conditions in the Cotton Belt, as bad as they were, were growing worse owing to the onslaught of the boll weevil, which had invaded Texas in 1892 and quickly spread through the cotton-producing areas of the South. This ravaging insect had caused production to drop and cost the industry, Skaggs reported, between $400 and $700 million in 1921 alone. Quoting the *New York Tribune*, he wrote: "There is nothing in the history of American agriculture to compare with the Mexican boll weevil as a costly blight upon those engaged in farming, neither has any other living thing made its pernicious influence felt by such a large proportion of the human race." Skaggs saw the migration of African Americans from the region to the industrial North, searching for an improved life, as a threat to the labor supply of the Cotton Belt. This combination of circumstances, the worst since the Civil War, he believed, meant "grave concern" for the twenty-five million people who depended on cotton, but "also to the industrial and commercial interests of the whole country." Worst of all, from his standpoint, however, an ancient regime ruled the cotton South, practicing and perpetuating a self-imposed state of mind that rejected modernity and kept cotton's laborers toiling in the fields and strapped to misery. For Skaggs, these degenerating conditions boded ill for the greater social and economic well-being of the United States.[17]

Tannenbaum and Skaggs were part of a growing number of analysts critical of the South. "The South," wrote one scholar, "had suddenly become an object of concern to every publicist in the country."[18] These writers meant to encapsulate all facets of southern life in need of reform, and nearly all wrote about cotton or some aspect of it. Poverty and the deprivation associated with sharecropping were their common denominator, the one aspect they almost habitually reviewed with zealous conviction. It could not be denied, however, that among cotton's laboring class there were people who had no shoes or adequate clothing, who lived on farms but suffered from the nutritional diseases of rickets, pellagra, and a general state of ill health. However brutal and mocking the critics' language might be, they wrote with the power of revelation and made the staggering conditions in the South more widely known.

Novelists reinforced the social and economic analysts and through the medium of fiction contributed to the perception, or realization, of a despairing and still worsening cotton industry. Southern fiction entered its well-known era of the Southern Renaissance, producing works that belonged to the realist school of literature and acknowledged the shortcomings of life below the Mason-Dixon Line. Gone were the romantic musings about the Old South or the Confederate spirit of defiance. Now the harsh realities of disease and degeneration, violence and racism, class resentment and conflict, characterized southern letters. Some of the finest literary art of the twentieth century exposed the hardship and shortcomings of the cotton culture, and like the writers of sociological and economic critique, novelists varied in their degree of harshness, but they too used the southern way of life as their backdrop.

In 1923 Dorothy Scarborough published *In the Land of Cotton*, a story of a girl's sunny childhood on her father's plantation, her growing knowledge of black folklore, her emerging awareness of the suffering endured by tenants, the ravages of the boll weevil, and the subjugation of all to the production of cotton. As was typical of Southern Renaissance writers, Scarborough did not back away from pain and sorrow; in the instance of the death of a small child in a cotton gin, she wrote that Rena Llewellyn, the novel's protagonist, painfully acquired a new perspective: "Of a sudden the realization swept over her that cotton was cruel. She had seen only the beauty of it before, its romance, its comedy, its poetry, but now she felt its tragedy."[19] Like Tannenbaum, Scarborough saw the devotion to single-crop farming as an obstacle to progress: "The one-crop idea is a curse to the state," says Rena's father, "and to the South as a whole."[20] Scarborough also bore witness to the attraction that cotton had for farmers, the closeness to the soil, the smells of spring, and the beauty of a field of cotton blooms. Her overall portrayal, however, was of the sacrifice and toil that farmers and their families endured year after year. She did not achieve major status as a Renaissance writer, but she demonstrated how the "way of life" meant impoverishment for many.

Literature was a principal means of molding public perception during the 1920s because radio was only beginning to grow in the consumer market. Writers critical of the South dealt with more than the cotton culture, but they were never far away. In the case of Tannenbaum and Skaggs, cotton was a major topic. Such treatment, wrote George Tindall, led to the identity of the region as the "Benighted South," a "neo-abolitionist image . . . that has strongly influenced the outlook of the twentieth century."[21] The social environment of the Jazz Age made the Cotton Belt a target of social pundits and critics of convention and orthodoxy. To be sure, the South could not deny its need for reform and innovation, and the assertions of the South baiters were not groundless, but their analyses and assessments often came with ridicule and mockery, written in the language of elitism and snobbery. They did not offer comfort and empathy, nor did they offer solutions. Even when novelists such as Scarborough wrote with compassion and understanding of the farmer and sharecropper, they nonetheless made clear the impoverishment and despair of the cotton culture. In the words of one writer, "The modernist era of the flapper, Freudianism, bootleg gin, and jazz had begun," so the ills and woes of the cotton industry were caught in the fashionable critiques of the era.[22]

With the onslaught of the Depression, the literature about cotton, both fiction and nonfiction, exploded, and some of it constituted the best sociological analyses and literary achievements of the period. Novelists continued to use farmers and sharecroppers to present universal themes of humanity, and no work surpassed Erskine Caldwell's *Tobacco Road* (1932) for impressing on the general public the dark side of the cotton culture. Caldwell's novel told the story of the family of a small Georgia farmer whose lives steadily deteriorate with the downward spiral of cotton farming. Unable to get credit to make another crop, they become degenerate and broken characters, preoccupied with hunger and enslaved by their own stubbornness, epitomizing the ignorance and backwardness that was becoming synonymous with the southern way of life. Like Scarborough, Caldwell expressed the optimism among farmers that the next season would bring a good crop, that tilling the soil and picking the snowy fiber gave a sense of fulfillment and satisfaction unavailable in other livelihoods. Despite the impoverishment guaranteed by stubbornly staying on small plots, Caldwell's central character Jeeter and his son Dude remained forever optimistic. *Tobacco Road* was made into a Broadway play in 1933 and ran for over seven years. In 1941 Twentieth-Century Fox produced a movie, directed by John Ford, based on the novel.[23]

A similar story about the declining cotton culture appeared in John Faulkner's *Dollar Cotton* (1942). His protagonist, old man Towne, hacks an empire out of the swamp and cypress of the Mississippi Delta. He devotes his life to his land and its lucrative yield, relying on black laborers, who continuously reward him with enormous profits that peak when the price of raw cotton reaches one dollar per pound, "dollar cotton." He acquires the wealth and recognition reserved for only the largest landowners. Along the way, however, he neglects family. "His cotton and his niggers," says his wife. "He has ever placed them before his flesh and blood." Cotton dominates old man Towne. "It is born in him to labor in the fields," says a friend. "Poetry, to him, is the growing of a seed into full fruit under his patient hand."[24] But when prices begin to plummet, old man Towne's greed and belief in the magic of cotton refuse to let him adjust, and he loses his land and fortune. So tragedy overcomes him and his kin just as it befell Jeter and Dude. Among the last of the Depression-era novels, *Dollar Cotton* appeared to be giving last rites to a dying way of life.

Southern letters depicted stagnation and revelation that effectively incorporated the universal themes of success and failure, justice and injustice, tolerance and bigotry. Novelists wrote of human shortcomings, of the disdain of one individual for another, all set in a rural culture where exploitation and weaknesses of character overpowered all else. Many writers exhibited a missionary zeal, some intent on exposing the sins of a culture, others sympathizing with southerners entangled in a net of feudalistic farming. Cotton was the common denominator among characters who played out their lives in the annual routine of producing a crop. In one fashion or another, the white tide touched nearly everyone as it swept over the region each year at "pickin' time."

MUSIC AND DRAMA

The cotton culture had long been portrayed in music and drama, much of which went far back into the nineteenth century and helped shape the image of cotton in 1945. Harriet Beecher Stowe's famous novel *Uncle Tom's Cabin* (1852) depicts growers in a negative light through the character Simon Legree, the plantation owner who beats Uncle Tom to death. Written on the eve of the Civil War, the novel effectively portrayed the brutality and savagery of slave labor in the antebellum era. The well-known song "Dixie," a minstrel song of the 1850s associated with the South during the Civil War, asserted the connection of cotton with slavery. In 1895 John Philip Sousa composed the uplifting "King Cotton March" for the Cotton States and International Exposition in Atlanta. It was a catchy and bouncy tune that imparted celebration. But the dark side of cotton culture reappeared often after World War I, as seen in 1926 when Edna Ferber published *Show Boat*, about a paddle wheeler that transported bales of cotton with the use of slave labor. The racist social environment of the antebellum period stands out, and Ferber's novel was made into a Broadway musical and movie. The association of cotton with African Americans appeared in George Gershwin's musical *Porgy and Bess* (1935). The female player, Clara, sings "Summertime," in which she describes how "the living is easy" and the "cotton is high." But in the land where King Cotton reigns, the characters in the folk opera are tormented by love and frustration, danger and murder.

But the most widely known example of music connected with cotton came with the blues that originated among blacks working in the fields. Rooted in the "hollers" chanted by workers plowing with mules, hoeing or picking cotton, the blues evolved into a popular musical form that spread across the United States. Southern blacks migrating out of the deltas and river bottoms took the music with them to cities like Memphis, St. Louis, Kansas City, Chicago, Detroit, and New York. Blues was the music of pain and suffering, an expression of the strenuous life as a sharecropper, "the personal emotion," stated Paul Oliver, "of the individual finding through music a vehicle for self-expression."[25] The music became associated with the mid-South, since many blues artists were born and raised there, but prominent artists also came from other cotton-growing areas, such as Blind Lemon Jefferson from east Texas. Blues was essentially black folk music, played and enjoyed by black people, until after World War II, when whites began to show enthusiasm for the form. Not all blues, however, originated in cotton fields. Songs came from levee builders, workers in turpentine camps, and prisoners, but life in the cotton culture stood as the overwhelming source of this purely American music. The development of musical records and studios in the 1920s enabled artists to record their songs and reach listeners on a large scale. Along with radio, the new recording industry spread this "cotton music" throughout the United States. Blues thus gave a musical form to the image of sharecropping as a harsh way of life.

NONFICTION STUDIES

Nonfiction studies of the cotton culture did not overshadow popular fiction, particularly in view of the success of *Tobacco Road*, but they provided a vast amount of data and analyses about the subject and went a long way in shaping public perceptions. Such studies dealt at length with tenancy, generally describing it as a new kind of slavery and warning that the cotton industry stood on the verge of collapse. In 1935 a team of three authors well versed in the conditions of the South published *The Collapse of Cotton Tenancy: Summary of Field Studies and Statistical Surveys, 1933–35*. The trio emphasized that sharecroppers survived hand to mouth, living on the edge of starvation and under the subjugation of landlords. They saw sharecroppers living constantly in debt, objects of ridicule and exploitation who had no hope for escape from the plantation economy. For a book of its type, it received unusual publicity because a press agent promoted it. "It was one of the most effective pieces of publicity ever done," stated William Alexander, one of the authors.[26]

When in 1936 Arthur Raper published *Preface to Peasantry: A Tale of Two Black Belt Counties*, he provided a thorough and perceptive analysis of plantation agriculture. In his study of two counties in Georgia, Raper documented the suffering of sharecroppers and small farmers, both black and white, implying that the conditions in these counties were common wherever tenancy was practiced. With the increasingly low return on investment in cotton farming during the 1920s, and the onslaught of the boll weevil, Raper believed the region's plantation economy had disintegrated to such a sorrowful state that it could not be salvaged, even with the development of a mechanical cotton picker. He saw share tenants doomed by the inherent structure of the plantation system, partly because the infamous crop lien required that maximum acreage be devoted to cotton. Even crop diversification had become impossible owing to the dependence on credit. "For the rank and file of plantation workers," he wrote, "there is little hope either in the rejuvenation of the present Black Belt plantation or in its collapse."[27]

In the foreword, civil rights activist and head of the Farm Security Administration William Alexander briefly but poignantly reinforced Raper's lack of faith in the plantation economy. "The decadence of this civilization is far advanced," and in the Southeast, "cotton farming . . . is doomed."[28] Agriculture in the South "has been crumbling for decades" because of misuse of land and exploitation of tenants, and it was now made untenable by the arrival of boll weevils. "So Negroes and poor whites live out the unenlightened years in the half stupor induced by malnutrition and neglect."[29] For the South to climb out of its sordid state of affairs, he recommended soil restoration and more emphasis on food production.

In a small portion of the Southwest, in south Texas below San Antonio and particularly below the Nueces River, Mexican Americans lived as sharecroppers. Their racial identity seemed more complex to Anglos than for African Americans in the older southern states. Latinos nonetheless has similar experiences, living in poverty and treated as second-class citizens. They faced the discrimination and exploitation of the cotton culture in the mixed Anglo-Latino society in the border area near Mexico. A special feature of cotton farming occurred in West Texas, the use of migrant workers to pick the crop each year, a practice that started when farming began there in the early twentieth century. Mexican pickers from south Texas would make their swing through the Rio Grande Valley in June, move to the Coastal Bend area near Corpus Christi in midsummer, and finish the cotton harvest in September through December on the plains in west Texas. They went back to south Texas and picked citrus and vegetable crops until the cotton season returned. Growers depended on migrants for labor and generally provided them with a form of housing, though it might be small shacks or abandoned buildings.

Santana Morales, a Latino migrant and tenant farmer, composed ballads, or *corridos*, that reflected the hardship and uncertainty of Mexican cotton pickers, *pizacordes de algodon*, as they traveled from town to town. Low wages and discrimination characterized their swing, and he put their travail into song. Other Latino composers writing from personal experiences were Justo Soto and Andres Garcia.[30]

However effectively nonfiction writers exposed the shortcomings of cotton farming, none surpassed Howard Odum's study *Southern Regions of the United States* (1936). Odum was the principal figure in the Southern Regional Committee of the Social Science Research Council, and his book "enjoyed such wide popularity as a textbook," according to one assessment, "that a large part of a generation of college students became familiar with the general outline of his ideas."[31] Considered in academic and policy-making circles as a first-rate research analyst with a grasp of the broader implications of education for improving southern life, Odum exerted great influence. In *Southern Regions* he dealt with industry and agriculture and naturally devoted a large part to cotton.

Southern Regions furnished the scientific backup for his assertions that cotton farming was defunct. The study had none of the sarcasm and ridicule of the 1920s literature, nor did it emphasize exploitation of sharecroppers. In a matter-of-fact fashion, it furnished narrative explanations with supporting data about the sociological characteristics of the cotton culture. Odum recognized that rising commercial farming placed the small farmer at a disadvantage, that underneath cotton's numerous deficiencies was surplus population on the land, that tenancy had increased, particularly among whites, and that "social advance will no longer countenance such low standards of living as now emerge from cotton culture."[32]

Novelists and analysts generally agreed that these conditions did not start with the Depression but resulted from practices that had been under way for several generations. They emphasized that much of the soil of the Cotton

Belt was worn out, driving up the cost of production to unprofitable levels, which accentuated the common indebtedness. Because of the low prices of the decade and the worsening depredations imposed by the arrival of the boll weevil, they insisted that cotton farming had become impractical in the Southeast. Some writers thought planters ought to shoulder the blame for the dire circumstances of the tenant class, while others felt that croppers were often responsible for their own troubles because of foolhardy practices and poor judgment.

The literature written between the two world wars made clear in irrefutable language that cotton was gasping its last breath, that the Cotton Belt had hung too long on to practices of cultivation and a system of labor outdated by the steady forces of technology and modernization. Through different genres, writers demonstrated how the South's infatuation with cotton was forcing significant numbers, whether serfs or small landowning yeomen, to find a new way of life. Writers of all fashion reached the same conclusion: the practice of one-crop farming with its complex and interrelated features fostered a declining environment. As the author Rupert Vance wrote, "The farmer is isolated from patterns of social culture by his life in open country, by his economic status, by his lack of educational advantages, and by the doctrine of native inferiority." The final product was "low economic status and rift in culture."[33]

HARDSHIP IN THE COTTON SOUTH

Hardship in the South impelled efforts to ameliorate cotton's low prices. In 1931 Louisiana governor Huey Long proposed a Cotton Holiday and "transformed the holiday from a rough idea," as one writer described, "into a South-wide movement."[34] The proposal asked farmers not to produce a crop for one year as a step toward reducing the mounting surplus and raising prices of the staple. Despite much effort and publicity, the movement failed.[35]

An incident that enhanced the image of cruel and oppressive planters occurred when in 1935 the Southern Tenant Farmers Union (STFU) staged a strike in the eastern Arkansas Delta because landlords there, particularly on the Norcross plantation, were not sharing the payments coming from the Agricultural Adjustment Administration (AAA) crop reduction program. Complaints about this practice were heard throughout the Cotton Belt, and evictions of sharecroppers were not uncommon as planters reduced their acreage through the AAA program. The strike led to gunfire, arrests of union members, tenant evictions, and police blocking roadways. Norman Thomas, a socialist, spoke on national radio: "There is a reign of terror in the cotton country of eastern Arkansas."³⁶ When no relief came to the union members, they staged a second strike in 1936, but still no concessions came. After joining the Congress of Industrial Organizations (CIO) in 1937, the STFU withered away.

In historical context, the strikes in eastern Arkansas, site of some of the nation's worst rural poverty, had significance. They engendered publicity in the national press over the plight of the rural poor and contributed to the famous 1935 "purge" in the AAA over the question of extending benefit payments to sharecroppers. A *March of Time* newsreel, "Trouble in the Cotton Country," was made about the unrest in the Arkansas Delta. "Largely because of the headlines, newsreels, and radio broadcasts stimulated by the STFU's struggles in the Arkansas Delta," according to a thorough examination of the incident, "the problem of farm tenancy had become a leading topic of discussion across the country."³⁷ Whether it became a topic so greatly discussed, the episode encouraged President Franklin D. Roosevelt, already aware of the impoverishment in the Cotton Belt, to take action on behalf of the rural poor, which led to the final events that shaped the image of a stagnant industry.

Partly because of the STFU, but also the growing recognition of the desperate situation among cotton's tenant class in the South, the Roosevelt administration, acting in conjunction with powerful southern members of Congress, took steps to bring relief to tenants. During the 1936 presidential campaign, Roosevelt spoke about the importance of farm ownership; after the election he created the President's Committee on Farm Tenancy with Secretary of Agriculture Henry A. Wallace as chair. In its report the committee lamented the squalor in rural America, especially in the cotton South, and recommended the creation of a new agency, the Farm Security Administration (FSA), to carry out a program to help tenants acquire land. In 1937 Congress passed legislation creating the FSA, which represented the most meaningful step taken on behalf of the cotton tenant.³⁸

Despite best intentions and some real innovations such as rural health clinics, and although most of the FSA's resources went into southern states, the new agency touched only a small percentage of farmers and tenants. Standard textbook explanations for its lack of impact emphasize resistance by landlords and the power of commercial agriculture. "It was eyed with suspicion and hostility by the agricultural establishment: the Department of Agriculture, the Extension Service, the land-grant colleges, the Farm Bureau," stated one authority.³⁹ Some studies judged the FSA to be too uncompromising with critics wielding political power.

The conditions responsible for the distress overwhelmed the FSA. Cotton farming stood near collapse owing to a multitude of difficulties ranging from outworn practices of cultivation to the behavioral characteristics of each group within its sociological structure. As repeated over and over by various analysts familiar with cotton farming and its humanist ramifications, too many people were involved in an activity with an extremely low rate of productivity, and whether restricted by technological barriers or human prejudices, they could not produce a standard of living sufficient for lively and active members of a rising industrial society. Because the cultivation of cotton took about six to seven months per year, small farmers and the laboring class were underemployed. Commercial operators could continue to succeed only if tenant labor remained available. All classes that engaged in the production of cotton experienced sacrifice and hardship during the Depression, a point poignantly expressed by STFU founder H. L. Mitchell: cotton, he remarked, was "disorganized, pauperized and kept alive only by Government subsidy."[40] It was unrealistic for the FSA to preserve a dying culture of tenancy and small agriculture unable to sustain itself.

THE NATION'S NUMBER ONE ECONOMIC PROBLEM

The image of cotton on the verge of collapse continued to grow, and the final blow came in 1938 when the Roosevelt administration declared the South to be the "nation's number one economic problem." A special study of the South ordered by the president and published as a brief but clearly written pamphlet reiterated the points that research analysts such as Vance, Odum, Raper, and Alexander had presented over the last decade. Prepared under the auspices of the National Emergency Council, the report, *Economic Conditions of the South*, covered both agriculture and industry, but its comments on farming, while offering no new information, drew public attention to the plight of cotton far beyond that achieved by previous studies.

The report pointed out that the southeastern states had the smallest farms in the United States, averaging only seventy-one acres, and that one-fourth of them were smaller than twenty acres. Because of their small size and exhausted soil, these farms yielded too little for a reasonable standard of living, and "a farmer with so little land is forced to plant every foot of it in cash crops; he cannot spare an acre of soil for restoring crops or pasture."[41] The number of farms under twenty acres had doubled since 1880, and over half of the farmers in the Southeast depended on cotton alone: "All their eggs are in one basket—a basket which can be upset, and often is, by the weather, the boll weevil, or the cotton market." Too many livelihoods relied on this single crop exposed to a world market: "No other similar area in the world gambles its welfare and the destinies of so many people on a single-crop market year after year."[42] The number of sharecroppers had increased in the last twenty years, and about two-thirds of them were white.

Most of this information had been known for a decade or longer, but the report's brief and direct statement caught national attention in a manner far superior to the weighty tomes by social analysts and policy makers. As stated in one text, "Never before had the failures of the cotton-tenant farmer tradition been presented so nakedly and graphically."[43] In a national radio address from Nashville, Works Progress Administrator Harry Hopkins reviewed the document and commented that the South's "dependence on cotton brought with it serious drawbacks."[44] The image of the cotton culture near death was complete.

The cotton industry acknowledged its status. In 1939 the *Cotton Trade Journal*, the industry's most widely read publication, featured an article by V. G. Martin, head of the Agricultural Education Department at Mississippi State College, and J. N. Lipscomb, head of the college's Agricultural Economics Department. They went beyond the report by the Committee on Farm Tenancy and reaffirmed the conditions resembling serfdom on plantations and farms by quoting a statement of Secretary of Agriculture Henry A. Wallace from 1936: "I have never seen among the peasants of Europe poverty so abject as that which I saw in this favorable cotton year from Arkansas on to the east coast." Martin and Lipscomb simply stated: "The cotton farmer is now the national example of poverty" and is "poor by every measure of economic well-being."[45]

Assertions and descriptions of the substandard living conditions among small farmers, tenants, and sharecroppers continued but captured little attention once war broke out in 1939. Further examples nonetheless bolstered the image of a way of life slipping away. In 1941 Herman Nixon, one of the Vanderbilt Agrarians, a group of writers sympathetic to the South, published *Possum Trot: Rural Community, South*, and closely echoed Tannenbaum with his insistence that "King Cotton is guilty of having caused many of the South's ills, and that king dies hard."[46] In 1941 George Sessions Perry published *Hold Autumn in Your Hand*, a novel that received the National Book Award. It portrayed the life of a cotton tenant in the Texas Brazos River bottomlands. Made into a movie in 1945, the work depicted the struggles of Sam Tucker, whose crop is destroyed by flood. Tucker is forced to leave the farm for a wage-earning job in the city because the repeated struggle of making a crop will not furnish enough income to provide for his wife and children. In 1941 W. J. Cash published *The Mind of the South*, a highly acclaimed interpretation of southern history and culture. Cash believed King Cotton "was growing continually sicker" and recalled how in the early 1930s all classes of cotton growers, including some planters, either abandoned their land or stayed and eked out only a marginal existence. According to Cash, the people at the bottom of the ladder, sharecroppers, whether white or black, had experienced real hunger, and many had drifted into towns and cities seeking relief. William Faulkner's *Go Down, Moses* (1942), a novel of interlinked stories, depicted the harsh lives of African Americans trapped in the South. The characterizations reinforced the perception of the cotton culture as stagnant and near collapse.

In the vast amount of literature written between the two world wars, ranging from journalists' accounts to the reports of presidential committees, including the emotional fiction by renowned novelists and socioeconomic studies by highly reputable university faculty, several themes emerged. Life on a cotton farm conjured up images of the most abject living conditions in the United States, shameful for any country, but particularly disturbing for the economic leader of the world. Erosion and depletion of the soil were easily identified as causes of low yields. Deeper causes of the substandard living included a single-minded devotion to one-crop farming, which owed much to ignorance as well as the fact that cotton was the only source of cash readily available for many farmers. Nearly all analysts concluded that self-sufficient farming and plantation agriculture dependent on subservient labor could no longer succeed in an industrialized mass consumer society.

Exploitation also emerged as a theme because writers often accused large landowners of acting as paternalistic caretakers of their tenants, of keeping them in economic straits through debt, or by not sharing the AAA benefit payments. Planters defended themselves, however, and characterized both white and black sharecroppers as a childlike people dependent on landowners for care and supervision, justifying their supervision as necessary in a socioeconomic system that depended on cotton to raise any money at all, particularly with few alternative sources of income. "The failure is in human nature," wrote the Mississippi planter William Alexander Percy, and "if the white planter happens to be a crook, the sharecropper system on that plantation is bad."[47] An inherent conflict existed between planter and sharecropper, the age-old quarrel between management and labor, but in this case one that constituted much of the social fabric of the South. By World War II, this division of labor had become intolerable regardless of its history or the conditions responsible for its existence. Incorporated into this maze were race relations, making it impossible to separate them from the paternalistic practices of cotton farming. Central to understanding this culture was the crop-lien system, which contributed to the hierarchy of dependence that began with the lowly croppers, rose through the planters, and reached the southern banks that were themselves dependent on outside sources of capital. Even the devotion, or bondage, to cotton owed much to the system of credit. "Inertia and ignorance had a part in establishing the one-crop system," wrote an astute observer who grew up on the banks of the Yazoo River, "but the fundamental cause was the lack of farm credit."[48]

COTTON IS SICK

However complex it might be, cotton farming no longer offered a reasonable livelihood for landowners with small acreage and certainly not for anyone farming on a share basis. Whether from the perspective of sharecroppers or planters, cotton farming as practiced in the South had become too archaic for a nation rapidly becoming highly industrialized with major cosmopolitan centers. The general public saw cotton as too inefficient, locked into an inertia that prevented its own progress and held back thousands of landless farmers from self-improvement. The planters, "the Bourbon politicians and economic royalists of the South," wrote H. L. Mitchell, bore the burden of responsibility, rightly or wrongly, and were targeted by social critics.[49] Small landowners, who came closest to fitting the image of family farmers, were respected as representatives of the agrarian ideal but were nonetheless thought to be entombed in a substrata offering no escape.

This recognition owed something to the coincidental development of media imagery and improved transportation. Mass communications began in the interwar era, with the rise of commercial radio broadcasting, the popularization of movies, the widespread use of automobiles, the new and improved roads and highways, and the greater circulation of newspapers and magazines. A penchant for sensationalism paralleled this development, fed by media exposés of the Ku Klux Klan and lynchings, which also included reports of peonage on cotton plantations. Advances in communications technology and improvements in the infrastructure of transportation broke down the isolation of rural areas nationwide, enabling news reporters to furnish next-day coverage of events from any remote pocket of the country. These developments made the conditions in the South more widely known.

Popular plays and movies reinforced the image. Along with *Tobacco Road,* Steinbeck's *Grapes of Wrath,* another famous Depression novel of 1939 later made into a movie, expanded the image of the poor southern whites, called Okies, trying to leave their lives of destitution for a better home in California. Cotton alone did not cause the South's woes, but it was the focal point of the image of backwardness by virtue of its overwhelming importance in the economic and social strata of the region. In *Requiem for a Nun,* William Faulkner, whose novels so effectively bared the soul of the South, saw cotton as an "omnipotent and omnipresent" king, the "white tide" that swept people along, the "white surf... altering not just the face of the land, but the complexion of the town too."⁵⁰ The "white tide" defined the South in American culture and accounted for the perception of the region by World War II as broken in spirit as well as pocketbook.

Within the context of America's social consciousness during the era between 1918 and 1945, the nation began recognizing its own racism, particularly the suppression of African Americans. In the tumultuous atmosphere when race riots, lynchings, and the resurrection of the Ku Klux Klan figured prominently, the question of racism came into sharp focus and evoked protests from black intellectuals such as W. E. B. Du Bois through the National Association for the Advancement of Colored People (NAACP) and white intellectuals as William Alexander of the Commission on Interracial Cooperation. An important contribution came from the studies of sharecropping and the discovery of the appalling conditions of servitude that tenant farmers endured. Certainly the climax of studies on racism occurred in 1944 with Gunnar Myrdal's *An American Dilemma*, the instant classic that in the words of one reviewer "played a significant role in changing the thought patterns and feelings of a people."⁵¹ Examinations of racism flourished because "of the extent of the social ills from which the body politic suffered," wrote John Hope Franklin, "and was one more indication of the tremendous frustration that characterized the lot of many a Negro, and some whites as well."⁵²

Because a large portion of the black population lived as sharecroppers on commercial farms, the new awareness of racism cast plantation agriculture and landlords in a negative light. Each new study of tenancy added to the number of Americans who saw cotton as an industry beset not only by poverty and economic stagnation but also with some of the worst characteristics of racism. Exploitation and discrimination of blacks happened in all regions, but their plight in the lower United States drew particular attention, so as the region carried the historical burden of slavery, the South bore the onus as the most racist sector of America.

Cotton's cultural image included more than a malaise caused by tenancy and economic depression. It incorporated a dimension that had only recently appeared: the reliance on government subsidies. Without the helping hand of Uncle Sam, cotton farming and the other segments of the economy that processed and handled the fiber, such as ginning and textiles, would be jeopardized. Price supports kept the industry from collapsing during the Depression. Other commodity staples such as wheat had a similar experience, so from the perspective of trying to restore prosperity to agriculture and the need to maintain the country's self-sufficiency with food and fiber, the New Deal's inauguration of subsidies in 1933 gave cotton more than a badly needed boost: it became a life-support system for the 1.5 million producers of the fiber. Subsidization altered the highly individualistic nature of agriculture, explaining a statement by a president of the Mississippi Valley Historical Association: "The New Deal hastened the dawn of another era in the cotton patch."[53] Meant to be only a temporary measure, subsidies had become a regular feature of cotton culture. But for independent southerners, historically the champions of states' rights, reliance on federal assistance violated the precepts of independence they so dearly cherished, so there was a desire to overcome the causes of cotton's illness and free themselves from government assistance.

Cotton farming was arguably the overpowering characteristic of the South's social fabric, but a large body of literature and general public awareness left no doubt that the cotton industry sorely needed improvement and reform. Proposals for lifting farmers and tenants to a prosperous level focused on soil conservation, cultivation practices, and credit and marketing. Much coverage dealt with substandard housing, improper diets, insufficient public health facilities, inadequate education, and a lifeless sense of civil community. The literature expressed the assumption that the lives of families would get better if advances in production occurred, crop diversification were undertaken, and landlords treated tenants equitably. But the dilemma went further than inequity, for as the *March of Time* newsreel concluded: "It is plain today that planter and sharecropper alike are the economic slaves of the South's one-crop system and that only basic change can restore the one time peace and prosperity of the Kingdom of Cotton."[54]

THE FAILURE OF ANALYSIS

The literature and analyses of the interwar years, so impressive and straightforward, fell nonetheless short. Writers explained the illness with heartrending examinations of institutions, geography, and social behavior. Showing how cotton was the defining feature of southern poverty, they established the wretchedness of a generation trapped in destitution and left no room to question the pain of a sharecropper's life. One writer considered the cotton tenant to be "the American peasant."[55] But writers paid scant attention to new economic forces responsible for the conditions they described. They cast the low incomes and stagnation in local or regional terms, taking a myopic look at a culture that in reality depended on world conditions. Cotton was too woven into the increasingly interconnected global economy for its condition to be resolved through adjustments in social behavior or modifications made within the parameters of an outdated culture. Too little attention went to the protectionist policies that restricted trade, and writers did not question whether the federal crop reduction program stimulated foreign production and retarded domestic prices. For all their penetrating studies of southern life, analysts and critics reflected the isolationism of the interwar years; the economic nationalism of the era gave them a self-contained mindset. In 1940 T. J. Woofter and A. E. Fisher published a pamphlet as part of the Federal Works Agency Series on Social Problems titled "The Plantation South Today" in which they listed a dozen recommendations for improving plantation life. They included the usual points: soil conservation, crop diversification, fair treatment of tenants, credit reform, and even tax reform and improved legal status for tenants. But they overlooked the loss of foreign trade and the rise of synthetic fibers and did not discuss the merits of the federal crop reduction program as a factor in keeping inefficient small farmers in operation.[56] Rising and shifting circumstances in the world economy, driven in no small way by militarist regimes in Europe and Asia, plus the autarkic trade blocs erected by the United States and its trading partners, demanded that a return to prosperity take into account broader developments already under way: the new competition from foreign producers of cotton, the intrusion into traditional markets by synthetic fibers, the erosion of foreign markets, and the need for mechanization, all of which meant the economics of scale had finally hit cotton. Self-sufficient farming was already doomed, and if small farmers and planters alike intended to remain on the land, they would have to invest in equipment, expand their acreage if necessary, and improve their cultivation practices. A scientific breakthrough in insect control had to occur to overcome the pestilence of boll weevils. Small landowners would have to become more akin to commercial operators, and all classes of farmers could remain in operation only by adopting the business ethic in which there was no room for the paternalism of tenancy, either as landlords or as sharecroppers.

As bad as economic conditions were, romanticism remained among cotton growers, especially small plot owners, who felt that transforming a field of raw earth into a sea of white bestowed a connection with nature and creation. They were part of the rhythm of the seasons, seeing themselves as stalwarts of a high calling. Southern folklore was indeed rooted in the cultivation of cotton, from which came music and poetry, the acceptance of hard labor, self-sufficiency and sacrifice, and an outlook on life as an enduring struggle. While sharecroppers wished to escape the toil and penury of chopping and picking cotton, landowners wanted to preserve their life on the soil, wishing only to enjoy better prices for the fiber they carefully nurtured. But even this romanticism had begun to lose to the forces of urbanization and industrialization as youth fled to the cities. There was a haunting feeling that growers of cotton were endangered.

For the United States, more was at stake than rescuing the southern way of life. Cotton ranked along with food in importance to human welfare for survival, since wool, synthetics, and silk made up only 25 percent of the fabric used to make clothing. It was self-evident that the country needed its own supply of cotton. On America's cotton farms there lived nine million people, and when combined with the thousands employed in ginning, textiles, and apparels, those people comprised a significant portion of society. From the perspective of America's common welfare, cotton farming had to become profitable. So embedded and long-standing were the causes of the decline, however, that the past seemed to overpower the future and erase all prospects that cotton could be reestablished as a viable part of U.S. agriculture. Keeping cotton out of the grave would require leadership able to grasp the realities of a shifting world economy and the common sense to adapt to modernity.

CHAPTER 2

THE NEW POLITICS OF COTTON

In 1945 the cotton industry had a new political dynamic based on the rejection of an order in which compliance and acceptance of existing practices were deemed insufficient. To restore prosperity to cotton farming, industry leaders knew that new rationales were needed, and they began with the insistence that poverty was not inherent in the cotton culture, that by infusing it with a renewed intellectual vigor and giving it consistent and sharply defined goals, practices and policies could be devised that would free farmers from hardship, sharecroppers from serfdom, and planters from fear of bankruptcy. This new energy and enthusiasm had been responsible for the creation of the National Cotton Council (NCC) in 1938.

ORIGIN OF THE NCC

The NCC grew out of the frustration over the persistently low price of cotton. Much hope had rested on the New Deal efforts to bring prices to higher levels, but after the 1936 Supreme Court decision *U.S. vs. Butler*, which overturned the AAA planting restrictions, there was a record crop in 1937, 18,946,000 bales, that drove prices down again. Unusually good weather had contributed. Warehouses of the Commodity Credit Corporation (CCC) bulged with cotton, and the price dropped to 8.7 cents per pound for 1937 and 1938. Throughout the 1930s, however, the price never climbed higher than 12.7 cents in 1936, below the meek average of 17.4 cents during the 1920s.

The NCC also arose from suspicion that federal support was in jeopardy. In 1936 Michigan senator Arthur Vandenberg, with the assistance of Senator Ellison "Cotton Ed" Smith of South Carolina, chair of the Senate Agriculture Committee, demanded that the USDA disclose the recipients of the largest cotton subsidy payments. After some initial resistance, Secretary of Agriculture Henry A. Wallace caved when pressure mounted from the press. Embarrassment fell on the Roosevelt administration with the revelation that the South's largest plantation, the British-owned Delta and Pine Land Company (D&PL), had received a hefty $396,000 between 1933 and 1935.[1] Oscar Johnston, manger of the plantation, had served as the finance director in the AAA and had liquidated the 2.5 million bales acquired through the Federal Farm Board during Herbert Hoover's presidency. Johnston had also served as Roosevelt's special emissary to Europe on a mission in 1935 to stimulate cotton exports. In 1936 Johnston managed the disposal of the large surplus of government-owned cotton, the "cotton pool" acquired through the CCC. Earlier in 1935, South Carolina senator Ellison Smith, a populist supporter of small farmers, had grilled Johnston in a special hearing on the cotton futures market, a hearing that ran sporadically for two years and kept the spotlight on the large landowners and brokers.

In 1937 *Fortune* magazine featured D&PL, consisting of thirty-eight thousand acres in Scott, Mississippi, as an example of a progressive plantation achieving the best results likely under the dire circumstances of the Depression. *Fortune* credited Johnston for the achievements, claiming that his management supervised "sharecropping at its best or, its least objectionable."[2]

But negative publicity, with Johnston sometimes the object of attention, reinforced the image of ruthless planters and brokers willing to exploit sharecroppers or manipulate the futures market for personal profit. Such criticism aroused fear in the South that Congress and the president might take action to reduce or even stop support programs. Should that happen, the industry would almost certainly collapse. Unkind characterizations of commercial growers were not uncommon in the Depression, but the wholesale indictment of the industry offered no solutions for its complex and deep-seated problems. This set of circumstances—prices in the doldrums, hostility in Congress, and suspicion of planters in popular culture—caused the progressive minded among cotton interests to reassess the condition of the industry and to seek means outside the government to ensure cotton's future.

With these developments beginning in 1936, the conviction grew that cotton had to act on its own behalf, to unite its various interests for action. Conventional opinion held, however, that unification would be impossible because too many segments—farmers, ginners, merchants, warehousemen, crushers, and shippers—made up the industry. Each segment, except for growers, had its own organization, such as the American Cotton Shippers Association, and several national farm organizations already existed, such as the American Farm Bureau Federation and Farmers Union. "The industry had no record of working in unison," wrote one author.[3] With the static practice of sharecropping and rejection of challenge to the established order, as seen in the crushing of the Southern Tenant Farmers Union, cotton growers seemed oblivious to change. Within cotton circles there arose a determination to challenge orthodoxy, however, driven by the failure of the New Deal to bring prosperity to the farm.

Agitation for a national organization began as informal discussions, sometimes among "friends meeting for morning coffee at a roadside café."[4] Cotton growers in the mid-South drove the concept forward, but a key organization was the Delta Chamber of Commerce in Stoneville, Mississippi. It originated in 1935 to lobby for improved roads and highways, but when the Mississippi state government passed legislation in January 1936 for improving roads, the Delta Chamber no longer needed to worry about its original purpose and questioned its future. With only one employee, a stenographer, the organization nonetheless remained determined to advance the needs of the eighteen counties in the Mississippi Delta. In 1936, short of funds, the organization hired Rhea Blake, a young sales representative of the Appalachian Power Company in Virginia, and instructed him to "get out and raise money."[5] No issue disturbed the Delta's planters and farmers more than the deteriorating condition of the economy, so on behalf of his new employer, Blake lobbied or assisted other lobbyists on matters pertaining to cotton, the lifeblood of the area. In January 1937 he presented testimony at the hearing of the U.S. House Committee on Agriculture on tenancy held at Montgomery, Alabama. Rupert Vance, the highly respected authority on the rural South, served as adviser to the committee.

Blake worked with some of the best-known and most prominent citizens in the Delta. W. T. (Billy) Wynn, attorney and planter from Greenville, Mississippi, an original founder of the Delta Chamber, had put Blake on the payroll. James Hand, a young and progressive-minded planter from Rolling Fork in the lower Delta, often shared his thoughts on rescuing cotton, and Oscar Johnston prepared the testimony that Blake read before the House committee hearing. Among these men and their acquaintances, mostly in the Delta, though not entirely, the idea moved forward that federal price supports and reduced production were not enough; at its annual meeting in mid-1937, the Delta Chamber went on record to recommend the following measures: promotion of cotton consumption with advertising, research to find more uses for cotton, lowering international tariff barriers, and research on lowering farming costs.[6] But no mechanism for carrying out these proposals came forth.

The plantation owner James Hand has received special consideration from chroniclers of the formation of the NCC. Described by Read Dunn, a longtime figure in the organization, as an "unusually capable farmer in the

South Delta and an original thinker," Hand ranked among the earliest proponents of mechanization owing to the shortage of labor in his area of operations; he was also an implement dealer, which gave him a propensity for using modern equipment.⁷ Hand had a precise mind and regarded research to further mechanization as essential for cotton to become steadily profitable. He helped design the proposals set forth by the Delta Chamber, but he also engaged in discussions with farm interests around the Delta, as all wanted a solution for the afflictions threatening cotton. He and Blake repeatedly went over each stage of production, from field to fabric, beginning with planting and following through ginning, storing, spinning, apparels, and finally the retail end of the fiber.

During a car ride into the northern Delta counties in the late summer of 1937 to stir up support for the Delta Chamber's resolutions, Blake and Hand hit upon the idea of creating a national organization. Blake later recalled that they spoke after the USDA August forecast for a bumper crop: "We have this mountain of cotton now staring us in the face," Blake told Hand. "Well," he continued, "you know the thing that bugs me about the whole business is that why in the hell aren't we doing something?" They focused on disposing of the surplus. "I remember that session with Jimmy Hand in the car vividly," Blake said, "and I would put that as probably the starting point of anything, so far as I know." They agreed that Blake would contact Billy Wynn upon their return. Blake went to Wynn's house, and they "took a drink or two and talked more on that first session." It quickly became obvious that the Delta Chamber was too small to lead the nineteen states in the Cotton Belt, so in further discussions among Blake, Wynn, Hand, and another young planter, Tol Thomas from Cruger, Mississippi, the notion advanced that a trade organization uniting the Cotton Belt should be created. But the task would be so overwhelming that it might be impossible, Wynn noted, and a new organization would have to be built around a person with a grasp of cotton's ramifications from the level of the sharecropper through the national and international levels. "We've got the man," he said. "We've got Oscar Johnston."⁸

Through private conversations, group discussions, and formal gatherings, the plan to unify the industry began. At its annual meeting in July 1937, the Delta Chamber passed a resolution drafted by Johnston, Blake, and Delta planters, including William Alexander Percy. It endorsed government restriction of production but called for steps to encourage consumption, lower tariffs, and conduct research to reduce production costs and develop new uses for cotton. Although Johnston was the ranking figure in the industry, Hand initiated some of the fundamental principles associated with the NCC, particularly the importance of research for better cultivation practices and expanding uses of cotton. Foremost in their thinking was the notion, stated Read Dunn, that "cotton farmers would have to help themselves."⁹

The "mountain of cotton" beginning to come from the fields led to still more action. Johnston spoke in early September 1937 to the annual meeting of the Mississippi State Commissioners of Agriculture held in Memphis. A sense of disenchantment with the New Deal cotton program prevailed, and Johnston, no longer an official with the AAA, had the floor as the keynote speaker. He again endorsed reduced production by government fiat, urged the end of the Hawley-Smoot tariff, recommended that the government instigate international trade agreements, and encouraged the industry to establish more political power in Congress. "The proposal also involved creation of an industry organization to achieve the above and generally help cotton," according to Dunn, but Johnston's plans for implementing the proposals were too "sketchy and almost incidental," and no action for such a body came out of the meeting.¹⁰

Johnston's persuasive speech in Memphis nonetheless accelerated the move for a national organization. Hodding Carter II, news editor for the *Greenville Delta Star*, wrote an editorial commending Johnston for proposing measures less dependent on the federal government. Johnston's stature began to rise as the person who best understood cotton and what needed to be done for it, but his own thinking remained cloudy about organizing the industry, and he turned to the American Farm Bureau Federation (AFBF).¹¹

No farm group in the United States equaled the Farm Bureau in political clout and degree of organization at the local level, which accounted for Johnston's interest in it. From his perspective, the bureau could go to work immediately on behalf of cotton, since it was deeply embedded in Washington lobbying. But the AFBF's constituency lay mainly in the Grain Belt, though it had a strong presence in Arkansas and Alabama. In three southern states, it had no chapters. Johnston stubbornly tried to shore up the bureau's involvement with cotton and managed to get the Mississippi Farm Bureau to change its name to the Mississippi Agricultural Council with himself as vice president.

Encouraged by this accomplishment, Johnston persuaded the Delta Chamber to meet in November 1937 in special session and consider the creation of a national association for cotton. He "was still thinking of an organization of producers only," said Dunn, and intended to "capitalize on the existing state Farm Bureau organizations—simply restructure them to make them more independent of the American Farm Bureau and rename them."¹² Johnston received a hearty approval from the Delta Chamber, but the AFBF understandably had suspicions. Johnston's concept at this point, basing his proposal solely on growers, was too narrow. The Delta Chamber and the promoters connected with him came principally from planters, who thus far reflected mostly themselves in the plans, although growers recognized the importance of unity and a lobbying arm in Washington. From other corners there arose a plan to spread the proposed organization across a wider spectrum of cotton interests.

A heightened sense of urgency developed when President Roosevelt called a special session of Congress on November 15 to consider new legislation for agriculture. Wheat and other commodities needed help, but the mid-South interests had their lives and fortunes tied up in cotton. The abundant harvest of 1937 yielded more than a storehouse of plenty; prices left growers struggling with debt and knowing that sacrifices would have to continue. Despite providing assistance on an unprecedented level, the New Deal had not pulled cotton from its abyss, and each class of farmers, from the lowly sharecroppers to the planter elites, now felt an even deeper anxiety over the future of their livelihoods. As Wynn told Johnston, "Everything the Delta's got, is tied right up with cotton."[13] Blake went to Washington on behalf of the Delta Chamber, but Johnston stayed in Mississippi to avoid agitating his adversaries in Congress. He needed to work toward a plan that would enable the industry to present a stronger front, and he believed that he should stay home.

In Memphis two well-known businessmen, Norris Blackburn and Douglas Brooks, operators of cotton warehouses, shared a plan for organization with members of the Memphis Cotton Exchange, but they wanted to include "all segments of the industry dependent on cotton . . . and [let them] pay part of the cost of operating such an organization."[14] Blackburn shared his view with friends at the Cotton Exchange, who liked the idea but pessimistically thought "it would be almost impossible to bring all segments of the industry into one organization."[15] As stated in one account, "There was skepticism, but no cold water was on the idea."[16] Blackburn took his proposal to New Orleans and reviewed it with interests from warehousing, shipping, compresses, and milling. They quickly became enthusiastic about creating an umbrella organization and took the idea to Richard Leche, governor of Louisiana, who liked the unconventional approach and shared it with reporters. In December 1937 the "Leche plan for a Council of cotton interests soon hit the press," wrote the NCC historian Albert Russell.[17] Leche managed to get the Conference of Southeastern Governors to endorse the concept. In Memphis and New Orleans, Johnston commanded much respect as the natural leader for erecting a national organization, which prompted Blackburn to state that such an organization "would take an outstanding cotton leader, and I felt that man was Mr. Oscar Johnston."[18]

Blackburn and his associates rallied business interests and brought them into the fold. Only if warehousemen, compress owners, bankers, and brokers were also committed could the industry exert enough lobbying force to obtain redress on factors affecting cotton, such as tariffs, that were not directly part of farm legislation. The changing nature of cotton's global market, which accounted for the falling exports, meant that international trade would have to be radically improved or else life on 1.5 million mostly southern farms would slide closer to collapse. Besides facilitating trade, the mid-South promoters wished to expand and redefine the genesis of farm assistance; they wanted to emphasize increased consumption rather than crop reduction. Such a shift would require research on improving the characteristics of cotton so that it could compete with synthetics. That was expected to lead to commodity advertising, which in the latter 1930s was unknown in American agriculture. "At the time the Council started," stated Ed Lipscomb, a NCC pioneer, "the words 'cotton house dress' just meant a cheap something. . . . Cotton was your cheap fabric."[19]

In the context of these perspectives, the push by business interests broadened the concept of the proposed organization so that it would encompass more than planters and farmers. Inherent in their thinking was the principle of equality among the various parties engaged in a major industry, the principle that unity based on cooperation should be the fundamental basis in promulgating new policies. "The concept was there," wrote Dunn, "and in a broader framework for the entire industry."[20] Thanks to Blackburn and his colleagues in Louisiana, it became clear, too, that Johnston stood as the recognized leader who should bear the burden of uniting the various segments into a single and cohesive organization.

By mid-1938 the clamor for action encouraged Johnston to move further. He now realized that cotton had a constituency consisting of more than farmers and planters, and he knew that the crown of leadership had been conferred on him. But his commitment to include all segments of the industry came only after others had worked to convince him, such as ginners, warehousemen, and oilseed crushers. Hand continued to stress the need for mechanization and research. But it was Billy Wynn and Rhea Blake, probably Johnston's two closest associates, who nagged him into accepting the role of leader.

In June 1938 the Delta Chamber of Commerce held its annual meeting in Cleveland, Mississippi, and the notion of creating a national organization held the spotlight. With the help of Blake and Read Dunn, Johnston had sent invitation letters to prominent leaders of the different segments of the industry. His theme for the gathering was international trade, and Johnston had asked Secretary of State Cordell Hull to speak, but Hull sent Dr. Francis B. Sayre of the State Department to speak in his place. Attendance was excellent because word had spread that a new organization to speak for cotton was in the making. Only one attendee came from outside the Mississippi Valley, George Payne, head of the Southwest Irrigated Cotton Growers Association, headquartered in El Paso.[21] Delegates included Mississippi governor Hugh White; his support proved beneficial because he promised to get a $9,000 appropriation from the state legislature to help start the proposed organization. Governors Richard Leche of Louisiana and Bibb Graves of Alabama also attended. Few questioned the wisdom of creating the proposed organization, and in the euphoria of the moment, doubts over unification of cotton's interests seemed inconsequential. In the words of a Johnston biographer, "the whole affair looked like an express train racing through the station."[22] The group passed a resolution for a meeting in Memphis in November to create a national organization. Most of the excitement centered on the beginning of such a body, but Sayre's warmly received speech promoting free trade indicated how the Cotton Belt looked glowingly on international trade.

No detractors attended the Cleveland meeting, but doubters remained in the Cotton Belt. No delegate officially representing the Farm Bureau went to Cleveland, though some attendees such as Harold Young, president of the Arkansas Farm Bureau, were active in their state chapters and supported the call for a commodity-centered organization. Edward O'Neal, national president of the Farm Bureau, stayed away. So did Will Clayton of Houston, cofounder of the powerful Anderson, Clayton and Company (ACCO), but R. C. Gregg from the ACCO office in Memphis attended. O'Neal and Clayton had expressed doubts about the concept of an inclusive body to represent all interests in the Cotton Belt, each for different reasons. O'Neal desired to protect AFBF turf as the spokesman for farmers in the South and saw the plan drawn up in Cleveland as a threat. Johnston might succeed without O'Neal, but he had to recruit Clayton, whose intellectual ability and respect in agricultural circles equaled Johnston's own.

With the success of the Cleveland meeting, Johnston proceeded to establish committees of the NCC on the state level in order for them to create the national organization. He worked hand in hand with Billy Wynn, Rhea Blake, and Read Dunn, and they arranged for organizations of other cotton interests, such as the American Cotton Shippers Association and National Cottonseed Products Association, to send delegates. To the delegates they mailed copies of a special study of commodity organizations conducted by four professors at Tulane University.[23] In some instances the state committees came together easily, while in other cases the process was difficult. Johnston and Blake had to tread carefully, always being respectful of differences of opinion and acknowledging territorial prerogatives. Because so many of the trade associations in the cotton industry were centered in the mid-South, the region was represented disproportionately among the delegates forming state committees, which made the western states resentful. In Texas, a state with powerful figures, there were doubts over the success of a national organization trying to increase cotton consumption; some interests there had reservations about Johnston, who had served in Roosevelt's New Deal. As stated keenly in one account, "Texas was the land of opportunity and booby traps."[24]

Lone Star cotton interests were not from the same mold as the Mississippi Valley growers and businesses; geography alone had made them independent. The Texans raised mostly short staple cotton, particularly on the High Plains. Too far from the textile mills in the Southeast, they exported much of their crop through Texas Gulf ports, and their capital came from western creditors. Distance from the mid-South also held down their personal acquaintances there. These circumstances gave them a different perspective and made them fiercely independent. But Clayton ranked as one of the best-known and most persuasive proponents of international trade, and for cotton to overcome its long history of division and reestablish itself as a viable partner in world commerce, he would have to come into the fold. Texas would throw its weight behind the proposed organization only if Clayton went first.

Johnston met with Clayton in Dallas to win him over. Both agreed on the importance of trade and that government price supports hurt cotton in the competitive global market. Johnston agreed with Clayton that if the government wanted to assist farmers, it should pay them with cash handouts rather than manipulating the market with subsidies and keeping U.S. cotton prices above the world price. But Johnston won Clayton's endorsement only when he explained that each segment member of the organization would have veto power over the others. This proviso, which Johnston, Blake, and others had developed in view of the divisive nature of the industry, meant that merchant companies such as ACCO would not be forced to follow NCC decisions if they, reaching a two-thirds agreement among themselves, chose to exercise their veto. This made membership acceptable to the Texas recalcitrants. Once Clayton came on board, Texas followed. Even the most conservative saw the merit of a central body unifying the industry: "I think ... the problems that the ginners from the cotton growing states encountered when they went to Washington," stated one cautious Texan, "caused them to feel that there was a great need for a national organization."[25] Johnston moved forward to hold the meeting for the formal creation.

The site for the historic meeting was the Peabody Hotel in Memphis, where, according to the southern writer David Cohn, "the Mississippi Delta begins."²⁶ Sitting at the northern tip of the Delta, where the rich mixture of white southern patriarchy and black culture fused into a steamy city of warehouses, wharves, and blues joints, Memphis pulled together the elements of the cotton South. From across the river and westward into Arkansas Delta, north from the Missouri Bootheel, and south from the Yazoo River, all manner of people flowed into the city, sharecroppers seeking temporary relief from their lives of drudgery and the landed gentry enjoying the city's lavishness. Cotton was the city's mainstay, and cotton was everywhere, stored in warehouses, loaded and shipped on barge, rail, or truck. At the heart of downtown on Second and Union streets, one block from the Memphis Cotton Exchange, sat the Peabody in its splendor and grandeur, the premier hotel of the mid-South. Popular big bands, the legendary likes of Tommy Dorsey and Paul Whiteman, played at the hotel's Plantation Roof Club.

In the Continental Ballroom on the mezzanine floor of the Peabody, approximately 150 delegates and attendees answered Johnston's call to meet on November 21, 1938. Designed in the style of Italian Villa Revival, the large rectangular room had a vaulted ceiling covered with plaster fretwork and large medallions. Along the side of the room overlooking the Peabody's lobby were French doors, each set within a Renaissance arch. A raised entry for grand arrivals stood at one end; it had ten Corinthian columns and twin staircases descending to the ballroom floor. From here belles of the city's elite families gracefully presented themselves to society. Across from the French doors, a small stage provided space for bands and small orchestras. It was the site for debutante balls and highbrow charity fund-raisers.

The elegance of the Peabody did not camouflage the seriousness of the meeting. Worried about the future prospects of cotton, the delegates eagerly waited to hear Johnston. After the opening remarks, Johnston took the stage, and a sense of historical significance filled the air as he began to speak. "The animals you see now in museums are animals that couldn't survive as conditions changed. The cotton business, if does not meet the changed conditions, will soon be as extinct as the dodo."²⁷ Johnston blamed only cotton for its pitiful condition, castigating the industry for depending on government to come to the rescue. "We have been too prone in the last few years to say, let's pass a law and look to Washington for help."²⁸ Continuing to wait for better conditions would be futile, he thought, because only bootstrap action could pull the industry up and move it forward to meet the challenge from synthetics and paper, which could be expected to shove the nation's major crop into oblivion if they were ignored. For anyone questioning the importance of the moment, he reminded them that "cotton, not the South, is the No. 1 problem of the nation."²⁹ He reiterated the need for unification and recommended the adoption of the resolutions that would be reported that evening.

It was a dramatic delivery that created an uplifting excitement, coming from a man known for his speaking ability, giving the most important speech of his career.

The rest of the meeting seemed anticlimactic. Delegates passed the resolutions, pledging to strive for increased domestic consumption and international trade. They saw research as essential to answering the challenge from competing fibers. To fund the organization, an assessment of two cents per bale would be levied on producers who joined, and members from other segments had assessments at different levels. They hoped to raise $240,000 annually. The *Commercial Appeal* reported the meeting "would go down in cotton history."[30]

The need for unification drove the founders. Skepticism about the future dependability of Congress and questions about the effectiveness of the federal support structure motivated leaders to organize and overcome their overriding weakness: lack of unity. For too long the various segments had preyed on one another, each defining itself by its own niche in the transformation of raw fiber into finished goods. Creation of a national organization, they agreed, would correct that flaw because now producers, ginners, warehousemen, crushers, and shippers agreed to work together for their common good. Through the NCC they achieved consensus, an agreement recognizing the value of serving each other's interests. A traditionally fragmented industry, if organized to speak in unison, could become an effective lobby, particularly since growers offered political clout by the sheer size of their numbers. A unified Cotton Belt could yield power because the different segments working as one organization with the backing of 1.5 million farmers would present a formidable front. The segments could also pool their resources and fight synthetics as a team rather than as splintered interests.

The actions taken in the Peabody's Continental Ballroom had been in development for about two years, and much of the credit went to Johnston, whose "driving spirit, blunt logic, and sharp interpretation of the southern problem," the *Commercial Appeal* reported, enabled him to articulate and make plausible a philosophy that had been denigrated or ignored for a generation or longer.[31] He did not achieve this feat alone, because the movement gained strength from the diverse elements of the industry, which acted against traditional and tired images of a defunct southern agrarian culture. Fear and necessity drove the founders as much as vision, but it was obvious to them that a new approach was required to fight for the economic livelihood of those southerners whose well-being was governed by cotton.

The advantages of cotton's interests working in unison through a central organization appealed to the textile mills, and when in 1941 textiles joined as the sixth member, unification from field to fabric became complete. Textile spinners had naturally focused on promoting their own finished products instead of raw fiber, but owing to the salesmanship of Johnston and Lewis T. Barringer, a cotton merchant in Memphis, C. A. Cannon of Cannon Mills became one of the textile industry's first believers in the new organization. "Cannon recognized the political clout of farmers," wrote Albert Russell, later executive director of NCC.[32] Early support also came from Bibb Manufacturing, Union-Buffalo, and W. N. Banks of the American Cotton Manufacturing Association. Securing their membership represented a major feat because "there were a lot of suspicions involved," recollected Robert Jackson, the longest-surviving attendee of the Peabody meeting. "The cotton farmers generally looked on the textile industry as this great big industry that was trying to buy the cotton as cheaply as possible," Jackson continued, and textile leaders suspected that cotton's leaders were resorting to legislation "to get their prices at abnormally high levels."[33] Nowhere had the industry's misunderstandings been more glaring than in the mutual suspicion of farmers and textile makers. Their joining together indicated the overthrow of the old order.

A VOICE FOR COTTON

Bad luck struck the NCC from the beginning. World War II broke out in 1939 and challenged the new organization to establish its legitimacy under wartime conditions. Original objectives had to be postponed or reduced to cope with the stringent restrictions imposed by wartime agencies on farmers' and cotton's related interests. Staff members such as Read Dunn, who enlisted in the U.S. Navy in 1942, were lost to military service. Others included MacDonald (Mac) Horne, who was in charge of economic research; Ernest Stewart, who was responsible for sales promotion and publicity; and Calvin Johnson, Clark Waring, and Gene Holcomb.[34] Advertising and sales promotion had to proceed on a minimal basis, since the NCC had to devote considerable energy to lobbying federal agencies for producers and processors needing materials in short supply: steel for making bands for bales, jute for wrapping bales, and insecticides for combating pests. The organization cooperated with the Quartermaster General in advertising cotton as a vital wartime commodity; it boasted that cotton was second only to steel in strategic importance.[35] Advertising cotton as a strategic item for the military enhanced its essentialness, but the benefit was temporary, and the need remained to increase consumption in the civilian market. Lobbying grew spontaneously as NCC personnel conversed privately with members of Congress or testified before committees, and the immediacy of the issues required the organization to keep personnel in Washington on a regular basis. Sam Bledsoe was the principal contact with Congress; when he left the NCC in 1946, Robert (Bob) Jackson, director of production and marketing, took over. Jackson assisted Bledsoe during the war and always stayed at the Mayflower Hotel.[36] Rhea Blake, the first executive director, frequently traveled from Memphis to assist with lobbying.

Lobbying proved indeed to be the most gainful labor of the new association during the war, and the most challenging and successful example dealt with the passage of the Bankhead-Brown bill. The Office of Price Administration (OPA), a temporary wartime agency created to fight inflation and price gouging, had placed ceiling prices on cotton textile goods. But the price of raw cotton had no restriction, which allowed it to climb, though only modestly. In 1943 the textile manufacturers asked the OPA to adjust textile ceiling prices, but it refused. Textiles went to the NCC for help. The budding organization accepted the challenge because textiles had become a segment member in 1941, but the OPA practice also lowered the price of raw fiber. From this perspective, the NCC felt emboldened to act on behalf of the textile manufacturers' request, and Blake, Bledsoe, and Jackson took over as the taskmasters. Bledsoe offered a compromise to Leon Henderson and Paul Porter of the OPA, but they "just sneered and laughed at me."[37] Senator John H. Bankhead of Mississippi and Representative Paul Brown of Georgia, whose constituencies had much at stake in the predicament, sponsored a bill to establish a mechanism for adjusting textile ceiling prices. When Bankhead threatened Jimmy Byrnes, director of economic stabilization, with public meetings across the Cotton Belt to explain how OPA held down the price of cotton, OPA backed off, and Congress passed the measure in 1943 over opposition from the White House.[38] Support came from the Farm Bureau and the powerful southern delegation in Congress. "It was a major breakthrough," Jackson later recalled, that kept textile companies solvent.[39]

Legislative favors and kind treatment from wartime agencies owed much to the South's powerful delegation in Congress, which sought to protect the "fabric of history." Leadership for obtaining higher parity rates went to Senator Bankhead, but he drew on the support of his fellow southerners in Congress. Some congressional districts relied on cotton as the principal commercial crop, and portions of some states such as the Mississippi Delta or eastern Arkansas depended on it almost exclusively. The NCC kept members of the so-called Cotton Bloc informed of its efforts and recruited their assistance whenever action taken by agencies offended cotton interests. For obvious reasons, one economist wrote, "congressional leaders from the Cotton Belt attempted by legislation to give cotton growers price advantages similar to those of farmers growing products needed in increased quantities during the war."[40]

Although these wartime efforts gave the ailing King Cotton new strength, measures taken on cotton's behalf did not always succeed. An issue arose in early 1942 when the War Production Board (WPB) approved the use of high-tenacity rayon over cotton for the manufacture of tire cord for military vehicles and aircraft; the board granted a second request before the end of the year. When in 1943 the WPB prepared to grant a third request, which would raise the rayon allotment from a prewar level of 25 million pounds to 200 million, the NCC squawked. It saw the allotment as an unfair incursion into cotton's right to open-market prices, a belief reinforced in 1943 when the Truman War Investigation Committee reported that based on trial tests, cotton proved to be as durable as rayon for use in military truck tires. Using this information, Johnston protested strongly to Donald Nelson, head of the WPB, and "demanded that cotton be given a fair hearing before it is condemned and executed in the tire cord market by the Government of the United States."[41] Johnston sent copies of his letter, with reproductions of the Truman committee's report, to members of the Senate, asking them to take "this matter up immediately with Mr. Nelson and Mr. Byrnes."[42] Despite its efforts, the NCC could not prevent the expanded use of rayon tire cord, which accounted for a portion of the total production of all synthetics, which doubled during the war.

A key ingredient in the ability of the NCC to speak persuasively in political circles rested in its knowledge of the needs and conditions of the nation's cotton producers, thanks in great part to the organization's Field Service. Generally overlooked owing to the natural tendency to focus on the glamour and intrigue of political lobbying, the Field Service from its beginning served as the council's lifeline with growers throughout the Cotton Belt. Acting as agents of the new organization, the employees of this branch, usually two per state, maintained contact with commercial planters and small-plot farmers as well as ginners. Agents recruited members for the NCC, who agreed to pay bale assessment dues. Gins served as the collecting point, a practice that Johnston and Blake had arranged in 1938. As they traveled their territory, Field Service agents learned about local weather conditions, insect infestations, labor supply, and other items related to farming. They took careful note of any political development that might impact the industry. Their information flowed back to headquarters in Memphis, where it was incorporated into crop forecasts, economic analysis, and proposed legislation. These agents were the glue and shoe leather of the NCC and enabled it to have updated information, so that its lobbyists and representatives spoke with accuracy and commanded respect for their knowledge of conditions of cotton farming. During the war the Field Service operated with a reduced staff, but Blake knew that it had to be brought up to full strength as soon as possible.

During the war, the price of raw fiber varied little, from 18 cents in 1942 to 22 cents in 1945, but the Cotton Belt benefited from the war. Approximately three million southerners migrated into other regions of the United States, and a sizable number moved into cities. Sharecroppers and tenants began leaving the fields to take advantage of new employment in cities, reaching such a level that growers, large and small, experienced labor shortages. The South had a 25 percent drop in farm population, and in October 1944 the *Cotton Trade Journal* reported that the "labor shortage retarded cotton picking."[43] Oscar Johnston had a loss of tenants on D&PL, but other planters lost labor, too. In some instances, German prisoners of war were put into the fields. Empty tenant houses littered the Delta, Johnston wrote, and he "despaired of ever being able to fill these houses with farm labor."[44] Despite this hardship on growers, the migration eased the burdensome oversupply of farm labor.

Regardless of their acreage, growers diverted land to food and feed crops. Johnston started running cattle on D&PL, devoting five thousand acres of plantation land to pasture eleven hundred head of beef cattle. He began switching acreage to feed and hay crops—oats, corn, and alfalfa—to feed out calves for the beef market. A spike occurred in the price of peanuts, rice, and wheat that encouraged shifting. Cotton nonetheless remained profitable, since Congress instigated a strong market through the CCC. Shortages of labor caused a reduction in planted acreage, but annual production remained steady from 1941 to 1944, rising from 10.74 million bales to 12.23 million.[45] Availability of fertilizer, better varieties of seed, use of fertile land, and generally favorable weather accounted for the strong yields. From a social perspective, the war eased the pain and drudgery of life for thousands of sharecroppers and prodded landowners to plant other crops.[46]

CONCLUSION

By the war's end, the industry stood poised to exert itself in a manner unknown since the antebellum era of the nineteenth century. A new breed of planters who did not fit the stereotype depicted in the literature of the Depression started pushing their interests in a new direction. They espoused some of agriculture's most progressive principles: the importance of scientific research, an emphasis on consumption, and a renewed international trade order that rejected the protectionism of the past. They also emphasized a new concept: commodity advertising. No longer would the cotton community follow the ancient practice of passively depending on an unfettered market to determine its well-being. Now it intended to disregard old and burdensome practices and seek to overcome the impotency of the past by looking only in a forward direction. Suspicion and mistrust among the industry's segments had begun to fade away, and they hoped the outreach of the NCC would provide the belated fulfillment of the South's long quest to bring prosperity and dignity back to the cotton kingdom. This commitment to self-reliance seemed hypocritical because of the dependence on government support, but there was an expectation that at the war's end, conditions would begin to improve and support could be phased out.

Despite the wartime gains made by the NCC as a budding new organization, King Cotton remained weak. Across the Cotton Belt, skepticism persisted about the future because the conditions responsible for the unhealthy state of affairs had not been overcome: production of artificial fibers had grown during the war; overseas markets were drastically reduced or thought to be forever lost; much farmland remained marginal; mechanization had lagged when shortages of labor seemed an omen of further migration off the land; and boll weevils persisted as a costly nuisance. Demobilization of the military was expected to send young men back to the farm, which would increase acreage plantings and make the surplus worse. Diversification into other crops was seen to be a short-lived undertaking for the sake of the war. By mid-1944 a feeling of unease began seeping through the Cotton Belt, for though the creation of the NCC marked the assertion of the industry to fight back, threatening circumstances persisted. Unity and organization did not appear to be enough to overcome the mounting dangers. In December 1944 an extraordinary gathering of cotton interests and influential government figures looked for solutions for the ailing King Cotton and its fiefdom.

CHAPTER 3

THE COTTON CONFERENCE

"The American Cotton Shippers Association appreciates the opportunity of participating in this meeting, whether it is an autopsy into the corpse of King Cotton, or will lead to a clear diagnosis of his ills." This brutal statement succinctly described the atmosphere of the Cotton Conference, hosted by the Subcommittee of the House Committee on Agriculture. It occurred in December 1944, about two weeks before the German offensive on the western front, the Battle of the Bulge. No meeting of such magnitude had ever occurred on behalf of cotton. It brought together all segments of the industry from field to fabric, as well as railroads, banking, labor, consumer groups like the National Consumers League, and officials from several agencies, including the USDA and the Department of State. With the prospects for the United States' chief global commodity growing dimmer in 1944, the conference participants hoped to identify the causes of the malaise and work out a policy that would fit all the pieces of the cotton puzzle into a coherent and feasible solution. The list of witnesses represented the best minds dealing with cotton and bestowed on the conference a level of stature uncommon for a congressional hearing on agriculture. Subcommittee chair Stephen Pace had arranged the meeting, and witnesses and interest groups came with their own prepared analyses and recommendations. He had asked them to examine every aspect of the industry, from soil conditions to the finished product that went into retail shops.[1]

From the federal government came Secretary of Agriculture Claude Wickard, Assistant Secretary of State Dean Acheson, O. V. Wells from the USDA Bureau of Agricultural Economics, and C. C. Smith from the Commodity Credit Corporation. Spokesmen for powerful planters attended, and the Farmers Union testified on behalf of small farmers. Pace had arranged for H. L. Mitchell of the Southern Tenant Farmers Union to speak, but Mitchell did not appear when Pace invited him to the podium. Several observers from foreign countries were there. Almost 225 persons came from over twenty-five trade organizations and government agencies. About sixty spoke as witnesses or submitted statements. Vice president elect Harry S. Truman, still chair of the Senate's War Investigating Committee, temporarily relinquished his hearing room to the subcommittee to accommodate the large gathering. Attendees had the privilege of asking witnesses questions as part of the general plan "to exhaust the subject and develop those proposals which are sound and expose the weaknesses of any which are unsound or impracticable."[2] The chairman of the full House Agriculture Committee, Democrat John Flannagan of Virginia, best described the objective of the gathering. "We are glad to look at the ills of old man King Cotton, who we all know is pretty sick. . . . If you will work out the right prescription then the House Committee on Agriculture will do everything within its power to fill that prescription."[3]

Pace represented the Third District in southwest Georgia and had practiced law and served in the state legislature before being elected to Congress in 1937. Like many southerners, he farmed the land and wished to see prosperity restored to agriculture. Macon County, the subject of Arthur Raper's revealing 1936 study of black poverty in the Cotton Belt, abutted Pace's district, which had many of the debilitating characteristics exposed in Raper's book. Pace thought it imperative for the United States to devise a way to rescue cotton, and he conveyed a sense of urgency to the attendees. Cotton faced an "impending peril," and "the conference is destined to be historic, either by reason of our accomplishments or by reason of our failures."[4] His plea for remedial action expressed the anxiety over global developments that were changing the international cotton market, but the most worrisome concern was the twenty-two million bales of world stocks, despite the importance of cotton as a strategic commodity in the war. Witnesses and subcommittee members alike understood the magnitude of their task and the ramifications of their discussions. As stated by Secretary of Agriculture Wickard, "What we often refer to as the cotton problem has a much broader base than that single commodity. We must think in terms of southern agriculture and the welfare of southern farm people."[5] Dr. Claudius T. Murchison, president of the Cotton Textile Institute, hoped that "historians a century hence will be able to refer to this week's cotton congress in Washington as a turning point in the life of the Nation's greatest agricultural commodity."[6]

The Pace Committee faced an almost insurmountable task because "cotton is engaged in a fight for its life," wrote the Memphis *Commercial Appeal*.[7] It was for this reason that leaders of the industry along with representatives of Congress and the executive branch assembled on Capitol Hill and took a broad perspective in their approach. They knew that some of the nation's most profound social and economic problems were rooted in the Cotton Belt, stretching from the Atlantic to the Pacific over a variety of topographical and climatic conditions. Some problems of the industry went back several generations, and to devise a remedy satisfactory and effective for all vested interests was impossible in a short time. "It is generally agreed," Pace had written a short time earlier to Will Clayton, administrator of the War Property Surplus Administration, "that the future of cotton is probably the most serious and complicated of all the commodities."[8] Clayton had opposed federal price supports since their inception in 1933 and was not optimistic that the conference would generate a solution. His voice and sense of authority extended through the South, and he encouraged Pace but added a warning: "I can only continue to warn of the disaster which is fast overtaking us and point to the right road. Time runs out fast now."[9]

First to testify after the opening remarks was O. V. Wells of the Bureau of Agricultural Economics. He presented an assortment of figures compiled by researchers in his agency. His data were noncontroversial and served as an accepted base of knowledge in the discussions that followed. Along with his statement, Wells used charts, graphs, and maps to describe the conditions in the raw cotton industry, with information on production, prices, acreage, imports and exports, foreign competition, the growth of synthetic fibers, and cottonseed consumption.

He demonstrated that acreage had declined since 1929 by slightly more than 50 percent, from 44.45 million acres to 20.47 million in 1944, but yields had improved about 60 percent, or about 100 pounds of lint per acre. Favorable weather, greater use of fertilizer, better success in combating boll weevils, and the use of improved varieties of seed accounted for the greater yields. Average staple length, a chief factor in determining quality and price, had gone from 7/8 to 15/16 inch since World War I. But exports had fallen as foreign production went up, perhaps the change with the most impact. Wells avoided the accusation that the New Deal program of reduced production had caused a surge in foreign competition. He demonstrated that in 1943 per capita farm income in ten southern states averaged less than half the average farm income in the remaining states of the Union. That evidence alone showed how the penurious cotton economy encumbered the general welfare of the United States. One fact offered by Wells illustrated the long-standing frustration of all farmers: the value of the raw product as a percentage of the final cost to the consumer. In 1944 the value of raw lint was 7.5 cents of each retail dollar spent on finished cotton goods.[10] This array of facts and figures, making up forty-four pages in the official record, substantiated common knowledge and expressed the growing conviction that new policies and approaches were needed to restore prosperity to cotton farming.

THE RISE OF SYNTHETICS

Cotton now had a new and unique slot in American agriculture that explained much of the attention it received for the rest of the century: synthetic fibers threatened to replace America's chief commercial crop. No other staple, wheat, corn, or rice, faced a loss of market. Invented in 1884, rayon had come onto the retail market in the 1920s. It quickly gained popularity with its lightweight sheer fabrics with satin qualities and gave extra strength to carpets, draperies, and tire cord. In 1924 rayon consumption amounted to only 2 percent of the domestic cotton market, but it reached 20 percent by 1945, equal to approximately two million bales. Consumption was double that amount in foreign countries.[11] The flowing characteristics of synthetics had made them ideal for the new and sleeker fashions of the twenties, and consumers associated cotton with less-fashionable and inexpensive clothes. This development caused P. K. Norris of the USDA Foreign Agricultural Relations to remark: "I, like you, get quite a chill out of the thought of the future. I think we all get chills from it."[12]

Synthetics had begun eroding cotton's market. The rising popularity of artificial fibers seemed destined to challenge the supremacy of cotton as the fabric of American life. This threat, growing more dangerous each year, explained why activists sought to overcome the indifference of the past and find new rationales that would maintain cotton as a viable force in the economy. C. K. Everett of the Cotton Textile Institute expressed the unease: "I think we must anticipate that the application of rayon fabric and other fibers is going to grow, and we have got to take those things into account now on an occasion of this sort."[13]

FOREIGN TRADE

Foreign trade was a major topic of the conference because cotton had been the country's chief export since the early nineteenth century. Until World War I, the South provided about 75 percent of the world supply, but the percentage started falling soon after the war owing to new foreign producers, until by 1939 the United States accounted for approximately 47 percent. The international market had long been essential to the South and gave it a global perspective not common in American business. For good reason, the Cotton Belt had shown strong support for the Roosevelt administration's reciprocal trade program in the 1930s, but now with American cotton losing its competitive edge in the world, the potential losses for the economy were significant. Accordingly, the witnesses at the Cotton Conference devoted much attention to exports. Oscar Johnston spoke principally for cotton but thought trade in all commodities and products was "fundamental and the most important step" in raising living standards around the world. "Because of cotton's tremendous importance in world markets," he said, "we strongly feel that the United States cotton industry should be represented at any and all conferences dealing with the international economic questions of the future."[14]

The structured and predictable world of cotton underwent fundamental changes after World War I, however, exacerbated by the Depression and government policy. During the 1920s the world remained on the gold standard, and since the United States had acquired creditor status during World War I, gold naturally flowed into the U.S. Treasury. This flow had an insidious effect: it enhanced the financial stance of the economy and strengthened the dollar but steadily eroded the buying power of other countries. At the same time, our protective tariffs further reduced the ability of other countries to sell in the American market, thereby limiting their ability to obtain dollars for buying American goods. The U.S. practice of extending loans to Europe, as well as the large expenditure of funds through private portfolios, pumped money into foreign governments and kept the system afloat. Cotton remained a popular item among foreign buyers, and the industry enjoyed hefty overseas sales, peaking at 10.93 million bales in 1926. The principal overseas markets were located mostly in Liverpool, Bremen, Le Havre, Milan, and Osaka.[15] The United States shipped 6.69 million bales overseas, or 53 percent of the world cotton exports in 1929.

The international financial system dependent on U.S. loans finally collapsed that year, and the Great Depression began worldwide. "Conditions were in the making," wrote one historian, "which led to the almost complete debacle of international trade at the end of the decade."[16] Exports of cotton continued, but as the Depression lasted, the U.S. percentage of world exports slipped. By 1937 it stood at approximately 47 percent compared with 66 percent in 1918, though the total number of U.S. export bales varied little, 5,598,000 and 5,592,000 respectively. New competitors had appeared, particularly Brazil, which produced two million bales for export annually. Other countries had begun supplying their own needs, and the United States' share of the international market shrank. In 1940, U.S. exports fell to a record low of 1.11 million bales because of the war.[17]

As cotton production climbed around the world, the carryover became larger until a surplus appeared. By 1944 the carryover expanded to 10.74

million bales, with the world total production hitting 25.4 million bales when world mill consumption was 22.2 million.[18] The surplus had begun growing like a malignant tumor in the 1930s, and its ramifications were staggering.

It killed the price. The glut on the international market ended the climate of predictable prices and assurance that cotton farming would remain a way of life in the South. As the United States increasingly developed a consumer economy, the fall of cotton prices, with only a modest recovery during World War II, placed the rural South on a dangerous plateau. Growers of all shapes and sizes found themselves competing with countries that had lower production costs, though the practice of sharecropping kept U.S. labor costs at a minimum.

James Hand, representing the Delta Council, testified that exports had fallen because of the inability of foreigners to sell in the American market due to measures such as the Smoot-Hawley tariff, and unless they were able to sell their own products and acquire dollars for foreign exchange, overseas buyers could not purchase U.S. cotton. It behooved the country to develop arrangements that would increase international exchange because "it is self-evident that we cannot sell abroad if our foreign customers cannot sell here."[19]

Among the attendees of the Cotton Conference was a determination not to build a system that would be just as vulnerable as before. There could be no viable industry without realignments and adjustments, and this thinking was especially strong toward foreign trade. War had worsened an untenable situation because two of the industry's principal markets, Germany and Japan, were lost, and sales to allies such as Britain and France had dropped dramatically. Bolder minds within the cotton community felt that a resumption of prewar practices would not suffice; they wanted trade policies to change because of the country's new status as a creditor nation. In cotton-growing areas, the trade barriers of the 1920s and 1930s were unpopular, and interests there wanted every effort made to promote free, or freer, trade to increase exports. This sentiment had begun to appear among other interests of the economy by late 1944, but no industry equaled cotton in this regard.

FEDERAL ASSISTANCE

In the course of testimonies and discussions over the five days of the conference, much attention focused on a new and compelling force that had entered the world of cotton, the federal support programs. The cotton farmer had a new ball to juggle along with weather, boll weevils, and foreign competition; he had to incorporate the complex set of rules and regulations associated with the support programs into his management decisions. This support had begun during the Hoover years with the inauguration of the Federal Farm Board and followed by the AAA crop reduction program. Other forms of assistance came during the 1930s from the Soil Conservation Service, the Farm Credit Administration, the Bankhead Act, and the controversial Farm Security Administration. State governments operated experiment stations, education programs, and various services pertaining to production and marketing. But the chief consideration from the growers' point of view was the manner and method of federal price support, which changed frequently. Farmers and related interests always criticized federal largess, but in this case they relied on it.

For those people dependent on cotton, over ten million, including the ginners, brokers, warehousemen, shippers, and spinners, price supports meant the difference between success and failure. Without a reasonable level of prices, farmers could be expected to abandon cotton, which would injure the American economy. Such a development would jeopardize the nation's supply of the principal fabric used by mankind, so by the time of the Pace Committee hearing, only a small minority questioned the need for continuing federal supports in some fashion or another. There was an undercurrent of resentment by southerners who philosophically opposed government intervention, but no outcries or vehement denunciations were heard at the conference. Witnesses hoped to see cotton return to prosperity and operate solely in a free market, but they typically recommended, allowing for variations, that government subsidization be maintained for a "transition period" of approximately five years.[20] Government assistance had become an integral part of the cotton culture, an unavoidable crutch until full health returned.

Opposition to subsidies was not new and represented a minority view at the gathering, but it nonetheless presented a compelling argument, that with the rise of artificial fibers and new suppliers of raw cotton around the globe, the price of U.S. raw fiber must stay competitive at world levels. Opponents pointed out how synthetic filament encroached on cotton's market whenever the U.S. support exceeded the world price by only a few cents. Nearly everyone at the hearing expected the world use of synthetic fibers to expand, but the speed of that development would depend "upon the price of cotton in relation to competing materials," according to the Interbureau Committee on Post-War Programs.[21] If the U.S. domestic price remained pegged at levels above world prices, exports should be expected to fall. From this perspective, price supports worsened the long-term outlook for cotton.

Subsidization reflected the dilemma. Price supports originated, of course, for the sake of agricultural relief during the Depression as a temporary measure until the world economy recovered, but since prices for raw cotton remained below parity, and because competition from synthetics and foreign producers was expected to continue, maintaining the subsidies appeared necessary to avert financial collapse of the country's major cash crop. Consumption of cotton had increased, but so had the world's annual supply; synthetics steadily grabbed more of the consumer dollar. To overcome this perplexing circumstance, conferees urged the following practices as steps to more efficient farming: soil conversation, crop diversification, and mechanization. Some wanted a shift of acreage to the West, where larger commercial operations were common and cotton could be grown at a price competitive on the world market. Secretary Wickard spoke for many when he proposed keeping supports during a transition period until such changes were accomplished. Even the most ardent opponents of price supports saw the wisdom of keeping them until farmers became more competitive with foreign producers.

COTTON POVERTY

Overshadowing the conference was the misery of sharecroppers and the poverty of small landowners. Many of the witnesses had direct experience with tenant farming, while the rest were familiar if not directly involved with it. Originally the perceptions of sharecropping were grounded in the conviction that, however mixed its character, there was more good than bad, because although it offered little advance for tenants, it gave them refuge in a world that offered no alternatives. No one at the conference suggested that sharecropping would thrive or that it was still sustainable, particularly after the past generation of deteriorating conditions. "Just as the Nation could not endure half slave and half free in the political sense," stated Russell Smith of the National Farmers Union, "so in the economic sense a mass-producing economy can only exist when the serfdom of poverty gives way to high mass-purchasing power."[22]

Conferees agreed that a surplus population lived on southern farms, but thought no amount of relief assistance or price support programs would overcome the poverty of the marginal landowners. Slightly over 50 percent of the cotton farmers, commonly relying on exhausted soil, produced four bales or less per year; such low volume would never generate enough revenue to sustain a household even if cotton prices reached parity.[23]

From the perspective of the conferees, sharecroppers and small-plot farmers could not remain on the land under present conditions, and if mechanization developed as anticipated, they could only expect to be displaced. Attendees agreed that only further industrialization of the South offered hope. Further migration of tenants into cities at the war's end appeared certain, a view reinforced by the growing number of people leaving farms. Small landowners remaining on the soil would need to diversify by concentrating on grains, livestock, and food crops. The Pace Committee witnesses regarded industrialization and crop diversification as the best future for the cotton peasant; Johnston regarded Ohio as the state with the best balance of agriculture and industry.[24]

THE GROWTH OF THE WEST

Conferees saw portents of change in the rising importance of cotton farming in the western states. Since 1910 acreage had been moving slowly westward, making the industry more complex and diverse than was generally known or established in social literature. Cotton had been regarded as a southern crop, and understandably so, but in California, Arizona, and New Mexico, production increased, and some entrepreneurs had established large-scale operations. From a small beginning, they had continued to expand their holdings of land; by 1943 westerners grew 563,722 bales, or about 5 percent of the U.S. total. They rarely used sharecroppers.

The economic and social characteristics of western cotton farming differed from the patterns established in the South. Small family plots producing one to five bales were practically nonexistent in the West. Owners there conducted their operations on a larger scale due to the availability of large tracts of land, where they formed long straight rows to facilitate the use of mechanized equipment. For this reason the use of machines had begun there first, but in 1944 growers depended on migrants, generally from Mexico, to handpick the crop. Weeds and grasses were comparably less troublesome, so the other labor-intensive activity associated with the cultivation of cotton, hoeing, was minimal in the West. Long staple predominated, and in California only one variety was permitted, Acala. It was premium cotton, but not strongly resistant to boll weevil depredation and thereby not appropriate for the southern states. Acala grew well in California, however, and in 1925 the state had inaugurated the "one-variety" requirement for the cotton counties in California's San Joaquin Valley. Riverside County, in the lower state, did not join the one-variety district.[25]

Without sharecropping, growers in the West practiced industrial agriculture. With large tracts of land, greater use of machinery, irrigation, wage labor, and no worry over boll weevils, they escaped many of the drawbacks of cotton farming. Westerners had insect marauders, particularly the pink bollworm, and their own marketing drawbacks, but they sought to maximize production by making their operations capital intensive. They saw themselves solely as investors because California was a prosperous and dynamic state, and cotton farming granted no particular esteem there as it did in the less-prosperous southern states. The "badge of poverty" associated with cotton did not apply to the western areas, nor did the West bear the cultural stigma of the cotton South. Westerners did not share the paternalistic views of their southern compatriots and did not practice sharecropping, so the human pathos commensurate with cotton farming escaped the West. Farming there had distinct advantages that could not be transferred to the South because of weather and geographic conditions, but southerners watched developments in the West for practices to implement into their own operations. In late 1944 the western states stood on the threshold of a larger role.

Few representatives of the far West went to the Cotton Conference, which caused it to reflect strongly the views of southern interests. David Davidson, a California grower and chair of the state's AAA committee, spoke on behalf of many individuals and groups at the county level. He concurred that cotton was in worse shape than wheat, rice, corn, and other commodities, and congratulated the Pace Committee for holding its special hearing.

Davidson felt that adjustments were inescapable if cotton was to have a future. He worried that the export market faced extinction, and wanted the United States to promote international trade agreements to stimulate overseas sales. Government restrictions on acreage should remain for the time being, along with supports for domestic prices, but he hoped the industry would overcome its dependence. Davidson recommended price guarantees for a larger number of crops to encourage small farmers to move away from cotton. Research into mechanization and consumer uses of cotton should be encouraged, but he doubted that improvements would affect the price of raw cotton. And while he observed the plight of the South's farm population from afar, he felt that the downtrodden should not be encouraged to remain on the land. They deserved help with relocation, Davidson testified, but he carefully avoided specific recommendations.[26]

MECHANIZATION

The conferees agreed that however burdensome the conditions of the past might be, a new force looming in the future held much promise: mechanization. Southern farmers had lagged behind their counterparts in other regions in the adaptation of, and conversion to, mechanical power, a condition attributed directly to the prevalence of cotton. By its nature, the cotton plant presented barriers to mechanization, especially picking the fiber from bolls. Removing weeds and "chopping cotton," the thinning of weaker or crowded plants, required hoeing by hand. No substitute for the backbreaking labor of picking had been found despite nearly a hundred years of efforts by inventors, agriculturalists, and farm shop tinkers. Not since Eli Whitney's invention of the cotton gin had a major mechanical breakthrough occurred. This technological lag, seemingly beyond anyone's control, was responsible for the extraordinary practice of sharecropping in the South. By late 1944, however, change was expected with the invention of mechanical pickers.

Mules remained the principal source of power in cotton cultivation, but the use of tractors had grown remarkably. During the late 1930s, rubber tires became readily available on tractors and made them less cumbersome and more convertible for various uses, so farmers throughout the South found the tractor quickly adaptable to crops other than cotton. During the war, southeastern farmers had begun a noticeable move away from cotton and concentrated on other crops because of the advantages that tractors offered. The Census Bureau reported that in 1945 the "ten cotton states" had 407,400 tractors, which was almost double the figure of 223,300 for the same area in 1940.[27] Much of the growth occurred, however, on large tracts of land in the Texas plains. And though tractors could be used for breaking ground and seeding, they could not overcome the chief reasons for the use of hand labor, hoeing and picking cotton.[28] An age-old dream of the thousands of persons, both black and white, who pulled long sacks along the cotton rows and stuffed them with the raw fiber taken from the bolls, toiling under a broiling sun and enduring the South's humidity, was a machine that would perform the task, a mechanical cotton picker.

Secretary of Agriculture Wickard stated that "mechanization of cotton production is inevitable," and Clarence Dorman, director of the Mississippi Agricultural Experiment Station, explained that research had already demonstrated the savings made possible with the use of tractors for preparing seedbeds, planting, and applying insecticides. A "pull-behind" picker, a forerunner of the first widely used picker, harvested one bale per hour and fifteen minutes during research trials. A few months earlier, a trial demonstration of a picker built by International Harvester near Clarksdale, Mississippi, showed much promise. For removing weeds, the other time-consuming chore, a device known as a "flame thrower" had been employed on larger plantations, but it had limitations. "These machines may or may not be improved to the stage at which they will be ready for widespread use," reported the Bureau of Agricultural Economics at the Pace hearing, so final judgment remained on their further development.[29]

These advances were expected to have a disturbing social effect, since they would reduce the manpower requirements for cotton farming. One study showed that 235 man-hours of labor were required per bale using mules and hand pickers in the sandy loam areas of the mid-South, compared with only 25 hours with tractors and experimental mechanical strippers on the Texas High Plains. A study conducted in the Mississippi Delta reported 150 man-hours per bale using conventional animal and hand labor, but the figure dropped to 30 man-hours when production was wholly mechanical. Projections of the reduction in manpower varied from area to area, depending on soil, farm size, weed growth, and type of cotton, but a dramatic drop was expected.[30] Tom Cheek of the National Grange foresaw thousands of people migrating into towns and cities with the advent of mechanization.[31] James Hand predicted a "national social problem" when tenants, sharecroppers, and laborers left farming. Oscar Johnston believed that a change for the anticipated displaced tenants and sharecroppers "is virtually certain to be a change for the better."[32] Use of mechanical pickers would obviously force workers out of the fields, so industrialization or migration away from the region offered their only recourse. In another respect, however, progress in mechanization would overcome the war-induced shortages of labor that were expected to continue afterward. Testimony before the Pace Committee indicated that revolutionary changes were expected in growing cotton, but that social life in the United States would be profoundly affected. "I fear and also really hope," one of the Vanderbilt Agrarians wrote about mechanization, "that more and more farmers are to be moved out of cotton production."[33]

REACTION TO THE COTTON CONFERENCE

The conference drew attention. The *Commercial Appeal* explained that Pace had convened the meeting because "some thoughtful leaders believe cotton has no future."[34] The *Cotton Trade Journal* reported that "no punches were pulled as industry's leaders gathered to discuss ways for helping King Cotton discard his crutches after the war."[35] The *Journal* reported the discussion of mechanization as a means of reducing production costs and related the general assumption at the conference that southern industrialization offered the best opportunity for the poverty-stricken thousands dependent on cotton farming.

An analysis of the hearing came from the Bureau of Agricultural Economics (BAE). It identified four areas of agreement among the conferees: incomes (cotton prices) were too low; further government price support would be necessary; efforts on behalf of soil conservation should be continued; and research should be expanded to make cotton more competitive with synthetics. From the BAE's point of view, however, disagreement more accurately characterized the conference because of the contrasting perspectives and solutions set forth to restore prosperity. Disagreements came not over goals but over the means to achieve them. "While no one objected to the development of mechanization," the BAE reported, "some doubt was expressed as to its probable significance in reducing the costs of production." In regard to generating trade through international agreements, "opinion was divided on the effectiveness of reciprocal trade agreements." The BAE reported that witnesses were divided over continuing price supports, with a few preferring to let "the price of cotton go to a free trade world market basis."[36]

Though he did not attend the conference, Will Clayton had long called for tariff reduction and a policy of moving away from price supports over a period of several years until the domestic price matched the foreign, but dumping the farmer into an unfettered market was not acceptable to him. Clayton encouraged mechanization and a shift of production away from the smaller plots in the Southeast to the large tracts of the West. Such a development, he thought, would enable U.S. growers to combat their foreign competition.

CREDIT STRUCTURE

As much as the conference encapsulated the ills of cotton into a single record, it seemed indifferent to the long-detested crop-lien system. Hardly any mention of the historic shortage of capital came forth, despite its place among the oft-mentioned causes of rural hardship. Testimonies and prepared statements by bankers and banking organizations dealt with credit only in general terms, tucking it away among their recommendations regarding tariffs, subsidies, marketing, and similar topics. This omission likely stemmed from two recent developments: the number of merchants who provided a "furnish" to producers, large and small, had declined because too many landowners had lost out during the Depression, making the old practice of extending credit based on next year's crop too risky; and the New Deal credit agencies had made borrowing easier. "The yoke of the archaic credit system began to be lifted at the same time that first steps were taken," wrote a congressman from the Yazoo Delta, "to stabilize the price of cotton above the poverty level."[37] Since the conference consisted of planters and related commercial interests, it spoke with the voice of the establishment, which had easier access to capital. To be sure, the Depression had touched commercial operators, driving some out of business, but the more adept, or lucky, had recovered by late 1944. Some merchandising houses "furnished" larger farmers after the war, but the expansive cotton economy of the era saw few foreclosures by merchants.[38]

CONCLUSION

The Cotton Conference intended to devise plans for the future rather than acknowledge weaknesses of the past, but since it did not generate a consensus for remedial action, the committee produced no recommendations for legislative action, and conferees saw social disorder about to explode with the migration of tenants and sharecroppers off the land. Despite the lack of agreement on future policy, there was "a clearing of the air . . . and the opportunity to get things off their chests," reported the *Commercial Appeal*.[39] Whatever the merits of the conference, an amazing range of grievances and aspirations was articulated in what amounted to the most comprehensive public forum on the cotton industry in modern times.

But the call for a prescription to salve the ills of King Cotton produced no medication, not even an ointment. Observers reported as much, indicating that the patient remained in a weakened condition with a bleak prognosis. The NCC had purposely not offered a slate of recommendations, hoping that a set of guidelines for making policy would develop. In this context, the days of testimony and questioning only reflected the division and special-interest selfishness that had characterized cotton since the Civil War. Pace had warned Clayton that "if this dissent and division continue the cotton industry may be destroyed."[40]

Although the conferees failed to reach an agreement, a bold spirit was engendered by the realization that the United States would emerge from the war as the world's economic leader. Cotton activists foresaw the United States expanding its hegemony and providing far-reaching opportunities for trade. Lamar Fleming, president of Anderson, Clayton and Company, wanted to redefine America's international role by eradicating protectionism and maximizing international trade through reduced tariffs. He also expressed the common objective to recapture the lost cotton markets of Japan and Germany, which had been two principal buyers of U.S. cotton before the war. An international broker, Fleming understandably had a preexisting tendency to focus on global trade, but he reflected the belief that the restrictive nature of trade in effect since World War I should be abandoned. As stated by Assistant Secretary of State Dean Acheson, the prewar uses of "trade barriers, quotas, exchange restrictions, and discriminations . . . brought about not only a severe shrinkage of international trade but also a wave of international friction and ill will."[41]

The subcommittee hearing did not amount to a meaningless series of statements venting the frustrations of the cotton industry; it was an expression of ideas and proposals to reinvigorate a major segment of the economy.

The South's blind faith in cotton had been shaken, and it was apparent that new attitudes and fresh energy were needed. Already the youth and vigor of the postwar era infected the industry as the burdens of the past were being shucked away and a desire for innovation began to appear. With the new year of 1945 only a few weeks away, the defenders of the cotton kingdom wanted the past relegated to the history books and the focus shifted to the future. "The widespread tendency to blame cotton for all the economic ills of the South is one of the nation's greatest mistakes," stated A. B. Cox, director of business research at the University of Texas, "in that it tends to prevent the adoption and aggressive development of constructive policies to save an industry."[42]

CHAPTER 4

A NEW ERA BEGINS

King Cotton remained weak and frail in 1945, but with a sense of vigilance its subjects intended to construct a new order to overcome cotton's stigma as the national example of poverty. A cold realization had set in that with the failure of the Pace Committee to develop legislative proposals, the roadway for the rehabilitation of cotton had yet to be constructed. Industry leaders resolved to remake cotton into a modern and vibrant part of the global economy, and they realized that with much of the world ripped apart by war, the United States had to take a leadership role in restoring the international economy. The industry knew that its welfare was connected with the renewal of trade, but it also realized that consumer demand had to be improved at home. Reformers recognized that they had to take action along a broad front: conduct research into more end uses of fabric; overcome the barrier to mechanization; compete with synthetics; develop improved seed varieties; and raise consumer demand. No task could be overlooked in this gargantuan endeavor because, as Herman Nixon wrote, cotton could no longer be "a cash economy without cash."[1]

It was apparent that customary plantation farming dependent on sharecroppers was fading and that small-plot agriculture was no longer viable. Survival required a new outlook infused by an awareness of the shortcomings of the past, but with the war winding down, cotton faced the challenge of moving forward without the artificial stimulation of defense spending. Could the industry restore its overseas markets, vital to keeping cotton the country's major cash crop? Would Congress be amenable to cotton's requests during normal peacetime conditions? Would the anticipated migration off the land by the tenant class jeopardize farming operations? How could the losses wrought by the boll weevil, the scourge of the South, be stopped? Throughout the Cotton Belt, from Georgia's small plots of sea island fiber to the irrigated fields stretching across California's San Joaquin Valley, each niche in the cotton culture knew its livelihood was at stake. For the sake of 1.5 million farmers, it was imperative to meet the challenge.²

Making cotton farming a way of life free of hardship and social stigma challenged even the most optimistic. Too many small landowners still attempted to eke out a living from exhausted soil, growing cotton inefficiently and producing so few bales that no program of assistance would lift them out of their straits. Cultivation practices still relied on hand labor and animal power long after mechanization had reached most of American agriculture. With the twenty million bales stored around the world, domestic prices remained shy of parity in 1945, though the United States had the lowest crop since 1921. In August 1945, excessive rains damaged the crop in the eastern belt, and bollworms inflicted heavy damage in Texas and Oklahoma. Synthetic fibers, particularly rayon and nylon, continued to erode the market for cotton. Production costs remained high because much of the industry's resources were locked into fixed costs: the reliance on hand labor for weeding and picking, marginal land, insect damage, animal power, and a conservative mind-set among some growers toward cultivation. And cotton remained synonymous with sharecropping, still perceived as the most deplorable socioeconomic practice in the United States. Failure by the Pace Committee to generate legislative proposals only enhanced the sense of futility. In May a special subcommittee of the Pace Committee met at the Peabody Hotel, attended by agricultural leaders such as H. R. Tolley of the Bureau of Agricultural Economics, and declared that cotton had a future as "black as midnight" if reforms and initiatives to meet the competition were not undertaken.³ Assessing the reasons for the South's rural poverty, the subcommittee resolved that "the existing unfavorable situation is further complicated by the fact that cotton . . . is facing serious new difficulties in addition to the old ones that have been plaguing cotton over a period of years."⁴

If cotton seemed beset with incurable illnesses, it also had powerful advantages. American values and folklore commonly known as Jeffersonian agrarianism were rooted in the tradition that farming was an expression of freedom and individualism. Agrarianism had lost much of its popular appeal by 1945 and would lose more after the war, but tillers of the soil still commanded respect, and even if cotton farmers had an image as a floundering and helpless lot, the natural sympathy toward agriculture provided a compassionate environment for helping at least small farmers. Not to be overlooked were the southern roots of the Democratic Party, which gave cotton's leaders easy access to the corridors of government power. Some congressional power brokers had spent their childhood chopping and picking cotton or else lived in the midst of the cotton culture. Speaker of the House Sam Rayburn had come from a small cotton farm in northeast Texas and always claimed that he "came within a gnat's heel of becoming a tenant farmer."[5] Cotton was still the most valuable commercial crop, with approximately thirteen million people involved in growing and processing it. The fiber was an essential ingredient in the daily lives of people and an important component in numerous products made by manufacturing firms. The nation's economic life required a strong agricultural base, and cotton ranked along with foodstuffs in priority, an inescapable fact that created a hospitable environment in Washington for reforms undertaken to improve the economic strength of this commodity fundamental to the welfare of the United States. So in spite of the shaky future in 1945, cotton had formidable assets.

A sense of a new beginning resonated through the country that year as Americans anticipated a fresh start in shaping their lives after enduring the Depression and the horrors of war. A distinctive period in history appeared to be unfolding with the introduction of atomic energy and the decline of Europe, leaving the United States as the supreme economic and military power. Fresh ideologies were bursting forth around the world that rejected the orthodoxy of past centuries; the ideology and rationale of privilege would soon come under fire as the three-hundred-year era of European colonialism quickly blew away in the face of demands for self-rule. In the United States, the isolationism and economic protectionism of the past generation were seen as mistakes that had enabled fascists and militarists to pull the world into war. History should not be allowed to repeat itself, so the reasoning went, and the country had to reassume its sense of mission and the lead the way in rebuilding the global economy. Free trade or at least the pursuit of more open avenues of international commerce must be obtained, not only for selfish economic reasons but also to encourage cooperation around the world, as seen in the support for the United Nations. The push for free trade and a spirit of mutual interaction among countries started the move toward an integrated world economy. "The central moment, the turning point," a historian later wrote, "was that of 1945."[6] In this atmosphere of anticipated change, the cotton industry sensed a future without the turmoil of the past, but recognized it had to formulate unprecedented rationales and strategies for its own well-being.

Through the NCC, a new principle of legitimacy had been established with the intention of launching cotton onto a trajectory leading to recovery. The organization had not been paralyzed by the war, but it had been unable to initiate action in areas such as foreign trade, advertising, and fashion apparel, where it hoped to stimulate consumption. Now it wanted to begin working on the matters that had propelled it into existence and to overcome the stagnation characterizing the cotton culture since the mid-1920s. Like a bundle of nerve cells concentrated to perform a particular function, the NCC planned to focus on the substantial issues and at every opportunity recited the mantra of unification, constantly reminding each segment of the industry that it was interconnected with the others, that economic viability would require cooperation with one another. The NCC's staff had worked enthusiastically during the war and had waited patiently for it to end, and now, during the closing months of the conflict, they began looking to the future.

A CADRE OF YOUNG AND OLD

Oscar Johnston remained at the top of the leadership pyramid as the first president of the NCC. He stayed at his home on Deer Creek in the middle of the D&PL plantation in Scott, Mississippi, but he frequently traveled to Memphis. He had recovered temporarily from the torment of his failing eyesight and maintained his correspondence, never hesitating to dictate or write lengthy letters in response to inquiries about the affairs of the NCC, or to push for funds from merchants, shippers, producers, and other parties active in the industry. He occasionally offered his views on mechanization, which he welcomed for mid-sized farms and commercial operations, but predicted production on small farms would continue as before, relying on animal power and hand labor. To one correspondent, he stated that "we will always, I think, produce three, four, or five million bales in the old fashioned way" because of the presence of so many small plots in bottomlands of rich soil.[7] Johnston remained in demand for speaking appearances and published sporadically. In the NCC's publication *Progress Bulletin*, he wrote that to prevent the recurrence of totalitarianism, "we should work to combat the forces which cause them by doing all in our power to revive world commerce."[8]

Rhea Blake had been made the executive vice president by NCC's board of directors in 1939 and took over responsibility for the daily operations of the organization. Few years were more hectic than 1945, when the organization launched specific programs for achieving the objectives announced at its inception in 1938. The organization now had to fulfill its promises to thousands of cotton farmers, and Blake had to provide the creative impulse for that to happen. He had the backing of the industry's most forward-looking leaders, but he carried the burden of achieving those goals. Blake faced a full array of tasks: recruiting high-caliber staff, launching initiatives to promote international trade, recruiting more members and raising funds, starting the economic utilization division, and increasing advertising and promoting cotton at the retail level. "You've got the responsibility of the thing on your shoulders," Blake later recalled. "You do whatever needs to be done at the moment, from sweeping the floor to kissing a pretty gal."[9]

A characteristic of cotton was the complexity of its economics, the labyrinth of data relating to staple length and grade of fiber, the conditions of both domestic and foreign markets affecting daily prices, government subsidies, futures trading, price competition from synthetics, and the need to forecast supply and demand. As was the case with other commodities, sellers, beginning with farmers, wanted to sell high whereas buyers wanted to buy low. From the raw product growing in fertile bottomlands and on the High Plains to the luxury garments in cosmopolitan boutiques, cotton went through many transactions before reaching its final destination, and each stage affected the price of the raw fiber. Accurate and reliable analyses of this puzzling maze of business operations had to be available for the NCC to carry out its mandate. The Pace Committee, reported the Memphis *Press Scimitar*, had been stymied in drafting a plan for postwar legislation in great part "because no authentic information had been available on which to base it."[10] Blake knew the man who could master the baffling economics of cotton.

In March 1945, Blake wrote to navy lieutenant M. K. Horne about rejoining the NCC. When the Cotton Conference failed to yield a consensus on future directions, Blake and Representative Pace wanted to pursue the subject further and make a thorough and scientific assessment of cotton's woes as a step toward developing a governmental policy for the postwar era. The NCC had encouraged Pace to conduct the original hearing, and now with the war expected to end soon, Blake, Johnston, and others hoped to reinvigorate the follow-up studies. Horne had worked for the NCC from 1939 to 1941, but soon after receiving his Ph.D. in 1940, he went to the University of Mississippi as director of business research for a short time before enlisting in the navy. Horne was stationed in New Orleans when Blake asked him to rejoin the NCC, but Horne had already discussed the new studies with Pace, and Blake wanted him to renew his work with the Pace Committee. "We trust that you are going to find it possible to apply for inactive duty," he wrote to Horne, "so that you can carry out the research work which was outlined in your recent meetings with Congressman Pace."[11] Horne would be a member of the NCC staff but would provide his services to the congressman. Blake's message had a sense of urgency about the need to carry on with the work, which Horne understood, but the letter encouraged him to tender his services for King Cotton.

Horne had the qualifications and background that Blake needed. He had a doctorate in economics from the University of North Carolina, he understood the economics of cotton as clearly as anyone, he had prior experiences with the NCC, he was acquainted with many of the leading figures of the industry, and he understood congressional lobbying. His youth and vigor made him the perfect choice to head the organization's Utilization Research Division. Horne eagerly accepted the offer, going back on the payroll in June 1945 with two jobs: chief economist and directing the research for the follow-up studies on behalf of the Pace Committee. Horne's appointment would enable the NCC to carry out one of its guiding principles and crucial functions: making sense of the complex and bewildering data about the industry. Horne once wrote: "The complexity of cotton economics can only be described as awesome."[12]

In August 1945, shortly before the Japanese surrender, Blake contacted Read Dunn, who had worked for the Delta Council in 1938 and 1939 and assisted Blake in establishing the NCC. In 1939 Dunn had joined the National Defense Advisory Commission for one year and then worked for two years as an assistant to the president of the CCC. He enlisted in the U.S. Navy in 1942, serving as commander of a Landing Craft Infantry (LCI) in the Pacific. Blake had kept Dunn in mind during the war, remembering his work when they strove to create the NCC, and he now hoped to persuade him to resume the crusade to save cotton. Dunn was in San Francisco when Blake wrote: "I have been doing an awful lot of thinking and hoping about your getting out of the Navy before long."[13] When Dunn mustered out and returned to NCC headquarters in Memphis, Blake offered him a job, asking, "What do you want to do?" With his infectious optimism, Dunn replied, "I want to organize a foreign trade program."[14] Since the NCC had no program for international trade, the responsibility fell to him.

Another important recruitment came with Claude Welch, who joined in March 1945 as director of production and marketing. He had worked as a county agent in Mississippi and then directed the state's War Food Administration. Johnston also asked General Percy Jones at the military hospital center in Battle Creek, Michigan, to release PFC Ernest B. Stewart, who had worked for NCC before the war.[15] Stewart returned to Memphis and assisted Edward Lipscomb in the Domestic Promotion Division. Navy lieutenant Philip Toker also came back after the Japanese surrender. Albert Russell, who later became executive director, returned from the navy in 1945.

An indication of the network established by the NCC was evident in the discharge of military personnel planning to return to the Memphis headquarters. Burris Jackson, a Texas cotton promoter, held the rank of major in the U.S. Army Reserve and worked in the office of Texas congressman Luther Johnson. Blake instructed his young recruits to keep Jackson informed of the progress of their discharges so that he could expedite them.[16] Blake's appointments and reappointments, along with personnel who remained during the war, such as Robert Jackson and Edward Lipscomb, made for a vibrant staff known for its youth, enthusiasm, and commitment to bringing prosperity and dignity back to cotton. "We had a team spirit in the thing," Blake later recalled, "and we had people that were dedicated to the job."[17] In his retirement, when he recalled NCC's earlier years, Horne said: "We had a strong sense of camaraderie and dedication to move cotton into the future."[18]

King Cotton had many friends in powerful places, but none surpassed the effectiveness of Lewis T. Barringer, a Memphis cotton merchant who enjoyed the confidence of Harry Truman. When Truman was elected president in 1945, cotton suddenly became well connected. Truman and Barringer had a common interest in the five Mississippi River counties of the Missouri Bootheel, with their rich sandy loam, abundant rainfall, and temperate climate. Textile companies liked the long staple produced there. It was an area where cotton reigned supreme with commercial operators and smaller family farms, but like other areas, it suffered the pains of sharecropping and the erosion of small-plot farming. Politically it belonged to the one-party South, and Roosevelt's New Deal was popular, so with his affinity for farming, Truman had established strong links with Bootheel Democratic party leaders such as the cotton broker Neal Helm of Caruthersville and Madrid County prosecuting attorney J. V. Conran, the "boss of the Bootheel." Other prominent Truman supporters included Judge Roy Harper and Bob Hannagan. The future president had done well there in his 1934 race for the Senate. In 1937 he met Barringer.

Barringer was a cotton merchant operating on Memphis's Front Street, the famous "Cotton Row," a center point of southern cotton culture. In 1918 he began his career as a clerk for Cannon Mills in Kannapolis, North Carolina, and moved ten years later to Memphis, where in 1933 he established L. T. Barringer and Company. He quickly became a powerful figure in cotton circles because he bought for Cannon Mills, taking each year almost the entire crop produced in Mississippi and Crittenden counties across the river in Arkansas, about two hundred thousand bales. When Barringer entered the market, other buyers temporarily withdrew because of his large orders.[19] Barringer had a fondness for the camaraderie of politics, for the intrigue of working with and through interest groups, and he remained devoted to the well-being of cotton throughout his life. He had been a central figure in the origins of the NCC, fully endorsing the concept of unification, noting that "I was in meetings ... many times before the Cotton Council was formally put together."[20]

Barringer interjected himself into Truman's affairs when the senator stood for reelection in 1940. There was a strong chance that Truman would lose the Democratic primary to Governor Lloyd C. Stark, so Barringer went to Truman's aid by taking half his office staff into the Bootheel on several occasions to help with the campaign. "Whenever I had a few spare days I would go to Missouri and join in on the general campaign," he recalled, "helping in any way possible to help create interest in Mr. Truman's behalf."[21] When the senator won his second term against Republican Manvel Davis, Barringer became more than a helpful volunteer. The cotton counties of the Bootheel always gave Truman strong support, and for no small reason he later told Secretary of State Dean Acheson that the "southeast corner of Missouri has always been in my corner politically."[22]

During World War II, Barringer had occasional contact with Truman when he chaired the Senate War Investigating Committee, but Barringer spent most of his time in Washington lobbying the various agencies that imposed wartime regulations on the cotton industry. This activity kept him in contact with the NCC's lobbyists, Bob Jackson and Sam Bledsoe, and frequently with Rhea Blake. Jackson, who spent many evenings conversing with Barringer in the Mayflower Hotel, recalled that Barringer made a sudden weekend trip to the Bootheel and upon his return told Jackson that along with others there, the decision was made to push Truman for vice president in 1944. Barringer would serve as campaign treasurer because of his experience in raising funds in the senator's 1940 race and his acquaintances in the cotton merchandising community. After Truman's inauguration, Barringer saw him "nearly every week" during his presidency.[23] On some occasions, Barringer sat on the edge of the basement pool in the White House while Truman took an exercise swim.[24] In his Memphis office and in suite 536 that he leased at the Mayflower Hotel in Washington, Barringer kept a direct telephone line to the White House. His "association with Mr. Truman ran along general lines as to how the public was receiving his Administration," Barringer later remarked, "and special matters in which Mr. Truman had a particular interest."[25]

It was not widely known that Barringer was a confidant of Truman, though Washington insiders knew him. A ghostlike figure who moved about quietly and went largely unrecognized, he relied heavily on the telephone, whether conducting his business or lobbying on behalf of cotton. He left almost no records of his activities and willed his estate to charity before his death. In 1976 he was the honored guest of a testimonial banquet hosted by Democratic power brokers at the Mayflower Hotel, where an impressive list of speakers honored him: Margaret Truman Daniel, Thomas "Tommy the Cork" Corcoran, Senators William Fulbright and James Eastland, former Secretary of Agriculture Orville Freeman, Truman's administrative assistant Charles Murphy, and others. Congressman Bob Poage served as the master of ceremonies. C. R. Sayre served as chairman. Corcoran described Barringer as a "catfish with a barb," referring to his well-known persistence as a cotton lobbyist.[26] The Memphis broker wielded influence in his ghostly manner, as indicated by Murphy, who credited Barringer for his move to the White House staff.²⁷

In 1972, shortly before Truman's last birthday, Barringer and Tony Vaccaro, a newspaper reporter and longtime friend of the president, visited him in the Truman home in Independence, Missouri. Truman's health was failing, but he had insisted they come. What was expected to be a short visit lasted well over an hour, and they reminisced about the past.²⁸

WILL CLAYTON

If cotton spoke to the president through the voice of Barringer, it had a direct connection at the highest level of policy making thanks to Will Clayton, the under secretary of state for economic affairs. Born on a cotton farm near Tupelo, Mississippi, in 1880, Clayton was a prodigy who went to work at the age of thirteen in the Madison County chancery court. He learned shorthand and with his near-genius knack for numbers became a deputy clerk the next year. He took a better job as a private secretary for a cotton merchant, Jerome Hill of the Texas Cotton Products Company. In 1902 Clayton became secretary-treasurer of the firm; he left two years later and formed a partnership with Frank Anderson. With additional backing from Anderson's brother, they created the Anderson, Clayton and Company (ACCO), based in Oklahoma City. A cotton brokerage firm, ACCO grew rapidly and proved lucrative for its partners. They moved the headquarters to Houston in 1915, and ACCO rose to become the predominant firm in the U.S. cotton industry. Clayton was the driving force in ACCO, "known as a one-man firm."²⁹ It built warehouses, cotton gins, and compresses and opened branch offices overseas. By 1933 ACCO was the world's largest cotton brokerage, making Clayton wealthy and influential in business circles, particularly as a spokesman on cotton affairs. He initially opposed the New Deal, principally for its price support program for cotton, and joined the anti-Roosevelt Liberty League in 1934. Clayton warned that subsidies would price U.S. cotton out of the international market and allow foreign competitors to capture more sales. Identified as a wealthy merchant in the eyes of the common farmer, and detested by cotton brokers in New York City for his campaign on behalf of Southern Delivery, he came under investigation during the 1930s by the Senate Agriculture Committee chaired by South Carolina senator Ellison "Cotton Ed" Smith. Despite much grilling by the committee, Clayton was charged with no unethical or illegal practices.³⁰

Clayton's experience in world trade propelled him into the Truman administration and gave him the opportunity to be an effective policy maker for the benefit of cotton. In 1936, after leaving the Liberty League, he supported Roosevelt because Secretary of State Cordell Hull championed free trade. In 1940, Jesse Jones, a Houston banker and head of the Reconstruction Finance Corporation (RFC), brought Clayton to Washington as vice president of the Export-Import Bank. During the war, Clayton moved into other positions, such as assistant secretary of commerce and surplus war property administrator. In 1944 he became assistant secretary of state for economic affairs. He accompanied Truman to the Potsdam Conference in July 1945, where he arranged a compromise with the Soviets over German reparations.[31] In 1946 he rose to under secretary of state for economic affairs. Free trade undertaken for the sake of economic growth and American leadership in the world had long been Clayton's fundamental belief. During his confirmation hearing for assistant secretary in December 1944, only a few days after the Cotton Conference, he expressed his views. Expansion of the world economy had to be encouraged because "the whole world is so closely knit together now economically that it is not going to be possible for any one area of the world to have prosperous conditions and a rising standard of living if the rest of the world is on the downgrade; it just cannot be done.... I think we should do everything we can to aid and assist an expansion of the world economy."[32]

The new political strength of cotton made clear that industry leaders would endeavor to exert influence in the remaking of trade treaties and postwar international negotiations. Clayton and Johnston were in full accord on the subject of international trade and agreed that cotton should have a primary place in new arrangements. When they spoke about the importance of overseas markets, the two giants appeared to mimic each other. One was a broker and the other a planter, each with experience in high levels of government. As such they had arrived at their convictions separately but thought as one on the matter of restoring exports. In 1943 Johnston planned to speak to the Cotton Research Congress held in Dallas, but illness prevented his appearance, so H. H. Williamson, director of the Texas Extension Service, read Johnston's speech. Johnston reminded the delegates that cotton had been the primary export of the United States, and it only made common sense that cotton be second to no other product in its "effect on post-war international economic relationships." Other interests would be represented at the conference tables, he continued, and the NCC would "seek direct representation among the delegates who will sit at the table" because it was the only organization that represented the whole Cotton Belt.[33] Johnston and Clayton regularly blamed restrictive trade policies for the growth of militarism, and from their perspective, cotton offered the best chance to rebuild the war-torn world.

ADVERTISING A COMMODITY

Only innovation, so went the thinking, would save cotton from becoming "as extinct as the dodo bird and the dinosaur," according to Johnston's oft-repeated line at the 1938 meeting in the Peabody Hotel. If consumers found cotton clothing not likable, all the efforts expended on political lobbying would be meaningless. In 1945 cotton fabric was still associated with low-cost clothing such as work clothes and the "cotton house dress." Finer clothing used wool, silk, and increasingly synthetics, and designers turned to them for upscale clothing. In 1938 the NCC had taken a self-imposed responsibility to change this popular perception of cotton, and now with the war winding down, Edward Lipscomb, director of sales promotion and public relations, began an advertising and promotional campaign on behalf of the "poor person's fabric."[34]

Lipscomb also continued a promotional campaign started in 1943 known as the "Governor's Lady Series," an idea that came from Wynn Richards, a photographer and sister of Billy Wynn. Lipscomb arranged for her to photograph thirty-five governors' wives dressed in stylish clothing specially designed for them and made of cotton fabric. Lipscomb's office would initiate contact with the first ladies and obtain their measurements and favorite colors. They included daughters "to get into the young people's and children's areas."[35] The pieces were shipped to the wives and daughters to make adjustments before Wynn arrived for the photo shoot. Lipscomb then sent the photographs, which included captions and quotes from the first ladies about the attractiveness of the cotton material, to popular fashion magazines such as *Harper's Bazaar* and *Vogue*. Lipscomb reported the program to be a success because "it wasn't long before whole sections of magazines were done on cotton."[36] When he returned from military duty in 1945, Ernest Stewart became public relations manager and assisted Lipscomb. They began the "American Designer" series, in which the NCC cooperated with clothes designers to feature all-cotton high-fashion outfits ranging from recreational pieces to evening wear. Placed in prestigious magazines, the special series portrayed cotton as glamorous and fashionable. Lipscomb's office arranged special displays and trunk shows with leading department stores such as Neiman Marcus and I. Magnin. "Out of the kitchen into the drawing room—that's cotton today" was the motto used in the advertisements.[37]

To associate cotton with glamour and fashion, the NCC had taken over the sponsorship of the Maid of Cotton program that originated before the war with the Memphis Cotton Carnival Association. For lack of funding and staff to promote the program, the association relinquished control to the NCC, which saw a chance to promote natural fiber. Lipscomb needed funding, too, and convinced Lever Brothers to become a sponsor. He also managed to gain sponsorship funding from the cotton exchanges in Memphis, New Orleans, and New York. In 1945 Jennie Erle Cox of West Point, Mississippi, was the Maid of Cotton, and she began her national tour in Miami, extending greetings from the cotton industry and promoting war bonds. That year the Maid of Cotton featured day dresses and frocks, as well as high-fashion evening gowns, all made of cotton. A stylist with Lux products, a division of Lever Brothers, traveled with her. The tour received enthusiastic attention as the public clamored for glamour at the end of the war; it offered a promising way to advertise cotton as a fabric of fashion.[38]

A less glamorous but valuable campaign dealt with cotton bags and sacks. Flour commonly went into large sacks ranging from 100 to 140 pounds for shipment to wholesalers. Retail-sized sacks ranged from five to ten pounds. Lipscomb's division promoted the "pretty print" feed sacks through the Textile Bag Manufacturers Association, which quickly became popular for making clothing for farm children. Depending on the size of the garment, farm wives would collect several feed sacks before they started sewing. Dairy feed sacks were particularly popular. Lipscomb's office developed a thirty-six page pattern booklet showing items that could be made from cotton sacks and received up to a million requests per year for the booklet, but the actual patterns were sold by Simplicity Patterns. The bag and sack market consumed about one million bales per year in the latter 1940s before paper began to take part of the market.

THE IDEOLOGY OF TRADE

Among its goals in the pivotal year of 1945, the NCC made overseas trade one of its priorities and took the position that new international organizations created to foster the exchange of goods and encourage commerce should be supported. At the organization's annual meeting, Everett Cook, a Memphis cotton merchant and advisor to the State Department, told the four hundred delegates that the Atlantic Charter was a live document and the statements regarding trade in the agreement of 1941 between Roosevelt and Churchill figured "prominently in all our relations with other countries concerning trade restrictions, quotas, subsidies and other matters entering into international trade."[39] NCC president Johnston recommended a downward revision of tariffs but warned that multilateral action on all aspects of trade would be necessary to foster global commerce. For immediate action, his organization endorsed the principles drawn up at the Bretton Woods Conference of 1944, and urged "the participation of this nation in the proposed International Monetary Fund (IMF) and International Bank for Reconstruction and Development," known as the World Bank.[40]

The Bretton Woods Conference grew out of the need to establish a foundation for world trade that had been upset by the Depression and war. Preliminary negotiations had begun in 1942 between Britain and the United States to avoid a return to the chaotic currency exchange rates of the Depression and create a stable economic environment for stimulating international commerce. In the context of the world's recent high unemployment, which had encouraged the growth of totalitarianism, trade was considered essential for peace, and stable currencies held a position of vital importance. Private trade had almost shut down during the war; global exchange consisted mostly of governments purchasing goods to fight their enemies. Both the Axis powers and America's allies had depleted their financial reserves, and large portions of their means of production—factories, farms, and transportation—had been destroyed. Trade was thought to be essential for world recovery.

Addressing these questions, delegates from forty-four countries met at Bretton Woods, New Hampshire, in July 1944 to hammer out a plan of international cooperation. The principal negotiators were Harry Dexter White of the United States Treasury Department and Britain's John Maynard Keynes. From this meeting came the IMF, responsible for maintaining a system of stable exchange rates for currencies and assisting individual countries that had temporary difficulties in meeting foreign payment requirements. Through the IMF, countries would peg their currencies to the U.S. dollar, the only sound currency at the time. Because the dollar was based on the gold standard, the fact that the United States held 75 percent of the world's gold gave it an advantage.[41] War-torn countries could obtain loans from the World Bank for rebuilding their economies, particularly infrastructure projects. Funding for both organizations would come from the member nations.

The White House, under both Roosevelt and Truman, lobbied Congress for approval, but in early 1945 confirmation was still pending although the proposals received general support across the country; only the banking profession opposed the IMF. Will Clayton saw the Bretton Woods proposals as essential for Europe's recovery and U.S. employment after the war.[42]

To help obtain congressional ratification, the NCC flexed its young muscles. To begin with, Johnston formally endorsed the proposed monetary pact; his service as a former member of the St. Louis Federal Reserve Bank's board of directors gave his opinion weight. He sharply defined the issue, stating that the Cotton Belt had much at stake in "the restoration of a flourishing intercourse between nations. Obviously, this is no time to be hesitant or timorous, or to be afraid to try something because it departs from familiar and often traveled paths." International traders of commodities had to rest assured that the currencies behind their transactions would be solvent and convertible, he continued, and the system proposed in the IMF represented "an effort to bring order out of disorder which has prevailed in foreign exchange for years."[43]

In April 1945 the NCC sent across the South a blitz of press releases, general letters, and statements endorsing the proposed agreement, often including copies of Johnston's urgent call. Copies of his statement went to all 151 Cotton Belt congressmen and senators along with a press release to 460 daily newspapers in the same states. Johnston's statement went to sixty cotton industry publications, and a general letter to 118 related organizations. A general letter also went to the 305 officers, delegates, and committeemen of the NCC. Sid West, a cotton merchant and chair of the organization's Export Committee, arranged a speakers' committee made up of leading cotton shippers to appear before civic clubs such as Rotary, Kiwanis, and Lions. Caffey Robertson contacted the Memphis Chamber of Commerce and asked it to recommend the Bretton Woods proposal to other chambers of commerce in the South. Resolutions by these civic groups were sent to members of Congress. Lobbyist Sam Bledsoe urged passage of the proposals among his contacts on Capitol Hill. And Frank Ahlgren, editor of the *Commercial Appeal*, made a speech on behalf of Bretton Woods and endorsed it in his editorials. Members of the NCC in other cities and towns personally contacted newspaper editors to get their support.[44] To Michigan congressman Fred C. Crawford, Blake wrote: "The opinion of the most thoughtful people in our group is that a start in this direction is the most important thing."[45] Blake timed this flurry of activity to stir enthusiasm for the monetary proposal as the congressional vote drew closer. In July 1945 the Bretton Woods proposals passed the House 348–18 and the Senate 61–16.[46]

No controversy erupted over the proposals drafted at Bretton Woods, and the NCC's stance conformed with the wider acceptance throughout the United States. But cotton's new lobby organization had effectively reached out on a large scale and demonstrated its ability to affect opinion. The NCC made clear that it stood heartily behind the principles of free trade by taking the historical course of the South on low tariffs and minimum restraints on trade, with stable currencies encouraged by the IMF and loans by the World Bank as vital steps toward a fully developed international economy.

An indication of a growing belief that tariffs hindered international trade occurred when the Carnegie Endowment for International Peace studied the impact of tariffs on agriculture. Commissioning prominent agricultural economists in 1945, the Carnegie Endowment arranged for John Van Sickle of Vanderbilt University to examine the importance of cotton in international trade and determine if a reduction or lessening of restrictions would facilitate prosperity, which the organization thought would promote world peace. Expressing the thinking in cotton circles, Van Sickle warned that the "postwar world will be very different from the world of the 1920s, or of the 1930s, or of that mythical base period, 1909–1914 when American farmers were supposed to be getting a 'square deal.'"[47] It would be wise, he recommended, to inaugurate trade policies that facilitated the flow of goods across borders, to encourage cotton to reach foreign markets, and to prepare farmers for living without federal price supports within a reasonable time of adjustment. Trade and efficient farming, Van Sickle thought, offered the best opportunity for the Cotton Belt to return to subsidy-free agriculture. He predicted, too, that the public would sooner or later grow weary of subsidizing growers. In this same spirit, Murray R. Benedict of the University of California wrote that tariffs for farm products were not suited to the economy as the country came out of the war. He urged their removal. The National Grange and Farm Bureau took the same position, and even Oscar B. Jesness, chief of the Division of Agricultural Economics at the University of Minnesota, thought that protecting dairy products, including the tax levied on oleomargarine, did not benefit the dairy industry. These writers for the Carnegie Endowment emphasized that freer trade would stimulate prosperity and encourage peace.[48]

The importance of trade in fostering prosperity and world peace received further recognition. In March 1945, the economist Herman Finer spoke to the International Labor Organization and stated that foreign trade encouraged a sense of friendly neighborhoods among countries. The world economy had reached such a level of interdependence, he said, that the prosperity of one country depended on the well-being of others: "There is woven a standard of living in which all form part of a pattern."[49] He urged postwar cooperation, claiming that an enhanced flow of international commerce would promote human welfare and combat unrest. He reminded his audience that diplomats now studied economics as a matter of course.

CONCLUSION

Nineteen forty-five, the year of transition to peace, saw cotton caught between past and future. The war had propped up the industry with better prices, but such props were only temporary, and with no assurance of foreign markets, the further growth of synthetics, along with the further migration of workers off the land, made the odds for a healthy future look slim. But a dynamic new world was unfolding, and it was apparent the future would tolerate no ineptness or failure to make the adjustments necessary for success.

As the war ended and the world entered the new era of nuclear power and the remaking of national alliances, the cotton industry was poised to exert influences effectively for itself. In its annual report for 1945, the NCC stated it had reached political maturity and proved its right to recognition as the representative and spokesman for cotton. Blake was apologetic about such bragging but felt the claim was justified.[50] Suspicion and mistrust among the industry's segments had begun to fade away as the NCC made its presence known in the corridors of Congress, the offices of the executive branch, and even the Oval Office. But the rapid changes of 1945 made cotton's leaders realize that action needed to be taken toward specific objectives.

CHAPTER 5

AMBASSADORS OF FOREIGN POLICY, 1945–1950

The South had long championed international trade, and the importance of cotton in the southern economy accounted for that attitude. Interests linked to growing and processing cotton depended on world commerce for their well-being, and like an axiom of mathematics, they stood against restraints on the flow of goods across borders. This faith in trade was driven by the demand for an abundant and low-cost spinning fiber in Britain and Europe that since the beginning of the Industrial Revolution had made cotton the principal export of the United States. The cotton industry had experienced a dramatic drop in overseas sales in the generation before 1945, which explains why the restoration of foreign markets became a pressing objective.

With the reshuffling of military and economic strength around the globe, realignments of international trade were expected to follow, and since cotton interests felt the "fabric of history" encouraged goodwill among nations by keeping open the doors of trade, they intended to influence postwar trade arrangements. In 1942 Peter Molyneaux, trustee of the Carnegie Endowment for International Peace, recommended that trade be encouraged after the war because it fostered peace; he believed that the cotton industry understood international trade better than "almost any other group."[1] Oscar Johnston reinforced the importance of trade when in 1943 he declared before the Cotton Research Conference in Dallas: "Cotton will be second to no other single product in its practical influence and effect on postwar international economic relationships." It was only reasonable, Johnston continued, that cotton's influence be felt on "those policies and agreements which together shall make up the terms of the international peace agreements."[2] Johnston articulated a general feeling that with the cessation of fighting, shipments of cotton should quickly be sent overseas as a matter of self-interest, but in a broader sense he believed the economies of war-torn countries should be rehabilitated to reestablish trade on a permanent basis. This last idea grew for the next two years within the Truman administration until it became a feature of postwar foreign policy, but within cotton circles it was advanced before the war ended. As early as February 1945, W. D. Anderson of Bibb Manufacturing, a textile company, recommended sending machinery to Europe to replenish factories damaged or destroyed in the war.[3]

THE AMERICAN COTTON SHIPPERS ASSOCIATION

An early step toward revitalizing trade with Europe came when the American Cotton Shippers Association (ACSA) assessed the extent of war damage to the European textile industry. In June 1945 officers of the association requested the support of Assistant Secretary of State Will Clayton in arranging transportation and hotel accommodations in the European cotton-consuming countries: England, France, Belgium, Holland, Switzerland, Italy, and Spain. Clayton well understood the importance of overseas trade and forwarded their request to Secretary of War Henry Stimson, since the U.S. military at the time was an occupying force in Europe. "Early reestablishment of the general trade rules under which this business is done and renewal of general trade contacts," Clayton wrote to Stimson, "is necessary to replace lend-lease shipments and to keep European markets for American cotton."[4] Such a trip, Clayton thought, was in the public interest, which justified assistance by the War Department. Stimson would not fulfill the request, however, because of the excessive demands imposed by demobilization, but a committee from the ACSA, consisting of Wolborn B. Davis, Marc Anthony, and R. C. Dickerson, traveled privately through Europe, relying on the U.S. embassy in London to help with arrangements. They left New York on a Pan American Clipper on July 16 for Croydon Field, near London. Bad weather forced several landings, but the trio reached their destination on July 18.

The committee found much damage in London with a "building or two" missing in every block. From St. Paul's Cathedral they "could see nothing but rubble for many blocks in every direction." In Liverpool, Britain's cotton importing center, damage was less extensive, but whole blocks were nonetheless demolished, and offices of the Cotton Exchange had taken hits from German bombers. Cotton merchants in Liverpool explained to the committee how they wished to resume trade quickly and to conduct business through private channels because it was easier to ensure accurate grades to British mills without government interference. Liverpool merchants asked about obtaining loans from the U.S. Export-Import Bank to restart the cotton trade. Officers in the Manchester Cotton Association, which included spinners along with merchants, similarly requested American assistance to spur trade. For the rest of their visits to Le Havre, Madrid, Zurich, Ghent, and Gothenburg, the Americans found a desire to return to prewar business without government controls. All the British and European interests expressed a need for U.S. aid to restart the manufacture of cotton clothing in their respective countries. Upon its return, the committee shared these observations with Rhea Blake and the ASCA membership, which included Lamar Fleming of ACCO, who remained in contact with Clayton.[5]

SOVIET THREATS

The ACSA committee's European trip and the interest they generated in the cotton press received the close attention of the USDA and State Department because of developments under way in central Europe. In June 1945 the U.S. government became aware that the Hungarian government had been negotiating an agreement for the Soviet Union to send "large amounts of cotton" to Hungary. The two countries had worked out an arrangement for the Hungarians to spin and weave the cotton and return a portion of the fabricated cloth to the Soviet Union as payment for the raw lint. Before the war, Austria and Hungary together had purchased about 123,000 bales of American cotton per year, compared with 53,000 supplied by the Soviet Union. It now appeared that the Soviet Union intended for the cotton transaction to be the "opening phase of a comprehensive trade agreement between Hungary and Russia."[6] E. D. White, assistant to the U.S. secretary of agriculture, forwarded this information to Clayton. At issue was the loss of sales to the Hungarians and Soviet intrusion into traditional U.S. markets in Europe.

Communists in Europe had already stirred concern among cotton leaders and the State Department. On May 28, 1945, not quite three weeks after the German surrender, Camillo Livi, the senior surviving member of Anderson, Clayton and Company's branch in Milan, sent a memorandum to Syndor Oden at company headquarters in Houston, requesting shipments of cotton to northern Italy as quickly as possible because when German troops withdrew, they did not destroy the spinning mills as had been anticipated. For lack of cotton, the mills had stopped in early 1945, while the weaving mills operated an average of twenty-eight to thirty hours per month. Unless raw fiber arrived soon, all mills would shut down, and the labor force of five hundred thousand would be idle. "In order to avoid serious discontent and hunger," Livi warned, "the immediate need is to receive raw material to keep the spindles active until normal trade can be resumed." Because of their large numbers, the Italian textile workers had "a deep and important influence on the local political life," and every day that passed without raw cotton to spin was a step closer to dissolution "and a great help to the destructive forces of communism which will find a good working-ground among the dissatisfied and starved labor of the cotton mills."[7] Despite the war damage and German confiscation, Livi added, port facilities and ground transportation remained operable, and the Italian spinners had saved enough foreign exchange and gold from the fascists and Germans to pay for the cotton.

Fleming forwarded Livi's message to C. C. Smith, head of the Commodity Credit Corporation (CCC), with his own recommendation for urgent action. But Fleming knew that postwar planning for Europe had only begun and that sending shipments quickly to Italy was unlikely. He quoted Livi, who feared that "too many are the problems to be solved by the Allied Commission and too much red tape is suffocating every initiative."[8]

Germany's ability to take cotton from the United States had more importance. As a major industrialized power, Germany had a large textile manufacturing capacity before the war, but the conversion to synthetics by the Third Reich had meant the loss of a major overseas market for American cotton. There now appeared to be an opportunity to restore the German market, but only if textile mills went back to natural fibers. An assessment by the U.S. Army in June 1945 showed that in the zones occupied by American, British, and French forces, the wartime losses in spindles and looms ran from 25 to 35 percent. Textiles in the northern sector depended on coal for fuel, which was far below the normal supply, so that the renewal of textile operations would be delayed. Southern Germany had hydroelectric power intact, however, and could quickly renew manufacturing, since it had over a million spindles and thirty thousand looms left undamaged. "The above situation suggests to me the advisability of moving some American cotton," White wrote to Clayton in a confidential letter, "into the southern zone during the next few months."[9] For payment, the Germans could supply the United States with a portion of the finished cloth as stipulated in the Soviet-Hungarian arrangement. White further recommended stockpiling cotton in the northern zone as a step to discourage the return to synthetics once coal supplies were replenished. Clayton encouraged White to pursue the shipments of cotton to Europe, and with that assurance, White went forward.

THE FIRST COTTON MAN IN EUROPE

In the meantime, the U.S. military faced the immediate task of furnishing food, clothing, medical care, and other necessities to the people of Germany. Living and working conditions were deplorable as civilians and returning military personnel scrounged for food and other essentials. General Dwight D. Eisenhower, commanding officer of the American occupying forces in Germany, had urged the Joint Chiefs of Staff to take action that would get the economy back into production, and to understand and promote the restoration of the various industries there, the War Department created a series of Technical Industrial Intelligence Committees (TIIC) to inspect various German facilities such as steel, transportation, chemicals, and electronics. For textiles, the Pentagon created a committee of thirty technical experts in textile production, but it asked the NCC to furnish a member from the raw cotton industry. This invitation presented an opportunity to learn more about German advancements in the production of synthetics because there was concern throughout the Cotton Belt that German "discoveries will result in considerable economy," Johnston told a member of the New Orleans Cotton Exchange, "in the production of synthetic fibers as well as marked improvement in the quality of these fibers for various uses in substitution for cotton."[10] Johnston's words expressed the fear that cotton stood to lose even greater shares of its markets. Moving quickly, Johnston and Blake arranged for Robert Jackson, who had left the NCC in 1944 and gone to work for Coker Pedigree Seed Company in South Carolina, to serve as cotton's representative on the committee. Robert Coker, vice president of the company, enthusiastically approved of Jackson as the choice for the committee and continued to pay his salary during his trip to Europe, which lasted over three months.[11] To Hugh Comer, president of Avondale Mills, Johnston explained that Jackson would pay particular "respect to the extent of the development of synthetic fibers as substitutes for cotton."[12] Jackson would become known as the first cotton man in Germany after the war, though he followed the threesome of the ACSA.

White learned about the upcoming trip by the Army Technical Committee and Jackson's role. In White's Washington office, the two men discussed the cotton situation in Europe and the plausibility of making arrangements for shipping cotton to mills there and taking finished cloth as payment. Jackson believed that White had "done more constructive thinking about the matter than anyone with whom I talked."[13] Whether White based the scheme on the Soviet proposal, he quickly sent a prophetic warning to Under Secretary of Agriculture J. B. Hutson: "Unless we remain aggressive regarding our cotton trade with central Europe I fear the loss of markets to Russia in the postwar period."[14]

Jackson prepared to debark from New York City, but the Japanese surrendered on the day before his flight left. With his wife Fan, he went to Times Square on V-J Day, where they were part of the throngs of people that engulfed the area. "There were literally hundreds of thousands of people there converging from every direction," he recalled, and only after some effort were they able to get back to the hotel. His flight left on schedule the next day, and when he reached London, his first contact was Will Clayton. Only because of the intervention by Fleming had Jackson managed to get an appointment with Clayton, who was in London conducting negotiations that led to the British Loan Agreement of 1946.[15]

Jackson went to Europe as an observer, but he also had responsibility as a messenger. His earlier discussion with White about disposing of the surplus and receiving fabric for payment became the "textile in payment" proposal. It was expected to reduce CCC holdings, which were forcing down the price of raw cotton, and encourage the German textile industry to concentrate on cotton rather than synthetics, at least for the immediate future. Eisenhower had requested fabric to make clothing for German civilians, and Germany would be able to start acquiring funds for foreign exchange. When he reached London, Jackson managed to catch Clayton in a barbershop and discussed White's idea there. The assistant secretary of state was reticent about cotton matters but liked the plan and urged Jackson to contact the members of the Bremen Cotton Exchange.[16]

Jackson first had to report to Frankfurt, the headquarters of the Field Information Agency, where he found chaotic conditions. American cigarettes were the only common means of barter among civilians, going for ten dollars a pack. Restaurants and overnight accommodations were scarce, making it difficult to bathe or wash clothes, but he had access to U.S. military facilities thanks to his civilian assimilated rank of lieutenant colonel. Jackson obtained the use of a jeep and the services of an army sergeant as a driver. He met Wily Kohler, an officer in the German Spinners Association before the war, who accompanied Jackson for about a month. Following Clayton's advice, Jackson went to Bremen and organized a meeting of a dozen leading mill executives in the boardroom of the Cotton Exchange, the only undamaged portion of the building. During the conference, German police came into the room and arrested the secretary of the exchange, who had been a prominent Nazi.[17] The rest of the meeting was emotional as the executives anxiously inquired about their American cohorts in the cotton business with whom they had been business friends before the war. Throughout his months-long trip, Jackson worked under stringent circumstances, often sneaking food to Kohler. He visited the German Textile Industry Association, a private organization, and the Reichstelle für Textilwerke, the official government agency that had controlled the industry during the war. Mostly he gained statistical information there. From Jackson's point of view, person-to-person connections with textile leaders offered the best chance to recapture the German cotton market, and he felt that "missions like that of the American Cotton

Shippers Committee are bound to be extremely helpful, for there is nothing like personal contact."[18]

While Jackson officially gathered information as a member of the TIIC, he sent a stream of reports to NCC headquarters in Memphis, providing information about shipping conditions in Europe. Loading facilities at Bremerhaven on the mouth of the Weser River, though damaged, provided "ample dock space, handling facilities, and storage space."[19] Thirty-five miles inland on the Weser lay Bremen, where much bombing had occurred owing to the location of submarine dry docks; "hundreds of blocks" had been flattened in the city. Reconstruction was progressing faster in Bremerhaven, where approximately 85 percent of prewar handling facilities were available. Jackson agreed with Clayton that Bremen merchants held the key to the German market and "would have to play an important part in any program developed by our military to import cotton to Germany."[20] Along with descriptions of physical facilities, to arrange future contacts, he provided an updated list of cotton merchants and textile executives living and dead since 1939.

Jackson understood why he had been sent. "Cotton has a lot at stake now and we certainly don't want to miss any bets."[21] In his reports, he explained that some sentiment existed in Germany for the renewal of synthetics, but he felt the cotton merchandising community and the desire of the textile industry to quickly restart operations worked in favor of cotton. Jackson pointed out that coal supplies were inadequate for consumer and industrial uses and that synthetics required four tons of coal to produce one ton of material, whereas cotton needed only one ton of coal. The railroad infrastructure was heavily damaged, which would delay even further the ability to haul coal to the mills; and the chemical and wood pulp industries, essential for synthetic manufacturing, could not in reasonable time supply their part for making artificial fibers. Textile production ranked, he reminded the NCC, among the top priorities throughout Europe, and the growing emergency arising from shortages of clothing lent an air of pending crisis. Cotton fabrics could be available in the least time, and for reasons other than selfish U.S. commercial benefits, he felt the plan to use CCC stocks had merit. Bremen merchants had warned him, however, that textile interests in their home country would attempt to reinvigorate synthetics if cotton did not quickly become available.[22]

A few months later when Jackson visited Italy, he reaffirmed the observations of Livi. He found 95 percent of the spinning and weaving mills intact and noted that thirty thousand bales had reached Italy thanks to the Allied Military Government, and more were expected; but he felt full resumption of trade should be reached as soon as possible because the labor force was a stabilizing factor on the region's economy. "Thus the most important step that can be taken in Italy to put people to work, create earnings, and cut down unrest," Jackson reported to the Pentagon, "is to reactivate the cotton textile industry on a full production program."[23]

Jackson shared his reports with Fleming, who along with Clayton knew most of the European merchants and had the most direct contact with the assistant secretary of state. Clayton had been extremely reluctant to "talk cotton" with anyone since his appointment to the Roosevelt administration in 1940. By sharing his reports with Fleming, however, Jackson ensured that his observations reached Clayton. The ASCA Shippers Committee sent a copy of its report, filed in September, to Jackson at his headquarters in Frankfurt, and the reports of both parties went to NCC headquarters in Memphis. Blake shared Jackson's reports with White and C. C. Smith, head of the USDA Cotton Loan and Utilization Division, so while the reports remained confidential, the NCC, USDA, and State Department knew the information in them.[24]

FRANCIS HICKMAN

As Jackson conducted his assessment of textiles, another figure well known in cotton circles traveled through Europe for his own purposes. Francis Hickman, owner and publisher of the *Cotton Trade Journal*, privately toured several European countries to study the postwar cotton market. Hickman went alone at his expense and with no assigned duties, though the War Department provided assistance for his travel accommodations. Originally Hickman planned to visit only France, but he expanded his trip to include Czechoslovakia, Austria, Italy, and Germany. Clayton formally requested assistance for Hickman from the War Department: "Practically all the cotton merchants in the United States are subscribers and readers of the *Cotton Trade Journal* and if Mr. Hickman could see the cotton situation first hand in Europe he would undoubtedly be able to give his readers the kind of information which they very urgently desire and need."[25] Hickman was a seasoned traveler, having last seen Europe in 1940, and, designated as a war correspondent by the U.S. Army, had access to American billets as did Jackson. Thanks to a friend in the United States, Hickman had an appointment for a personal conference with Eisenhower. Hickman did not meet Jackson, but when he visited the Bremen Cotton Exchange, he learned that Jackson had been there garnering support for the proposed textile payment plan.

Upon his return, Hickman published a lengthy article in *Cotton Trade Journal* detailing his experiences and reporting that European brokers anxiously wanted southern cotton so that textile production could get under way immediately. In Liverpool the merchants gave a large part of their business to

Brazil or other competitors, and though they wanted American cotton, merchants complained that the new Labour government provided no direction for renewing trade through private channels. Approximately five hundred thousand bales of U.S. cotton had reached France through Lend-Lease, and textile production was under way there, but further imports were stymied by prohibitions on private trade. In Holland only a small shipment of cotton had reached Enschede, and a little spinning had begun, but Dutch textiles still remained in disarray. In Austria Hickman found no cotton and one small plant spinning artificial fiber. Germany had the worst circumstances; not only were clothing and raw cotton for spinning severely short, but food and coal were only partially available. Everywhere he traveled, Hickman reported much distress and fear, warranted by the shortages of food, clothing, coal, and other necessities of life. The appearance of recovery on the streets of Paris was only a charade, he wrote, and cities such as Antwerp, Rotterdam, Vienna, and Pilsen, Czechoslovakia, were visibly damaged and badly in need of help. Only Brussels seemed to be recovering, mostly because the Germans had not been so destructive, and "cotton manufacturing is very active again."[26] Like Jackson, Hickman described the plight of the people and their yearning for American aid, how he saw people with "sick and tired hearts" scouring the countryside for food and firewood. A black market prevailed that kept goods out of the hands of the needy. He warned of a small but dedicated cadre of communists in each country whose influence was small but expected to grow.

In this atmosphere of fear and uncertainty, with refugees commonly wandering the roads and highways, Hickman considered it essential for the United States to extend aid of various sorts quickly. Cotton could obviously alleviate the clothing shortage if a way to reactivate Europe's mills could be found, but cotton alone would not overcome the effects of the war, when "the average European is staggering around in a sense of hopelessness."[27] Soon after his return in December 1945, addressing the Memphis International Center, Hickman urged, "We must do our utmost to help Europe regain her economic stability, even if we are required to give her the initial boost."[28] From the standpoint of Jackson and Hickman, the United States had an obligation to go to Europe's rescue, and they saw the cotton industry's effort to reestablish its international market as part of a larger mission to restore the European economy. An advertisement placed by the Memphis-based shippers Hohenberg Bros. Company in the *Cotton Trade Journal* expressed the attitude in the Cotton Belt: "The problems of reconversion in this country, and the problems of rehabilitation abroad make good shipments [of cotton] more vital than ever."[29]

Suggestions for sending cotton to Europe continued to arise. In October 1945, Clayton had lunch with three Texas congressmen, W. R. Poage, Luther A. Johnson, and George H. Mahon, at Washington's Metropolitan Club. Stephen Pace joined them, along with Willard Thorp of the State Department. The congressmen explained the textile payment plan as a feasible system for clothing the people of Europe and furnishing Britain with raw fiber while reducing the stockpile of cotton in the United States. Payment remained the central difficulty, and they suggested that Britain could cede territory in the Caribbean as payment for its portion. Clayton was aware of the textile payment plan but saw little chance that countries would sacrifice territorial possessions for a raw commodity. The congressmen also suggested the United States should consider destroying Germany's cellulose plants as part of the demilitarization plan, which would prevent the manufacture of synthetic fiber. This idea had been presented earlier in the Morgenthau Plan for postwar Europe because cellulose facilities could easily be converted for making gunpowder. The War Department had told the congressmen to take their suggestion to the Department of State, but Clayton showed no enthusiasm for the idea, and Thorp felt that careful allocations of coal could be made to favor cotton spinning. No plan emerged for eradicating German cellulose plants, nor did the idea of taking territories for payment move forward, but the textile payment proposal, which had originated with White, Clayton, and Jackson, remained under consideration.[30]

THE READ DUNN REPORTS

These early initiatives for extending aid to Europe were reinforced in 1946 by Read Dunn. When he took over the NCC's newly created Foreign Trade Division in November 1945, Dunn had the prime responsibility for the organization's efforts to improve foreign sales. He had an excellent grasp of the mid-South cotton culture: he was born in the Mississippi Delta in Greenville and spent his boyhood there. He earned money by selling milk from a pet cow, and during the famous 1927 flood, he delivered newspapers by boat. Dunn graduated from Millsap College in 1936 and went to work as a reporter for the *Democrat Times* of Cleveland, Mississippi, for which he covered the Delta Council. Dunn developed an interest and concern over the affairs of cotton and became convinced that the resurrection of cotton farming was imperative for the South to progress. He enrolled in graduate school at Columbia University, studying economics for one year, after which he landed a job as Rhea Blake's assistant in 1938, where he did much of the footwork in the creation of the NCC. Now out of the navy, Dunn was part of the organization's youthful staff, known for its esprit de corps and determined, in the words of Mac Horne, "to move cotton forward."[31] Instead of Memphis, Dunn lived in Washington, D.C., where the Foreign Trade Division was headquartered because its work entailed government operations. When Sam Bledsoe left as the NCC's Washington representative in 1946, Dunn took over the job until Robert Jackson became the representative in September.[32]

Dunn had to postpone some of his work on behalf of trade during the interim, but he nonetheless lobbied the Export-Import Bank to finance exports to Western Europe. While Germany had difficulties peculiar to itself, much of the textile industry in Western Europe was still intact, with a large percentage of its spindles still operable though textile goods were in short supply. Lack of raw material held back production, which in turn stemmed from a lack of funding to buy cotton. To address this drawback, the bank established a line of credit at $100 million for fifteen months at 2.5 percent interest. This offer helped some countries to buy cotton, but others could not afford the loan despite the low rate.[33]

Dunn got an unexpected opportunity in 1946 to promote shipments to Europe when the U.S. Department of Commerce undertook studies of various industries in Germany to learn about any new technologies that might be useful to American commercial interests. It asked him to go to Germany and investigate the "three-cylinder system" developed there during the war, a technique for spinning low-grade and coarse cotton into yarn. Dunn's experience resembled Jackson's earlier trip. Traveling as a member of an official Technical Investigation Committee, one of the "green uniform boys," he headquartered in Höchst, near Frankfurt; later he switched to Berlin. Like Jackson, he saw the devastation of Germany and the people's struggle for survival. Cigarettes still served as a black market currency, highly prized because they could be swapped for food. "It was miserable," he later reminisced, "to see these people really scrounging and groveling for potatoes, and all."[34] Dunn concluded that the three-cylinder system would not be practical for the United States because it was an "ersatz" program, a wartime emergency use of "off-grade and poor quality raw materials" for making low-grade yarn. But Dunn learned much about the condition of the textile industry there and recognized that the overriding considerations were the lack of capital, the compelling need to find a way to supply Germany with funding for restarting textiles and subsequently buy down CCC stocks, and to achieve his goal, "to move American cotton."[35]

Again like Jackson, Dunn sent reports directly to NCC headquarters in Memphis, providing firsthand knowledge of conditions in Germany, Britain, France, and Russia. His preliminary report arrived in the fall of 1946 and pictured a lagging conversion to a peacetime economy. In France, Italy, Belgium, Holland, and even Switzerland, factories ran at less than half the level of production of 1938 for lack of coal because mines in the Ruhr and Silesia valleys produced less than half the 1938 tonnage. Shortages of coal were crippling the reconstruction of economies in countries that historically depended on German supplies.[36] During the war, captive laborers had kept up coal production in the Third Reich, but they were now being repatriated, and the remaining German workers were underfed and overaged. The miners, moreover, often left the pits "to forage the countryside to find food for families."[37] Dunn pointed out that output lagged badly in all industries "due to social and political differences among the occupying powers, military muddling, unrepaired destruction, and shortages of food, transportation, skilled labor and a thousand and one items."[38]

Synthetic fiber production continued to remain below 1938 levels for lack of coal and other raw materials. Germany had ample forest reserves for cellulose wood pulp, but they were "being heavily cut at present for firewood and lumber for export."[39] Practically all synthetic plants in the Russian zone had been destroyed or relocated into the Soviet Union. In the British zone, the war's destruction of synthetic factories ran about 50 percent, and Dunn reported that "current operations are thought to be very small."[40] Dunn had no figures for the French zone but thought production was limited. In the American zone, output reached only 12 percent of prewar levels, and the shortage of coal was expected to restrict increases until a fundamental reconstruction of the economy occurred.

Regarding cotton, Dunn reported German mills running at 25 percent of capacity for lack of raw fiber. He felt confident that German textile output would increase to two-thirds of capacity in the American zone by the end of the year if contracts for coal were fulfilled and raw cotton were available. In the British and French zones, cotton textile production ran at one-fourth to one-third capacity. Again, lack of coal remained the culprit. Dunn complained that Russia would not furnish information, but he felt assured that production stood at thirty thousand bales, equal to the combined output in the British and French zones. He reported that the textile payment plan had been endorsed by General William Draper, commander of the Economics Division of the Office of Military Government for German recovery. Dunn knew the plan was "now being considered by U.S. government agencies in Washington."[41]

Jackson and Dunn paralleled each other in their trips to Europe. They had similar assignments: to report on the condition of textiles for the likelihood of restoring production, and to investigate new technologies that pertained to cotton and synthetics. Each furnished the NCC with straightforward information about the desperate conditions of the people there, especially Germans, and linked their plight with the greater question of recovery, seeing cotton as part of the larger concern. When Dunn sent his reports in late 1946, he detected a slight improvement as assistance trickled in from the United States because food rations had been raised in the American zone from 1,250 to 1,500 calories per day for each person. Both men discussed Europe's need for funding to jump-start cotton textiles, either through the textile payment plan or through direct lines of credit from the U.S. government. They acted as operatives of an industry seeking to improve its commercial self-interest, but they also demonstrated the dire conditions of Europeans struggling in the rubble of war. Both tied the resuscitation of textiles with the reconstruction of the German economy; they concluded that general rehabilitation had to be undertaken to restore the cotton trade.

COTTON'S AMBASSADOR

Assistant Secretary of State Clayton was familiar with the reports, thanks to Fleming at ACCO, but he had to tread carefully on the subject of cotton, since he still owned 40 percent of the company's stock. He could not take any action that would arouse suspicion and endanger his reputation for impeccable integrity. After his confirmation hearing in 1944, he wrote to a friend: "The most stubborn part of the hostility which the Senate demonstrated against me arose with the Southern Senators. Incidentally, cotton has always been political dynamite and I suppose always will be."[42] Clayton's reticence explained why Jackson complained of his reluctance to "talk cotton" when he met him in London over the textile payment plan. Clayton had a sense of cynicism toward cotton. He complained that "southern barefoot standards" perpetuated poverty, and he believed that only mechanized farming offered a future. In 1945 he told a young lieutenant who was thinking about a future in the cotton business: "I am sorry there is nothing I can tell you at present of an encouraging nature." Conditions are so uncertain, Clayton said, that "I cannot advise any young man to go into it."[43] Only a short time later, however, he told a colleague, "King Cotton is not dead" because he saw an opportunity to redesign America's trade policies for the benefit of the national economy and regain overseas markets for cotton. Clayton's presence in the State Department encouraged the assumption evident before the close of the war that efforts would be made to restore European markets. Starting in 1947, he had to work cautiously, since Republicans had captured control of Congress; he had complained earlier to one confidant that they were too isolationist.[44]

Clayton's stature as a cross-link for international affairs and the welfare of cotton became evident when in August 1945 he received an invitation from *Cotton Trade Journal* to prepare an article titled "A New Era in World Trade." He eagerly accepted the invitation, and his piece appeared in the journal's special international edition. He succinctly stated the position of the State Department's Economic Branch, which paralleled the view commonly found in cotton circles. "It is obvious that some kind of new era of world trade is upon us," he wrote, but "what kind of new era?" America should not let the prewar trading blocs and cartels reappear, but rather strive for "fair and open competition" with "lowered restrictions and orderly exchange rates." Autarkic practices must not impinge on the freedom to trade across borders, but "trade cannot prosper until the means of production have been recreated, people put to work again, trading connections reestablished, and a basis laid for business credit." Because so much of the world lay in ruins, he wanted the United States to promote agreements aggressively, sometimes on an emergency basis, to revive trade around the globe. Since our trading partners were seriously short of capital, he felt it essential to offer them financial assistance, which would enable them to reconstruct their economies and engage in trade. Barriers to trade or discriminatory practices should be removed. It was not clear, he admitted, if "other countries will be able to go with us," but the United States must act to help them. "The world is very sick," he warned, and the creation of a new trading era will not be easy, but it was essential to "raise standards of living at home and abroad for the benefit of everyone."[45]

Clayton was repeating similar calls made earlier in the Cotton Belt. "If the world perseveres in economic nationalism," wrote Lamar Fleming in 1943, "the suspension of military struggle will bring no peace, but only economic war in preparation for the inescapable next military war." And in 1944 Roger Dixon, a cotton merchant from Dallas, condemned the Fordney-McCumber and Hawley-Smoot tariffs for restricting trade between the United States and Europe. Restrictions must not be allowed to remain, he insisted, "while two and half million farmers starve."[46]

Clayton remained aloof from the NCC and other farm organizations for good reason. Congress held great respect for him and listened to him closely, so for the sake of securing new economic alliances for the United States, Clayton could afford no accusations of showing favoritism. But he kept up with cotton affairs principally through Lamar Fleming, who served on the NCC Foreign Trade Committee and whose knowledge of cotton on an international level was unsurpassed; and in October 1944 Clayton urged Fleming to dispose of ACCO's holdings in view of the large CCC stocks and his expectation that prices would be flat at the end of the war.[47] No one manipulated his thinking, but he was not insulated from the ebb and flow of thought in the Cotton Belt.

One of the clearest spokesmen connecting the need to rebuild Europe with the revival of trade was Lamar Fleming. In June 1945 he told Texas senator W. Lee O'Daniel that the destruction of Europe had never been so severe, that "industries are destroyed" and "homes are worn out."

> So industries everywhere will have to be rebuilt; dwellings, cities, and transportation will have to be rebuilt in the war-torn countries, and they will have to be repaired, renovated and extended in all countries. Diets will have to be built up again and people will have to be reclothed, and stores will have to be supplied with new inventories.[48]

THE BRITISH LOAN AGREEMENT OF 1946

An example of the desire by the cotton industry to help Europe occurred with the British Loan Agreement of 1946. Beginning in late 1945, Britain began negotiating with the United States for a large loan because it had expended a strikingly large portion of its financial reserves to fight the war, and now with its productive capacity reduced and in view of the pressing needs of its people in the home market, Britain could not allow the exports needed to finance imports. Yet the country had shortages that only imports could fill. When Britain requested a $6 billion loan from the United States, it did not come unexpectedly. Clayton and John Maynard Keynes conducted the negotiations, resulting in an agreement for $3.75 billion. In the United States, reservations about the loan were widespread, with some cotton interests seeing a short-term gain for the United States but questioning the long-term benefit, since Britain had a staggering debt. Opposition to the loan was strong, particularly in New England, and the Senate passed the measure with a margin of only twelve votes, indicating how Congress looked askance at loans to foreign powers, including America's political and cultural ancestors. As put by Secretary of the Treasury John W. Snyder, "The loan was very controversial."[49]

The respect for Clayton and the importance of his fellow cotton enthusiasts in the reconfiguration of American trade became clear when in March 1946 Oscar Johnston wrote the assistant secretary about the loan agreement as it was pending in Congress. Johnston expressed his disappointment that Mississippi senator Theodore G. Bilbo planned to oppose the measure, but Johnston understood that several "of the most powerful and influential" friends of the senator had undertaken to change his mind. Johnston was under pressure to announce publicly his personal support for the agreement, "acting purely in my individual capacity and in no way involving the National Cotton Council of America." He had refused because he was president of D&PL, owned by the Fine Cotton Spinners and Doublers Association Ltd. of England. Because of his employment with the English interests, he anticipated that his "own views would be distorted by opponents of the proposed legislation." Johnston cited an article in *P.M.* magazine, dated March 17, 1946, describing the NCC as a "greedy lobby" and implying that he would be "doing the bidding of a group of English capitalists and financiers."

Privately Johnston gave his support to the pending agreement and told Clayton that "it will make a very substantial contribution toward enabling Great Britain to recover from the terrific losses she has sustained as a result of World War II." Britain "bore the brunt of the war," he continued, before the United States "went into the fight," and "we should render every conceivable aid and assistance." Johnston recognized that the loan would enable the British to buy U.S. cotton, but he acknowledged they could spend the money on cotton produced in another country, a competitor. Still, the loan should be made to help Britain rebuild her economy and "to prevent the further development of Communism in England and throughout the British Empire." Johnston added the last comment because he had received warnings from associates in Britain that the Soviet Union opposed the loan. This last statement indicated how closely cotton interests remained in touch with conditions overseas, because at this point communist activity in Western Europe was only beginning to be recognized outside the confidential circles of government and military officials.[50]

Johnston shared this correspondence with Read Dunn with instructions that he was "at liberty to show this letter to anyone who may be interested, but I prefer that it not be handled as to fall into the hands of the Press."[51] Unlike its activity on behalf of the Bretton Woods Agreement, the NCC purposely refused to take a stand on the British loan, which accounted for the dearth of publicity among cotton activists on behalf of the proposal.

Johnston's encouragement to Clayton, while avoiding the limelight, demonstrated the dilemma of the State Department. Opposition to making foreign loans resulted from America's experience with loans made during World War I. Most of them had not been recovered, and the private investments in Europe during the 1920s had helped maintain its shaky postwar economy. With the onslaught of the Great Depression, these loans had come to be regarded as a mistake. Making loans or grants to Britain and Europe now seemed out of place because Americans had faced rationing during the war and continued to incur shortages. Refusing to extend assistance, however, endangered the crushing need to rebuild the European economy, on which the

United States depended. Time had become a factor, too, as evidence mounted of communist influence into the countries damaged during the war. To stand by idly meant greater threats. Clayton was reassured by the knowledge that Johnston's endorsement expressed the view of cotton interests. It also demonstrated how these two giants of the Cotton Belt were linked by the responsibility of the United States to exert leadership in restoring the world economy.

Clayton and Johnston were kindred spirits, each convinced that free trade, or at least a minimum of restraints on trade, was essential for a strong economy in the United States and provided the most effective means of preventing the rise of totalitarianism in the world. Such views had been common in the Cotton Belt before the war, but they now had special meaning after a generation of protectionism and world war and the rising influence of communism in Europe. And since international trade was important to further employment in the United States, the cotton community, which included the assistant secretary of state, thought the restoration of foreign markets should be an objective of public policy. As the United States began to establish its leadership in rebuilding the world economy, the trade philosophy of the cotton South offered much attraction. This combination of circumstances bestowed extraordinary influence on cotton interests.

KING COTTON AND THE MARSHALL PLAN

The momentum for extending aid to Europe continued, though it proceeded awkwardly and with no defined objective. As early as December 1944, when the Senate Committee on Foreign Relations conducted hearings on Clayton's nomination for assistant secretary of state for economic affairs, he testified that for the "first two or three" years after the war, the United States would have to extend credits to foreign trading partners.[52] In October 1945 he had stated during a radio interview with Barnett Nover of the *Washington Post* that the scarcity of food and clothing in Europe would require the United States to extend "gifts or loans" to those countries until they could acquire sufficient financial reserves to engage in trade.[53] In March 1946, Rhea Blake forwarded to the State Department the NCC's formal recommendation for an international conference on trade and employment as a step toward establishing a permanent international trade organization. From the State Department's Office of International Trade Policy, Blake received an explanation of the plans under way to initiate such a conference, reporting that the council's resolution "is entirely in harmony with the policy which this government is following . . . and [the State Department] hopes to have the continued support of NCC's members in a joint effort to expand world trade and employment."[54] In September 1946, a subcommittee of the United Nations Economic and Social Council reported "serious shortages of food, housing, domestic equipment, tools, clothes, footwear, and raw materials throughout Europe."[55] Truman had asked former president Herbert Hoover to make a study of economic conditions in Germany, and he responded with a dramatic and straightforward report summarizing the weak economy there and warning that our former enemy must receive aid. In the conclusion, Hoover wrote, "It may come as a great shock to American taxpayers that, having won the war over Germany, we are now faced for some years with large expenditures for relief for these people." Hoover justified his proposal on humanitarian grounds and thought it wise to "establish such a regime in Germany as will prevent forever again the rise of militarism and aggression with these people."[56]

A boost toward sales of cotton overseas came in January 1947 when Secretary of Agriculture Clinton P. Anderson addressed the NCC at its annual meeting in Galveston, Texas. Held in the Buccaneer Hotel on the city's bay front, this meeting again demonstrated the symbiotic relationship between policy makers in the Truman administration and cotton interests in the evolution of a rehabilitation program for countries devastated by the war. Anderson wanted further efforts to get cotton to foreign countries, with particular emphasis on Germany and Japan. His position meshed perfectly with the organization's official resolution, which urged the Departments of State and War to take prompt action "to supply American cotton . . . to the mills of occupied Germany and Japan." Johnston forwarded this resolution to Secretary of War Robert P. Patterson, who replied that in addition to shipments already sent to Germany via the textile payment program that went into effect in 1946, he recognized that the former enemies needed still more cotton, and the War Department intended to push for a program to that end.[57]

The trade philosophy of the cotton South paralleled the position taken by Cordell Hull, who had struggled against the restrictive practices of the 1930s; the Reciprocal Trade Act of 1934 had moved in his direction. During World War II support for open trade had manifested itself in the Bretton Woods Conference and the effort to promote cooperation among U.S. trading partners through the General Agreement on Tariffs and Trade (GATT). In January 1947 Truman sent a list of proposals to Congress that included a recommendation for erecting a system of open trade via the proposed International Trade Organization (ITO) that Hull had wanted. That same month, however, the Southern State Agriculture Commissioners opposed the announced reduction of tariffs to be made through GATT negotiations.[58] Tariffs still retained strong support, and Republicans, the historic champions of tariffs, controlled the Eightieth Congress, all of which caused British negotiators to drag their feet on forfeiting their preferential trading system for the Commonwealth, the sterling area, during the GATT negotiations because they saw little public support in the United States for Truman's stance on trade. When Congress passed a wool tariff even as Clayton fought for reductions at the GATT negotiations in Geneva, the momentum for promoting a greater flow of commerce stood in real danger. Clayton made a special trip back to the United States and persuaded Truman to veto the wool bill, which saved the Geneva conference.[59]

Truman emphatically embraced open trade when in March 1947 he spoke about foreign economic matters at Baylor University in Waco, Texas. His speech was broadcast on national radio, and Truman unequivocally stated the case for trade as necessary for promoting peace and freedom around the world. For too long we had acted as isolationists, he said, and the United States must now "join forces with other nations in a continuing effort to organize the world for peace.... The slogans of 1930 or of 1896 are sadly out of date." He denounced America's Hawley-Smoot tariff, Britain's system of imperial preferences, and the economic restrictions adopted by Nazi Germany. "The future is uncertain everywhere," he warned, and "economic policies are in a state of flux." Our mistakes must not be repeated, and "whether we like it or not, the future pattern of economic relations depends on us." He again urged the establishment of the proposed ITO. "In economics, as in international politics, this is the way to peace."[60]

The president and King Cotton reaffirmed each other, and each had a sense of urgency. Each saw trade in the totality of world recovery and thought it should include aiding nations recovering from war and stabilizing currencies as well as reducing tariffs and ending preferential trading blocs. Only a geopolitical perspective would work, and that objective required a commitment to multilateral cooperation. American prosperity was linked to international prosperity, and U.S. leadership was indispensable for providing the framework for global trade. It was this position that Clayton expressed during his Thursday luncheons at Washington's Metropolitan Club with staff members of the State Department.[61] From the perspective of cotton interests, the permanent reestablishment of foreign markets still looked uncertain because the postwar shipments of raw fiber to Europe and Asia had been undertaken through government auspices as temporary relief, and the private trading of cotton since the end of fighting had been minimal. By spring 1947 the outlook for restoring normal trade worsened as the European economy suffered setbacks from weather and flooding, pressing the case that massive aid for the rehabilitation of Europe was essential for a permanent peace and new trade order. To keep up the fight, the presidents of the affiliate organizations of the ACSA telegraphed President Truman and endorsed his call in Waco for free trade and lower tariffs.[62]

Despite the position of cotton interests and the Truman administration, there was no clear definition of, and certainly no consensus about, U.S. purposes in Europe. The populace was naturally inclined to concentrate on home life and domestic affairs after enduring the hardship and sacrifices of the Depression and war, though the public was not self-absorbed as it had been in the hedonistic Jazz Age, and knew that indifference to world affairs would be costly. Still, the midterm victory of the Republicans in 1946 left no assurance the country would abandon protectionism, which created within the State Department a natural skepticism over public acceptance of further trade alignments. It looked for support just as cotton interests fought for a new trade order.

By spring 1947 the cotton surplus disappeared, largely, though not entirely, thanks to the efforts by the NCC to stimulate overseas shipments. The surge in consumer spending accounted for the domestic consumption, with strong employment and the need to overcome wartime shortages in goods. In 1947 the price for raw fiber was 33.3 cents, slightly over the parity, or government support price, of 28.52 cents.[63] The textile swap plan for Germany had brought about the export of over one million bales, and the money received from the sale of cloth reimbursed the CCC, which had furnished the cotton. Through the British Loan Agreement, along with smaller shipments made to the Netherlands, Belgium, and France, the CCC sold 990,000 bales. E. D. White reported that 500,000 bales had been shipped through the Export-Import Bank by mid-1947. Shipments to Japan had been slower because only about one-sixth of its textile spindles had survived American air raids, whereas in Europe four-fifths of its spindles had remained intact. In 1946 a textile payment program established by the State and War Departments and the CCC sent 890,000 bales to Japan, with the proceeds going to the CCC. As reported by the *Cotton Trade Journal*, the NCC "has been of immense assistance in getting the program underway."[64] Plans for shipping cotton to Germany and Japan through private channels, the ultimate objective of the cotton industry, were under way, with private funding established at $150 million and $60 million respectively. Production of artificial fibers had not resumed overseas. White later told the Eighth Cotton Research Congress in mid-1947: "We are rid of our price-depressing surpluses."[65] Apprehension now grew about the carryover for 1947–48, which most estimates placed at slightly more than two million bales, considered to be the minimum for the needs of the United States. This development, however, did not deter the efforts to reestablish world markets and trade because as Cotton Belt interests acted on economic motives, they continued to recognize, as did the cotton merchant Burris Jackson of Hillsboro, Texas, that "cotton is important to the peace of the world for it is one of those strategic products essential to the economic security of every country on earth."[66]

Direct pressure for overseas cotton shipments with implications for a broader program to rehabilitate Europe came in April when senators from the Cotton Belt formed an ad hoc committee to study the advisability of legislation "to ensure retention of markets in Germany and Japan for American cotton."[67] The committee included California senator W. F. Knowland, who joined them in urging the State Department to devise a program to extend sales or else cotton interests would take steps. This group was likely responsible for Secretary of State George Marshall's concern that "others, particularly people in Congress, would start coming up with ideas of their own about what ought to be done for Europe."[68] By April 1947, however, "American aid to European reconstruction was in the air," so the cotton industry by this point was part of the growing momentum for establishing a European recovery plan.[69]

From the standpoint of cotton promoters, foreign trade took a historic leap forward in their backyard when Under Secretary of State Dean Acheson spoke on May 8 to the Delta Council in Cleveland, Mississippi, a speech described by Truman as "the prologue to the Marshall Plan." Acheson's trip originated from a promise Truman made to his personal friend Billy Wynne, one of the founders of the NCC, that he would address the Delta Council at its annual meeting. It is likely that Truman's confidant Barringer originated the visit because he was the viaduct that mid-South interests had to the White House.[70] The death of Senator Bilbo, however, had created some division among Mississippi Democrats over his successor, so to avoid becoming embroiled in the squabble, Truman withdrew his offer but promised to send a "heavy hitter" from the administration, who would make a major foreign policy speech. The president asked Acheson to speak, and Acheson suggested that he spell out the worsening economic crisis in Europe and indicate the need for the United States to extend large-scale aid. "If the Delta Council wanted an important foreign policy speech," Acheson later wrote, "here was one." Truman agreed, and Acheson put several staff members, including Joseph Jones, to work on the speech. Acheson flew to Mississippi on May 7, and the next day the Wynnes drove him through the area, showing off the lush crops and "cattle knee-deep in rich pastures." Acheson noted the "picturesque but ramshackle shanties giving way to neat, well-fenced farms and painted houses."[71] When he spoke after an outdoor barbeque, Acheson stood before a throng inside the gymnasium at Delta Teachers College, but a larger crowd sat outside. Children played while mothers cared for babies. Loudspeakers carried the remarks and speeches made inside the gymnasium. "The setting for the speech was as fine as anything Hollywood could have contrived," said one account.[72] Acheson wore a three-piece suit, but upon reaching the rostrum, he removed his coat and rolled up his sleeves. The scene was "thoroughly American," he wrote.

Acheson pointed out that because of the war, "Europe and Asia are today in a state of unbelievable physical destruction and economic dislocation" and that Germany and Japan, "two of the greatest workshops of Europe and Asia," had yet to begin reconstruction.[73] Conditions in Europe had worsened, he reminded them, because of blizzards and flooding in 1946. The United States should export foodstuffs and other badly needed commodities to Europe, but since the countries there had no financial reserves, Congress would have to furnish the financing for them to acquire American goods. "No other country is able to bridge the gap in commodities or dollars."[74] Acheson spoke in general terms and did not mention cotton, but his references to the agricultural surpluses and commodities needed to alleviate hunger and shortages of clothing made clear that the produce of the Delta was urgently needed.

His plea, made only twenty miles from the mud banks of the Mississippi River, resonated through the Cotton Belt, where the flow of commerce was perceived as essential for world prosperity and political stability, and restrictions on the exchange of goods were regarded as a threat to living standards in all countries. In 1943 Lamar Fleming had repeated the belief that cotton was tied to peace in grandiose terms before the Cotton Research Congress in Dallas: "The future of American production and exports of cotton happens to be at stake on the same decision as the future of peace, on our standard of living, and our civilization."[75] With the growing apprehension about communism by 1947, the link between trade and peace had significance.

For the *Commercial Appeal*, Acheson spoke "sensible realism" and presented a State Department policy that should be endorsed "whole heartedly and without counting dollars and cents."[76] He offered an opportunity for the resources of the Delta to be used for creating a new order, so the *Commercial Appeal* opined, which would simultaneously combat the spread of communism. Acheson's plea that American aid should move beyond traditional loans recalled the earlier effort by Hickman of the *Cotton Trade Journal* to raise $100 million of cotton fabric to be donated to France. And only a few days before Acheson's speech, a Dallas cotton shipper proposed huge loans to Germany so that it could buy U.S. goods.[77]

The State Department had anticipated the enthusiastic response to Acheson's proposal. Since 1945 correspondence and reports from the NCC, plus other cotton organizations, had made clear they saw large-scale support as essential for Europe's recovery. In August 1946 confidential surveys made by the State Department showed that the southern states believed more than all other regions that trade restrictions impeded economic growth and that trade fostered world peace. During the war, free-trade proponents such as Sumner Wells, Joseph Grew, and Nelson Rockefeller published in the *Cotton Trade Journal*.[78]

After Acheson's speech, the plan for reconstructing Europe moved rapidly ahead in the State Department. Clayton furnished much impetus with a famous memorandum of May 27 in which he expressed the worsening European crisis in severe terms. He had been in Europe conducting negotiations with the newly formed GATT and had seen the deteriorating conditions. "Millions of people in the cities are slowly starving," he warned, and the "grave economic crisis" was leading to a political crisis. If the standard of living fell further, "there will be revolution." He recommended shipping surplus commodities: "6 or 7 billion dollars worth of goods for three years . . . principally of coal, food, cotton, tobacco, shipping services and similar things."[79] After food, Europeans most needed clothing, and since the manufacture of synthetic fiber required an abundance of coal, which was not available, cotton textiles presented the only feasible means of clothing the destitute millions in western countries. Sending cotton to Europe for the sake of humanitarianism and recovery had never been a stated objective, but it was a common assumption in the evolution of policy.

With this message and Clayton's impelling reasoning within the inner circle of the State Department, the proposal for massive aid moved forward. "William Clayton was the most powerful and persuasive advocate to whom I have listened," Acheson later recalled.[80] Clayton worked on behalf of world recovery and American security because he felt the two were inseparable, and rather than seeking only to recapture markets for cotton, he knew, as did his friends scattered through the South, that U.S. cotton offered the quickest means for covering the poorly clad of Europe. The commercial benefits, so evident, were not the final objective but would be an effect of a larger and more worthy goal.

When Secretary of State Marshall announced the administration's proposal on June 7, however, he represented the contributions of policy makers besides Acheson and Clayton. Marshall had indeed set the State Department into decisive motion when in April he returned from a meeting in Moscow and put George Kennan to work; Kennan, with his policy planning staff, developed "thoroughly reasoned papers . . . all compiled for a report . . . that was given to Secretary Marshall."[81] Indeed, the contributions that led to the Marshall Plan had no single source, but as far back as mid-1945, the operatives for cotton had made known their position that the resumption of trade would require massive aid to rebuild the economy of Europe. With the help of cotton's ambassador of goodwill, who now held the position of under secretary for economic affairs, Oscar Johnston's wish that King Cotton sit at the table of postwar trade arrangements had come true.

After Marshall's announcement, doubts remained over the acceptance of the proposal. It faced a stingy Congress, and consumer shortages persisted. "The greatest fear was an adverse reaction from the Mid-West," Marshall recalled, "from Bert McCormick and the *Chicago Tribune*."[82] The *Kiplinger Newsletter* reported, however, that while the general public felt lukewarm about the plan, it was not particularly opposed.[83]

The looming shortage of raw lint did not reduce support for the proposal among cotton's vested interests. They knew that the overseas sales made since the end of the war were temporary and undertaken for the sake of sending relief to countries ravaged by the war. The economies of foreign trading partners remained in disrepair, and in view of the perceived political instability in Western Europe, so went the reasoning, the United States must seek to rehabilitate the European economy not only for King Cotton's health but also for the general welfare of the nation. Read Dunn had cast the question in the broadest terms, seeing the restoration of purchasing power in Europe as the only assurance of a reliable market for cotton, which would happen only when the economies there overcame the aftermath of the world's most devastating war. Repeating the stance taken by Oscar Johnston, Dunn wrote, "We cannot sell any large quantity of cotton or other goods unless we can buy more foreign goods."[84] In April Dunn had pointed out that unless a new assistance program became available, relief contributions, which were benefiting the whole U.S. economy, would "fall off" by the end of 1947. For these reasons, cotton promoters pushed for aid to Europe.

To overcome congressional doubts and resistance to the proposed rehabilitation program, the NCC launched a promotional and lobbying campaign across all levels of public opinion, from the grass roots to the most powerful figures in the Senate and House. It became apparent that the trade philosophy of the Cotton Belt had gained powerful friends when Secretary of State Marshall spoke at the NCC's annual meeting held in Atlanta's Biltmore Hotel in January 1948. Europe had to be rebuilt, stated Marshall, for America's own welfare, and the free intercourse of commodities and goods offered the best chance. The United States and Europe should strive to erect a "mutually profitable pattern of world trade . . . that holds the prospect of greatest benefits for American industry, commerce, and agriculture alike." He continued that only through such exchange, buying and selling in each other's markets, could Europe acquire the means, or dollars, to buy our produce and ensure its own well-being, too. Predicting that about ten million bales would be sent to Europe during the lifetime of the recovery effort, Marshall recognized that agriculture—cotton producers, in this case—stood to benefit, but unashamedly stated, "This is one of the major objectives of our foreign policy."[85]

The *Atlanta Constitution* saw the connection between cotton and the origin of the Marshall Plan. "There is a definite parallel between the National Cotton Council's program for rehabilitation of the cotton industry and the Marshall Plan for the recovery of Western Europe. The Cotton Council has performed an admirable service to the South and the world. It has done much to revive a sick industry, just as the Marshall Plan will do much to rescue the bedridden economy of sixteen Western European nations."[86]

The piecemeal shipments of cotton to Europe since the end of the war had been arranged through expedient channels, but here was an open call for a comprehensive commitment to fill the need for raw cotton in Europe along the lines that agents of the industry had begun urging just weeks after the German surrender. Marshall expected raw cotton to constitute a small percentage of the total assistance package, but because it was essential for European clothing, he believed that cotton and the rest of the four-year program went hand in hand and that neither could stand independently of the other. He warned that powerful forces, Soviet agents, were urging Europeans to abandon their political faith, and that if the United States did not go to their aid, the welfare of the nation would be affected. The fight against communism had, of course, already started via America's aid to Greece and Turkey, but Marshall's insistence that trade needed to be restructured was not so widely popular. His position that cotton played a critical role in European recovery echoed the sentiments expressed by the South and bonded the Truman administration with the region in respect to trade. On the next day after Marshall's speech, the organization passed a resolution fully endorsing the European Recovery Program (ERP) on the grounds that it was "vital to the peace of the world and of fundamental interest to the United States."[87] As it began a campaign to help win approval, the NCC reprinted Marshall's speech in the council's publications. Marshall later recalled that he had to sell the ERP to the cotton industry because of low stocks, but he misunderstood the overwhelming support for the proposed recovery plan.[88]

Following the same tactic it had employed on behalf of the Bretton Woods proposals, the NCC contacted all members of Congress from cotton states and reported their position on the enabling legislation, known as the Vandenberg Bill. According to the NCC, there was a "considerable lack of interest" and "outright opposition" in some cases.[89] Some confusion existed in Congress because of the number of bills introduced on the subject, which the NCC expected to disappear once the Senate Foreign Relations Committee approved the Vandenberg Bill. Some members had not heard from their constituents, so the "time has come for the people of the Cotton Belt,"

Johnston told the NCC membership, "to let them know how they feel about the Marshall Plan. We must move fast."[90] He urged that each NCC member write his senators and representatives. Johnston wrote a brief essay on the proposal for members to use when addressing meetings of growers, civic organizations, and the like. In addition, Johnston wrote a letter of endorsement that he wanted to be used freely. The NCC's Field Service agents promoted the pending legislation and reported the position of members of Congress to local interests, hoping they would apply pressure to their congressional representatives.[91] The NCC headquarters in Memphis kept its Field Service agents updated on the progress of the bill and the position of the congressional members from their area of operations. The organization worked through its Washington lobbyists, Read Dunn and Robert Jackson, to convert any member of Congress from a cotton state who had reservations. Clayton and Acheson had lobbied, too. "We called on individual men and discussed the matter with them," Clayton recalled.[92]

Congress approved the Marshall Plan in March 1948, and Truman quickly signed the legislation. In the course of the debate in Congress, House Minority Leader Rayburn of Texas declared: "It is as important to the cotton farmers of the Fourth Congressional District of the State of Texas that Western Europe be rehabilitated as it is for Western Europe itself."[93] The well-known Speaker of the House, temporarily ousted by the Republican victories in 1946, worked quietly and effectively, and Clayton later singled out Rayburn for his efforts on behalf of the Marshall Plan.[94]

Cotton exports soon picked up, jumping 143 percent over the previous year. Not all exports resulted from the ERP, but it accounted for the boost, with Britain, France, Italy, and Germany taking 2.5 million bales of the total, or 75 percent of the total sent to Europe in 1948 and 1949. In the previous year, all of Europe took 975,000 bales. The shortage of cotton anticipated in mid-1947 did not materialize because prices were strong and farmers responded with large crops, producing 16.1 million bales by 1949. Production tapered off in 1950 with a drop in prices but zoomed again in 1951 because of the outbreak of the Korean War. When shipments arranged through the ERP ended in 1952, ten million bales had been sent to Europe.[95]

During the postwar era, the United States initiated major economic realignments to put the world economy back into order. It accepted the role of leadership thrust upon it, and through a series of measures with global ramifications, whether to stabilize currencies or to rehabilitate wartime friends and enemies, the United States sought to make a new order based on open trade. In the development of this policy, which bounced along awkwardly, the cotton industry, acting as a commercial interest, saw its own well-being yoked to the general welfare of the country and supported the creation of trade alliances. The NCC drove most of this activity, acting as a young, self-interested organization seeking to establish credibility. Through its efforts, shipments of badly needed cotton went to foreign mills and ultimately clothed the refugees of war while removing the burdensome surplus and below-parity prices that had gripped the Cotton Belt for over a quarter century. Cotton organizations had moved quietly, like an ocean current rather than the waves of a pounding surf, which caused their contributions to remain out of sight.

In 1943, when Fleming spoke to the annual Cotton Research Congress in Dallas, he made one of the earliest calls for the United States to take the leading role in restoring world peace and the economy. He foresaw that the recovering countries would need resources and financing from the United States to rebuild and establish a lasting peace. "Peace will be just a recess if rehabilitation and reconstruction are denied the peoples who have suffered from the war."[96] In January 1945, Fleming wrote to Rhea Blake that "I will guess that we will assume right after the war a paternal relationship to many countries."[97] In his speech two years later to the Delta Council, Acheson made the same plea. "For it is generally agreed that until the various countries of the world get on their feet and become self-supporting there can be no political or economic stability in the world and no lasting peace or prosperity for any of us."[98]

CONCLUSION

In retrospect, the activity to restore foreign trade had more than a short-term commercial impact. It disposed of the last vestiges of the 1937 bumper crop that had depressed prices, and for the millions of Americans who depended on cotton, this accomplishment brought renewed hope that better prices for cotton would enable them to keep up with the rising social and economic mobility in the United States. Cotton farmers anticipated getting free of debt and modernizing their homes and farms. A surge of sales in appliances, particularly refrigerators and kitchen ranges, occurred in the rural South.[99] As the United States began the second half of the "American century," opportunities improved for the cotton farmer to cast aside the badge of poverty and enter a world in which nearly all things old had been ripped asunder. Restoring foreign markets had been only one effort among others, however, to regain health and stature for King Cotton.

CHAPTER 6

THE DINNER TABLE WAR
Postwar Struggles

Restoring foreign trade accounted for only a portion of the endeavors to push cotton forward. Industry leaders knew the postwar era was unsettling and threatening to the status quo but realized that it provided opportunity for the imaginative and bold hearted, and that they had to act fast to keep cotton afloat in the swift currents of fashion and consumer tastes. To keep pace in this challenging environment, cotton would have to be lifted out of the quagmire of small-plot inefficiency, boll weevils, and reliance on mules. Among the immediate goals was an end to the discriminatory advantage enjoyed by the dairy industry at the expense of cotton farmers. King Cotton began its battle on behalf of oleomargarine.

THE DINNER TABLE WAR

In 1868 Napoleon III of France offered a prize for the invention of a synthetic edible fat because Europe had a shortage of vegetable oil. The chemist Hippolyte Mège-Mouriés responded with a clear oil that when made into drops resembled pearls, so he called them margarine after *margarites*, the Greek word for pearls. In the United States production of margarine started in the 1880s, with cottonseed oil comprising about 65 percent of the basic ingredients. Manufacturers began offering margarine as a substitute for butter, but dairy interests strongly objected. They persuaded Congress to pass the Margarine Act of 1886, which prohibited the use of yellow coloring to make margarine resemble natural butter. Congress also prohibited the sale of margarine to the military services, stiffened the license for manufacturing it, and put a two-cents-per-pound tax on margarine. In 1905, when the new process of hydrogenation made it possible to harden margarine, consumer fondness for the substitute grew. But federal law did not require retailers to identify the product, so dairy lobbyists concentrated on the states. Thirty-two states slapped taxes and instituted a coloring ban, meaning that consumers could buy margarine but had to add coloring in their home kitchens and pay a food tax.

During the Depression, the issue of oleomargarine versus butter began to intensify. In 1931, when South Dakota imposed an excise tax on margarine with another five cents per pound on vegetable shortening or cooking oils, it provoked a reaction from the Alabama agriculture commissioner: "It is very deplorable that an attempt be made by the various states to erect artificial barriers which are thoroughly unsound."[1] In 1935 Wisconsin hiked its margarine tax from six to fifteen cents per pound, prompting Alabama governor Bibb Graves to declare that his state would retaliate by boycotting Wisconsin cheese. Wisconsin lost "considerable business" in road machinery in Alabama, though the Wisconsin Agriculture Commission denied it.[2]

The discriminatory measures against margarine struck a nerve in the South, where there had been outcries about the region's colonial economy. In *Divided We Stand* (1937), the Texas historian Walter P. Webb insisted that the South suffered under "economic imperial control by the North." Below the Mason-Dixon Line, one had to contend with "economic carpetbaggers and scalawags of America's new feudalism."[3] Restrictions on margarine and the sensitivity over the question of a colonial economy reinforced one another with the *Report on Economic Conditions of the South* of 1938. The report accepted the claim of imperial control over the region and blamed discriminatory freight rates, absentee ownership, and a faulty tariff structure. When the report singled out the treatment of margarine, it added great weight to the charge from the cotton fields. "The further development of cottonseed oil for oleomargarine and kindred products has been hampered by taxes, licenses, and other restrictive legislation not only by states outside the region but also by the Federal Government."[4]

In his speech at the Peabody Hotel in November 1938, Oscar Johnston referred to the discriminatory laws against cottonseed oil. For him such action helped justify the creation of an organization to fight on behalf of cotton farmers. Once the NCC got under way, it formed the Trade Barriers and Penalties Section with Philip Tocker in charge. The objective was to persuade Congress and various state legislatures to overturn the margarine restrictions, but the outbreak of war forced the new organization to mothball the endeavor. Tocker went into the navy. When in 1941 the *Dairy Record* called for "the complete extermination of oleomargarine" and urged dairy farmers not to rest "until the manufacture and sale of oleomargarine had been outlawed in this country," the Cotton Belt stiffened its determination to put cottonseed oil on an equal footing.[5] Agricultural interests split along the Mason-Dixon Line, but the cotton farmer had a powerful ally: the housewife. During the war, the price of butter rose, and American homemakers searched the grocery shelves for ways to stretch the family budget. The NCC tried to exploit the rising price of butter and lobbied for a repeal of all federal taxes and restrictions. In 1943 Congress dealt with a repeal bill introduced by South Carolina congressman Hampton Fulmer, chair of the House Agriculture Committee, but failed to pass it.[6] Dairymen held too firm a grip, and the servants and allies of King Cotton would have to gather their strength for a later fight.

From 1945 to 1947, the NCC conducted only a mild campaign, but it made some progress. In 1947 legislatures in sixteen states reviewed their restrictions affecting the manufacture and sale of margarine, and a few states enacted friendlier laws. Oregon repealed the prohibition against the state-owned institutional use of margarine, and Pennsylvania removed the requirement that retailers place a label on the carton in gothic type. Utah threw out the licensing and fee requirements for retailers. In 1947, however, four bills to repeal the restrictions on margarine died in Congress. Margarine continued to be the only food product taxed by the federal government, with colored margarine taxed at ten cents per pound and uncolored at four cents. Manufacturers still had to pay a $600 licensing fee, while wholesalers paid $480 per year unless they sold only uncolored margarine, which required a fee of $200. License fees for grocers remained at $48 and $6 respectively. "The fact that the housewife has to color her own margarine," the NCC summarized, "and pay the discriminatory taxes which have been imposed on a food product is adding momentum to the repeal movement."[7]

Sam Bledsoe, now with the National Association of Manufacturers (NAM), wanted the fight taken to housewives rather than "pussyfooting" around with the legislatures, so in the latter months of 1947, the NCC organized one of its most extensive lobbying and public relations campaigns to date. To begin with, Rhea Blake and his assistants devoted more time to the effort. The Office of Public Relations prepared to send messages, editorials, radio scripts, and other information to newspapers, radio stations, farm publications, and civic and women's organizations in the cotton-producing states. To further encourage public demand for repeal of discriminatory taxes, the Field Service likewise planned to contact retail grocers, women's clubs, and civic organizations, supplying material from the Margarine Manufacturers Association and providing up-to-date information on the positions of members of Congress. Bledsoe threw the weight of NAM into the campaign.

Arrangements were made to run advertisements in the *Washington Times-Herald* on behalf of repeal. Publicity alone, however, would likely not overpower the opposition, and political muscle would have to be used.[8]

Robert (Bob) Poage from Waco, Texas, took over as House leader in the struggle with dairymen. He identified with the young rural Democrats who went to Congress during the Depression with a determination to get help for farmers and ranchers. His district lay in the heart of the Texas blackland strip where dryland family farming dominated. Cotton was the principal crop, but corn, wheat, cattle, and forage were popular. The small city of Waco, which had a population of around fifty thousand, served as the urban center of Poage's district. A Texas populist when he went to Washington in 1937, he appropriately got a seat on the House Agriculture Committee. He served for fourteen years as vice-chairman until he took the chair's seat in 1967. For Poage, nothing was more important than helping farmers deal with an outbreak of bollworms, or smoothing the way for them to get electric service from the REA, and generally lobbying for federal assistance to farm families.

Bledsoe had recommended that a denunciation of the antimargarine laws be made in Congress. In January 1948, Poage made a powerful speech on the House floor on behalf of oleomargarine. The laws restricting its consumption had passed in the 1880s, he reminded his colleagues, when margarine was admittedly an inferior product, but improvements in refining and the addition of vitamins now made such restrictions obsolete. Discriminatory laws held down the value of cotton and soybeans, but the pleas to remove the legislation had been "drowned out by the clamor of the butter interests demanding protection and special privilege."[9] Poage blamed the "butter lobby," insisting it had been petted and pampered long enough at the expense of cotton and soybean farmers, as well as housewives juggling family budgets. Cotton interests had never wanted a tax levied on synthetic fiber, so why, he asked, should there be a penalty put on wholesome food? He attributed the perseverance of the protection to Republicans and urged Congress to "tear out these antimargarine laws by the roots and fling them where they belong—upon the scrap heap of oblivion."[10]

Poage introduced a bill for repeal, but Congress gave it little attention and squabbled instead over Truman's proposals for an antilynching bill, extension of the Fair Employment Practices Commission (FEPC), and abolition of the poll tax. But the NCC nonetheless saw progress and predicted a better sentiment when the new Congress convened in 1949. Not all had been lost that year, because the campaign had persuaded four states to repeal their prohibitions against the sale of yellow margarine, leaving nineteen states on which to concentrate. A large number of newspapers and magazines had recommended repeal, and Blake reported that no amount of money "could have purchased the publicity and good will" generated by the effort, which *Time* magazine called "the greatest publicity campaign of all times."[11] Blake seemed confident that some opponents would not return to Congress after the 1948 elections because the campaign had effectively stirred public sentiment. Bledsoe recalled that "it just caught fire."[12] Soon the dinner table battle would jump into mainstream politics.

This step occurred when the Democratic National Committee's platform for the 1948 election stated, "We favor repeal of the discriminatory taxes on the manufacture and sale of oleomargarine."[13] The matter now had a political dimension, though it was overshadowed in the election by other issues. Still, the defenders of the antimargarine laws, the dairy interests, found themselves having to defend an unpopular law. Truman's unexpected victory reinforced his party's commitment, and in his economic report to Congress in January 1949, he recommended dropping some excise taxes, "particularly on oleomargarine."[14] When Congress convened in 1949, the Democratic Party, the solid party of the South, had regained control. A misunderstanding soon arose in Congress, however, over whether the Democratic platform had accepted the proviso that margarine should not be colored yellow to resemble butter. Dairy interests were ready to concede the taxes but wanted to keep the restrictions on coloring. For the NCC, this concession nonetheless retained the discrimination, so the council maintained pressure for repeal of all restraints. Democratic representative Emanuel Cellar of New York ended the confusion when he explained to the House that he had helped prepare the party's platform, and "the plank meant outright repeal of the federal anti margarine laws—without reservations."[15]

If progress seemed slow, it was nonetheless apparent. The prevailing law required a license for restaurant operators who sold colored margarine, as well as the requirement that margarine had to be identified at the serving table. Butter sold for a dollar per pound at the time, and there was considerable suspicion that the licensing requirements were being violated. In 1948 the National Grange had endorsed repeal of all taxes as long as satisfactory means could be developed to differentiate butter and margarine at retail grocers and restaurants. That same year, the Farm Bureau refused to become embroiled in the fracas, but in 1949 it adopted the stance of the Grange.[16] More support came from the American Federation of Labor, National Council of Jewish Women, and the American Home Economics Association. In December 1948 the NCC sponsored meetings of margarine's vested interests in Dallas and Atlanta, again organizing the dissemination of information to the general public and arranging for letters and various resolutions for repeal to be sent to Congress. By 1949 the question moved to the forefront of the news, and a cartoon by Herb Block of the *Washington Post* featured an Easter bunny seeking permission from Congress to color its eggs.[17]

In 1949 members of the House introduced twenty-nine bills for repeal, with Poage leading the fight. Each side lined up supporters, witnesses at hearings, and statements to show public support. Within the butter lobby were impressive supporters such as the Cooperative Milk Producers Federation and the United Paperworkers of America, with over 250,000 members. But the momentum had begun to shift in favor of cotton, owing to the now overpowering consumer preference for buying margarine without restrictions. Only one hurdle remained: dairymen wanted to prohibit the sale of colored oleo, making sure there could be no confusion with butter. In the eyes of the Cotton Bloc, this reservation was a ruse to discourage sales, because margarine backers had conceded that distinct and visible labeling should be put on packages to distinguish them from butter. In 1949 legislation for ending all restrictions passed the House by a margin of three to one, but it bogged down in the Senate. Arkansas senator William Fulbright got a commitment from the Democratic Policy Committee, which set the schedule for legislation, that the repeal measure would retain its active status and be the first order of business in January 1950. C. G. Henry, chair of the NCC committee on margarine, expected the measure to pass and go to the White House. "We have the upper hand in the fight," Henry reported, "but overconfidence could result in disaster."[18] He noted that three more states had lifted their restrictions. Henry proudly reported, too, that Congress had authorized the army and air force to use margarine. "We want to continue our efforts," he concluded, "and win the big victory in 1950, the complete elimination of federal restrictions on white and yellow margarine."[19]

Success came in 1950. In January the Senate passed the Poage-Fulbright bill by another three-to-one majority, and in March the president signed the bill that "brought to an end one of the bitterest domestic struggles in the history of the country," according to the *Cotton Gin and Oil Mill Press*.[20] For the cotton community, this clash tested its resolve and capability to accomplish goals. Credit went to the NCC, which had changed "the complexion and direction of the battle" when it entered the fray. Will Clayton told one correspondent that the NCC "has acquired a kind of recognition as spokesman for cotton people."[21]

A handful of states retained their restrictions, but the drive peaked with the removal of the federal excise taxes and coloring qualifications. The issue was now identified with consumers' rights as much as a self-interest objective of the cotton industry. In July 1952, for example, New York assemblywoman Geneta M. Strong, a Long Island grandmother, took a mixing bowl onto the

Assembly floor and, wearing an apron, demonstrated "the messy chore of adding color to white margarine."[22] Her colleagues reacted quickly, and New York became the eighth state to remove the color ban since 1950. The handful of states holding out slowly conformed until Wisconsin, the last state, lifted its restrictions in 1967.

FIGHTS WITHIN THE USDA

Throughout its campaign on behalf of oleomargarine, the NCC juggled other tasks. To have the greatest effect in obtaining federal assistance for growers across the Cotton Belt, the organization had to be aware of the internecine affairs of agencies dealing with cotton. It faced a challenge when Truman appointed New Mexico representative Clinton P. Anderson as secretary of agriculture in June 1945, which raised a few eyebrows because Anderson had no political identity with agriculture. Cotton leaders accepted the reassurance of Congressman Pace, who had served with Anderson on a special House committee on food during the war. Soon after taking office, the new secretary announced plans to reorganize the department by creating a special branch for each commodity. Ten new branches were envisioned, so the head of the cotton branch would be able to influence and guide the federal program for the cotton industry. Loan activities for cotton centered in the CCC but were expected to be shifted to the new branch. "The head of the proposed new cotton branch," reported the NCC, "would be about the most important federal official where cotton is concerned."[23] Speculation had started on the selection of the new official, and the NCC favored E. D. White, assistant to Under Secretary of Agriculture J. B. Hutson. C. C. Smith, head of the CCC, also held favor in cotton circles.

White had a cotton background in Arkansas and began his career with the Department of Agriculture through the Extension Service. He moved to the Agricultural Adjustment Administration (AAA) in the 1930s and then served as head of the CCC. White identified with cotton and had developed the textile payment plan for Germany patterned after the Soviet arrangement made with Hungary. As Anderson now contemplated his appointments, he discovered that White and Smith had political baggage. Smith had alienated growers in the western states by placing a CCC loan discount on irrigated cotton. Though Smith had been following the common practice by textile companies of offering a lower price for irrigated cotton, his actions nonetheless had negative connotations for his appointment. Within Democratic circles there was opposition to White that the NCC could not explain, stating only that the "grounds for such protests have remained a mystery."[24] As a compromise, Anderson appointed C. C. Farrington of the USDA as temporary head of the new branch. Smith visited growers in Arizona for damage control, but he and White were "definitely in running," reported *Cotton's Weekly*. As winter approached in 1945, it became apparent that discord existed in the USDA. The trouble began upon Truman's selection of Anderson as secretary because Hutson expected Anderson to name him the department's "top-flight administrator." Hutson understood cotton, but his standing to become top administrator was not ensured. This jockeying had significance because wartime emergency agencies that had affected cotton such as the OPA and WPB were expected to end. New legislation for all commodities was anticipated since the war had ended, justifying the NCC's statement that "cotton has a big stake in the Department situation."[25] By the summer of 1946, however, no solution had come forth.

Then the ghost entered the fray. In August 1946, Barringer wrote Truman about the supply of raw cotton in the United States in view of the large exports that had gone to Europe and Japan since the war ended. Barringer praised Secretary Anderson's "keen interest and complete cooperation" but thought he needed a special assistant "to assist and coordinate all government cotton programs."[26] This proposed slot, Barringer explained, would outrank the head of a commodity branch. Truman forwarded the letter to Anderson with the suggestion that he discuss the matter with Barringer.[27] In January 1947, Anderson responded to the president and explained that after several discussions with Barringer, he planned to make White his special assistant for cotton affairs. Truman quickly informed his longtime friend and confidant Barringer of the decision; the NCC had maneuvered its first choice into the high echelons of the department. At its annual meeting held in Galveston, the organization passed a resolution commending White's appointment. This incident indicated the quiet and effective manner in which King Cotton's reach extended to the Oval Office.[28]

THE CLASH WITH LABOR

In the postwar struggle to resurrect King Cotton, the industry fought for a variety of objectives. Trying to tear down the barriers blocking the road to prosperity, it believed that every penny had to be squeezed out of the cotton dollar. Meeting the challenges from synthetics and competition from foreign growers meant that all costs, beginning with the preparation of seedbeds to the final sewing of apparel, must be kept to a minimum. Labor costs could not be allowed to spiral upward in the ginning and handling of cotton in warehouses, compresses, and shipping. This objective grew, too, from the general anticipation that inflation would occur as the country converted to a peacetime economy. During the war, the NCC had managed to block a request for an increase of twenty cents per hour for textile workers made by the Congress of Industrial Organizations (CIO), so when in late 1945 legislation was introduced in Congress to raise the minimum wage by twenty cents per hour, the NCC confidently joined the opposition.

The proposal had features directly affecting cotton. It would make workers in cottonseed oil mills eligible for time-and-a-half wages and eradicate the Fair Labor Standards Act's restrictions on wages and hours for cotton gins, compresses, and warehouses. And it would establish industry committees that would set minimum wage rates for skilled and semiskilled workers whose job classifications were above the basic minimum wage set for unskilled labor. As seen by the NCC, this last provision would leave the national wage structure solely to the discretion of the federal government, "a major step into national socialism," and would be "utterly incompatible with the maintenance of both free competitive enterprise and a democratic form of government."[29] Blake wrote Senator John Bankhead that "all of the progress which we hope to make in the next several years in reducing our costs and improving cotton's competitive position might easily be wiped out overnight by the passage of this legislation."[30] The proposals failed, including a hike in the minimum wage, but this result should not be attributed solely to opposition from cotton interests because opposition to labor was growing. After V-J Day, strikes increased in heavy industries such as automobiles and steel. In February 1946, three officials of the OPA branch in Georgia resigned their posts in protest of CIO influence. "We are sick and tired of government of the CIO, by the CIO and for the CIO," they told the press.[31] Labor's image suffered when Winston Churchill came to New York City in March that year and CIO and communist protesters surrounded his hotel and called him a warmonger. When the CIO announced the next day that it would soon start a drive to unionize the South with a war chest of one million dollars, it put the NCC's membership on alert.[32]

In June 1946, King Cotton and labor drew their weapons when the International Teamsters Union publicly declared, "The Cotton Bloc in Congress is responsible for most of our domestic troubles." It alleged that the bloc consisted of "reactionary southern Democrats" who had joined with "reactionary Republicans to destroy the living standards of the American people."[33] Southerners in Congress, so the Teamsters Union continued, were "interested primarily in perpetuating the plantation-slave economy of the South." America should strike back, went the accusation, by boycotting cotton goods. To "crack the Cotton Bloc," the union encouraged consumers to buy rayon and nylon clothing and "refuse to buy anything made of cotton." A boycott would presumably drive cotton prices down and force many Cotton Bloc members out of Congress. To promote full employment, social security, public health, and higher wages, the union had a simple solution: "DON'T BUY COTTON."[34]

Oscar Johnston, the "elder statesman of cotton," reacted with fiery passion. To "the pompous, boundless insolence of today's power-bloated union bosses," he replied, "the union has much yet to learn of the South." This effort to "starve America's cotton states into economic despair" would backfire because it would unite southern members of Congress to thwart the plan by "Teamster Dictator Dan Tobin" and his "un-American fellow tyrants to dominate this nation." Johnston met Tobin's vindictiveness with his own. "Our nation's pampering and appeasement of professional false gods of labor have reached the point where a Tobin has the impertinence to proclaim that we shall no longer be permitted to elect our congressmen of our choosing." Regarding elections, Johnston concluded, the challenge will be accepted "in every ballot box in every precinct in every county of the cotton states."[35]

The NCC deployed its standard mailing campaign. It sent a copy of the Teamsters statement to its members and to newspaper offices across the Cotton Belt with a request to give the matter editorial consideration. With an attached letter, Johnston repeated the assertions of his reply to the Teamsters but hurled a few new stones at them. In the upcoming election, Johnston asserted, the "candidates of the labor-lords" would be thrown out at the polls, while southern candidates would "have solid and unswerving support in their resistance" to labor's attempt to dominate the United States.[36]

From the Teamsters came a quick reply. Under the headline "The Cotton Monopoly Squeals," the union asserted that the NCC was "a subversive organization" acting in conjunction with members of Congress "to undermine the American standard of living." It interpreted the NCC's role in getting the OPA not to impose ceiling prices on cotton goods as a sinister plot to increase prices and take "the American people for a ride." To achieve its ends, so the Teamsters claimed, the NCC had killed congressional legislation on the following: minimum wage, full employment, unemployment compensation, veterans' housing, and public health. Only a boycott would drive the cotton monopoly out of power and let the South send new members to Congress "who are useful to the American people."[37]

This exchange showed deep anger on both sides. Each believed it was working in the best interests of its constituents, the cotton farmer and the working man. Both organizations attributed any attempt to oppose their constituents' well-being to greed and selfishness, and since both represented some of America's most downtrodden, they fought with zeal. Cotton interests saw wage increases as a step toward spiraling inflation and regarded further advances in labor rights as a threat to private enterprise. The Memphis-based NCC carried the flag for King Cotton and girded itself for battle.

This throwing of stones belonged to a broader movement that trimmed labor's power, the Taft-Hartley Act of 1947. When Republicans gained control of Congress in the midterm elections of 1946, unions expected an effort to scale back the gains they had made in the 1930s. In view of the Teamsters' call for a boycott of cotton, the NCC thought restrictions on labor would be appropriate. Shortly after the 1946 midterm election, Blake told Johnston, "The chief topic in Washington these days is, of course, what is going to be done by the new Congress to curb the labor unions."[38] Johnston wrote Georgia senator Richard Russell that his organization had been disturbed over labor practices, "particularly since the audacious attack made on cotton" by the Teamsters.[39]

At its annual meeting in January 1947, the NCC formally endorsed several resolutions against labor. It wanted to prohibit the closed shop and secondary boycotts; to prohibit strikes against government, utilities, and services necessary for health, safety, and general welfare; and to protect the right of a worker not to join a union. "While there are many evils that need to be corrected with respect to labor legislation," Johnston wrote to the membership, "there is one that stands out above all others in importance—the closed shop."[40] And to Senator Russell he opined that the issue of labor "transcended in importance everything before our people at the present time."[41]

To start combat, the NCC hosted an organizational meeting at the Peabody Hotel in March, where representatives from business groups along with cotton interests shared their views and committed themselves to fight for a revision of labor laws. Johnston wrote a speech that went to newspaper editors and radio stations for their use. And literature went to cotton businesses and NCC members, urging them to express their opinions to members of Congress. But the southern congressional delegation needed no prodding because it leaned heavily against labor, and coupled with the strength of the Republicans in Congress, the vote for the Taft-Hartley bill to trim labor passed easily. Truman was expected to veto the measure, which prompted Johnston to predict to a friend that such a step would make it reasonably certain that many solid Democrats would instruct delegates at the 1948 National Convention to oppose Truman's nomination.[42] Truman vetoed the bill, but Congress promptly overrode the veto with a surprising margin, 331 to 83 in the House and 68 to 25 in the Senate. Such a lopsided vote demonstrated how the issue provoked strong feelings across the United States. Lobbying by the NCC did not always succeed, however; for instance, Burris Jackson, who had a close friendship with House Minority Leader Sam Rayburn, made the wishes of the cotton constituency known to the powerful and highly respected Texan. Rayburn voted against Taft-Hartley and voted to sustain the veto.

In contrast to the fight on behalf of oleomargarine, in which the NCC stood conspicuously at the top of the leadership, the crusade for the Taft-Hartley Act had a broader base, and the Cotton Bloc had no prominent role, though it enthusiastically supported the move against unions. The South outranked all regions in animosity toward labor unions, and the newly organized cotton interests conformed with the prevailing attitude in the lower states. Whereas the NCC aligned with consumers and enjoyed popular support in fighting the butter lobby, its stance on labor gave it a Wall Street identity. Cotton growers, however, shed no tears over the passage of Taft-Hartley and saw it as a step toward a level playing field in their own endeavors.

Wages and prices moved upward despite all efforts to stop them. In 1949 a bill from the Truman administration sought to raise the minimum wage to one dollar, a hefty percentage above the forty cents in effect since 1945. The cost of living had climbed, and public sentiment favored a raise. Congress adjusted the bill to provide seventy-five cents per hour, however, and proceeded to consider it. Again the NCC fought the measure but found itself alone. For one thing, the economy was strong, and industrial interests voiced only minimal objections. Few strikes had occurred to stir public anger. Democrats outside the Cotton Belt supported the measure, and Republicans, who lost their majority in the 1948 election, offered little resistance. Despite the efforts of the cotton interests, the new wage went into effect in 1950, when the average price for raw cotton was forty cents per pound, up from the twenty-two cents of 1945.[43]

THE POSTWAR STUDY

As the NCC juggled balls in the air, whether to lobby for exports, to repeal laws on oleomargarine, or to hold down wages, it directed and contributed to the massive postwar study of cotton that had gotten under way through the House Subcommittee on Agriculture, the Pace Committee. No organization was better suited to supervise the undertaking, and the responsibility fell to Mac Horne; he went into the project with much enthusiasm. Acting on behalf of the House Subcommittee, he invited forty-one government agencies and farm organizations to a planning session to be held at the Peabody Hotel in May 1945. Representatives from a broad spectrum attended the Peabody meeting: H. R. Tolley of the Bureau of Agricultural Economics (BAE), James Patton of the Farmers Union, David Lilienthal of the Tennessee Valley Authority, and Edwin G. Nourse of the Brookings Institution. Horne cochaired two subcommittees. This was the same committee that had predicted a "future as black as midnight" for cotton if corrections were not made.

This series of studies lasted two years. Each subcommittee had professional leadership, and Frank J. Welch, dean of the School of Agriculture at Mississippi State College, served as overall chair. Various figures well-versed in cotton affairs and with a penchant for research headed the subcommittees; some came from the BAE, and others occupied academic posts. All eyes were on the future, and the study's raison d'être was to gather and analyze factual information, including trends, to be used in formulating policy. From the committee's perspective, such a goal required a full understanding of the rural culture in which cotton grew, explaining the presence of two tangential sections, Southern Education and the South's Health. The undertaking was ambitious, almost breathtaking, but in view of the depth and breadth of the socioeconomic problems of the cotton kingdom, anything less would be insufficient. Even before the subcommittees went to work, however, Horne anticipated shortcomings, particularly if the study questioned the use of federal supports to keep marginal producers in farming, a topic he thought "too hot for much handling."[44] Work commenced nonetheless with much optimism. Each subcommittee worked independently, and when they finished and presented the results before the House Subcommittee in July 1947, they had produced a lengthy and impressive document.

Printed in 1947, the 877-page document was crammed with facts, statistics, and analyses. Its rich narrative presented evidence in a straightforward manner with no hidden truths. It reinforced the observations and conclusions made earlier by the Southern Regional School at the University of North Carolina, citing on occasion the work by Rupert Vance and Howard Odum. But the updated data included Horne's study of cotton's competitive position in the world market and the drag that tariff walls had on the flow of cotton across borders. The researchers thought that modernization of cotton was imperative, that too many people lived on "horse and buggy farms," that mechanization should be encouraged, and that low-cost, efficient farming had to be achieved. Research offered hope, but fundamental adjustments leading to operations on a large scale had to be made if U.S. cotton intended to compete in the world. In the context of 1947, when 50 percent of all cotton farms produced fewer than eight bales per year, this recommendation applied to small-plot farmers, who would need to diversify with cattle or other crops if they intended to remain on the land.[45]

The researchers delved into the heady question of two-price cotton, meaning a domestic price set by government fiat for cotton grown on controlled acreage and an open-market price for cotton grown on "free acreage."[46] To the consternation of legislators, however, the spokesmen for the nine subcommittees would not recommend one or the other as a proposal for legislation. No two-price program had been used, and the idea had only minor support among growers. In gathering data and information for the sake of making policy, the committee applied current economic conditions as it forecast the next few years, an exercise the researchers recognized as nothing more than gazing into a crystal ball. To anticipate how the economy would develop in the future amounted to guesswork, though little else could be done. But the committee felt strongly about one point: the South needed further industrialization to absorb its surplus population. This advice had come from nearly every study of the South for the past generation.[47]

Whatever weaknesses appeared in the study, it was an accomplishment. From the National Planning Association, a high-powered group of policy thinkers and vested-interest leaders such as public-power enthusiast Morris L. Cooke and Philip Murray of the United Mine Workers came the recommendation that the report be published because "it is going to be a great landmark in the literature about the South."[48] Other than its printing as a House document, however, the report never went to a publisher.

As a step toward policy, the study had no impact. It clearly defined the woes and ills of cotton but provided no specific proposals, since it was meant to be a fact-finding endeavor chiefly for use by congressional policy makers, and the wealth of those facts was tucked deeply into the reports of the nine subcommittees. Pace requested that Horne analyze each report for information that could be considered in drafting an overall cotton program, but Blake suggested to Horne that he needed only to study the "Conclusions and Recommendations" sections. Horne reported after his review that "the summaries and conclusions do not do justice to the full reports." They were "vague, evasive, and dull." He found the full texts, however, to be a "wealth of excellent material" written by savvy cotton economists. But the undertaking had failed, in his opinion, when the researchers tried to move beyond fact finding and draft recommendations. "They demonstrated that no amount of facts can eliminate prejudice or resolve differences of fundamental conviction to make policy decisions." Picking through the reports and gathering the information requested by Pace would take much time. Horne was pessimistic and replied, "I lean toward letting that whole committee structure rest in peace," though it made him "a little sick to think how much good, expensive work lies buried in this mighty tome."[49] With this judgment from the country's recognized expert on cotton economics, the study reached its finale.

Mac Horne supervised other economic studies such as the annual *Cotton Counts Its Customers*. It demonstrated the numerous uses of cotton in the general economy, which Alan Greenspan, past chairman of the Federal Reserve Board, remembered in his memoir, *The Age of Turbulence*, as a particularly valuable work for understanding the cotton industry.[50]

THE BRANNAN PLAN

In 1949 a controversial fight broke out over the Brannan Plan, but cotton interests played only a small role. Truman's new secretary of agriculture, Charles F. Brannan, presented the plan. Secretary Anderson had left his cabinet post to run for senator in his home state of New Mexico and recommended Brannan to the president. Brannan, who identified with the small farmer, began his career with the Resettlement Administration in 1935 and became director of the FSA for Colorado, Wyoming, and Montana in 1941. In these posts, he assisted "agriculture's disinherited."[51] Brannan's philosophical bent resembled the liberal Farmers Union, partly because of his long and close friendship with its president, James Patton. Truman appointed Brannan to replace Anderson in May 1948, "partly to secure the support of the Farmers Union," according to one political analysis. In his 1948 campaign, Truman called for flexible price supports for agricultural commodities instead of the rigid supports then in operation. The president was trying to appeal to the midwestern farm voters who favored his opponent, Thomas F. Dewey, and wanted to distance himself from the Republican-controlled Eightieth Congress. To the surprise of many, Truman won in 1948 and partly owed his victory to a last-minute switch in the midwestern farm states. His new secretary of agriculture went to work on what became the Brannan Plan, described as a "postscript to the election."[52]

When Brannan introduced his proposal to Congress, the plan immediately drew fire. The Arkansas Farm Bureau and the Agricultural Council of Arkansas attacked the proposal on the grounds that it would end the concept of the parity price and cut supports. Such action, according to the Farm Bureau, would mean a return to peasantry for the smallest farmers.[53] Rhea Blake quickly concluded that Brannan would have no support in the cotton industry and expected Congress to give him the cold shoulder. Blake had discussed the plan with his contacts in Congress and based his prediction on their negative responses. Hence, from the beginning, the NCC saw little chance for Brannan. To W. B. Coberly, a cotton trader in Los Angeles, Blake reported "unanimous and bitter opposition" to the plan, and he told Lorenzo K. Wood of the Burley Leaf Tobacco Dealers Association that he saw "no possibility of this program receiving favorable action in Congress."[54] Blake found no support among agricultural organizations except for the Farmers Union, and he saw little enthusiasm among officials of the Truman administration. Owing to the overwhelming dislike that Blake found, the NCC kept in the background.[55]

The fight over the Brannan Plan, however, shaped and defined the power structure of U.S. agriculture for the rest of the century. Brannan faced the likelihood that surpluses would reappear, which would mean another wave of hardship for America's farm families. Surpluses would also place a burden on the U.S. Treasury if the government furnished more price supports and warehoused the glut of food and fiber. In the South there still remained a large portion of cotton farmers who produced only a few bales per year.[56] Brannan faced the task of ensuring reasonable incomes for farmers while not encouraging excessive planting that would bring back surpluses. He wished to further address the fact that large-scale commercial operators received the lion's share of government subsidies. To complicate his endeavors, he also wanted to provide consumers with low-cost food. As reported in one treatment of the subject, Brannan "would provoke a national debate over the farm problem, create deep political antagonisms, and catapult the Secretary of Agriculture into highest prominence both in the Administration and in the Democratic party."[57]

The Brannan Plan intended to furnish farm families with a reasonable livelihood, known as the Income Support Standard, by concentrating on income rather than prices. With his advisers, Brannan devised an income goal that would be applied to each farmer based on his receipts over the past ten years. Adjustments for inflation would be made. To reach the income goal, farm prices would have to be set and calculated into the formula, "so that in the end, his new formula merely substituted one set of support prices for another."[58] To limit government support for large operators, he proposed that subsidies be held to 1,800 "units" of a farmer's production, which amounted to a maximum of $26,100, based on each unit being valued at $14.50. This proviso engendered opposition, so Brannan said that he would not fight to retain it.[59] The plan also lengthened the list of commodities that would be eligible for price supports from the current list of six to ten: he added meat, eggs, chickens, and milk. He hoped to see increased production of these perishable foods and thus a lower price for them, which would benefit grocery shoppers while the farmers received benefits to keep their income at the "standard level." To prevent a surplus buildup, he wanted to reserve the right to impose restrictions on production. A major objection to the plan was understandably its "complex and technical" nature.[60]

The NCC launched no campaign against the proposal comparable to its battle for oleomargarine or the Taft-Hartley Act. So low-key did the organization remain that it took no formal position, and its dislike for the proposal went mostly unnoticed. Other cotton organizations did the same, and except for an occasional item in the local press, little discussion ensued. In a careful study of the plan's impact on cotton, one analyst reiterated the long-standing recommendation: the South's surplus rural population needed to find alternative employment, and the South should improve its educational facilities so that the needy could learn new skills and seek to change their lives. The proposal would have no impact, the study continued, for farmers growing only a few bales; "this perhaps is the major weakness of the Brannan Plan from the standpoint of many southern people."[61]

Although cotton interests felt confident the plan would perish without their attacks, they disliked it. For them, any plan to guarantee an income for each farm family smacked of socialism. Replacing the standard practice of price supports for a scheme to determine each farm's income, so went the common reasoning, threatened each grower's existence. Cutting off commercial operators from benefits after a certain point seemed discriminatory and would be injurious to their business. But in a deeper sense, the Brannan Plan went against cotton. To fight synthetic fabrics and compete in the world market, the cotton industry needed to encourage large-scale operations, that is, practice the economics of scale. Read Dunn recalled this last point as a fundamental reason for the objection to the proposal.[62]

The struggle over the plan centered in the midwestern farm states, which meant the Farm Bureau stood out in the campaign against it. Since Brannan had proposed major changes to the farm program, he drew attention from the national press, which generally opposed them. Labor supported Brannan's proposals because it hoped to lower food costs, and the Farmers Union enthusiastically endorsed them. In Congress the Cotton Bloc wanted the Gore bill, an alternative to the Brannan Plan. From the White House came only lukewarm support, and former secretary of agriculture Anderson would not stand behind the proposal. In February 1950, Brannan and Farm Bureau president Allan B. Kline conducted an open debate in Des Moines that the *Des Moines Register* described as an "ugly and partisan" event at which the audience both booed and applauded Brannan. Read Dunn considered it a "disgrace."[63] In the final outcome, Congress instead passed a bill that Anderson offered as an alternative, and Truman accepted it. The Anderson measure focused on flexible price supports with no provisions for guaranteed incomes and no restrictions on benefits to larger growers. Almost overnight, however, the anxiety over an anticipated surplus disappeared with the outbreak of the Korean War in June 1950. Suddenly the country faced possible shortages of foodstuffs and cotton, and the surge in farm prices ended the quandary about federal support.

Fighting over the Brannan Plan had ramifications. It marked the superior power of organizations that embraced large-scale farming. The Farm Bureau, which represented commercial agriculture, had clashed with the Farmers Union, where smaller farmers tended to congregate. The NAM joined the

Farm Bureau; leading newspapers and magazines opposed the proposal. Among this group stood the NCC and other cotton organizations whose silence indicated not a lack of passion about the plan but a strategy to stay out of the fray. This fight resolved, for a while, the issue of extending federal supports to large planters and reinforced the influence of those growers participating in the business-oriented agricultural organizations. Small-plot cotton farmers exited this struggle with little voice, and their only recourse continued to rest in new forms of employment. In 1949, when the Soviet Union exploded its first atomic bomb, the concern about communism grew, and Brannan's plan appeared to be too socialistic for many. Only a few months after Congress voted down Brannan's proposal, Senator Joseph McCarthy made his allegations that communists had infiltrated high levels of government, and in the ensuing climate of paranoia, any proposal to establish a standard income for farmers while restricting supports to commercial interests engendered fears of radicalism.

THE KOREAN WAR

With the outbreak of war in Korea, the United States suddenly faced possible shortages of cotton. This unexpected possibility had been exacerbated by the imposition of acreage restrictions for 1950, instituted by Secretary Brannan in view of an impending glut of cotton. Brannan had acted on the fact that farmers opted to put about 30 percent of the gigantic 1948 crop and about 20 percent of the 1949 crop into the CCC loan, indicating that the postwar domestic demand for cotton appeared to be slowing down. Exports through the Economic Cooperation Administration (ECA) were, furthermore, scheduled to end in 1952. Brannan had no recourse since the 1938 AAA measure made such action mandatory. With the reduced plantings in 1950, from 27.7 million acres to 21.7 million, as well as bad weather, the 1950 crop produced only ten million bales, one of the lowest annual yields since 1900. By contrast the United States had produced over sixteen million bales in 1949. Now when conflict started in Korea, the price of cotton jumped 30 percent, and farmers had fewer bales to market.

Both the NCC and USDA quickly foresaw a real threat of shortage as they measured consumption and exports against the existing carryover and small 1950 planting. Supplies outside the United States were low, too. In September 1950, just as picking began, the NCC issued a press release calling attention to the "emergency" and recommending that all restrictions on planting be removed. Throughout the industry there was confidence, however, that any shortage would quickly be alleviated. That same month, NCC leaders met with Brannan, USDA officials, and the Munitions Board at a special conference and recommended a program for the government to follow. Controls on planting must be removed, and "farmers must be assured of sufficient planting seed, fertilizer, insecticides, farm machinery, and labor." Ginners, warehousemen, and textile manufacturers would also need assurances of labor and equipment. No commitments came from Brannan or other federal officials.[64] The NCC felt time could not be wasted and became more worried when Barringer reported that Brannan was out of favor at the White House, which prompted some speculation that the organization might have to work around him.[65] However, Brannan formally removed controls for 1951, calling for a crop of sixteen million bales, the amount recommended to him by the NCC's Cotton Mobilization Committee at the September conference. Now the government wanted the Cotton Belt to plant from fencerow to fencerow. Congressman Pace worried about a "critical" lack of labor to pick such a large crop and urged Brannan to facilitate the use of Mexican nationals.[66]

In October 1950, Brannan placed export quotas on cotton to ensure a sufficient supply for civilian and military needs. He intended, furthermore, to hold down the price as part of a broader goal of preventing inflation. Cotton interests, particularly shippers, disliked the quotas and pressured the administration to ease the restriction. In its assessment of the cotton supply, the NCC had foreseen a possible need for such action, and Walter Randolf of the Farm Bureau had agreed that restraints on quotas were necessary.[67] Brannan raised the quota by 62 percent and thus eased some of the criticism. In view of the recent efforts to restore foreign sales, this move seemed unrealistic, which demonstrated how volatile market conditions could be.

But the more important step, one that angered growers, came in March 1951, when the director of price stabilization (OPS), Michael V. DiSalle, put a ceiling on the price of raw cotton. It had been climbing rapidly from the 30.3 cents of 1949. Afraid of the power of cotton to force hikes as far as the retail shop, DiSalle slapped the ceiling at 45.76 cents, which exceeded parity. His action sparked anger. The NCC accused DiSalle of seeking to please "consumers and the voting public." So unpopular were the export quotas and price ceiling that the congressional Joint Committee on Defense Production held a special hearing where it quizzed Brannan and DiSalle. At the urging of the NCC, members of Congress from Cotton Belt states attended the hearing and expressed their dislike for the actions, particularly the price ceiling. DiSalle indicated that he would not back down, which appeared to forecast a showdown between him and the Cotton Bloc. When the world price of cotton temporarily surpassed the U.S. domestic price, a rare event, farmers' exasperation worsened.[68] The conflict with DiSalle quickly evaporated, however, when prices began to decline as a huge crop appeared certain. In

May 1952, the OPS dropped controls. "He had earned the enmity of powerful cotton interests," concluded one writer, "on behalf of a cotton order that was never needed."[69] By 1953, when the Korean conflict stopped, cotton had fallen to 32 cents, and it continued falling through the rest of the decade.

CONCLUSION

In the postwar era, the servants of King Cotton sought to reach objectives established in the previous decade, and no cause energized them more than the discriminatory taxes and restrictions placed on oleomargarine. For the NCC, the new custodian of the regime, a fast-paced and competitive global economy had no place for fiefdoms whose well-being depended on archaic privileges. When an unanticipated conflict arose between the NCC and unions, each having its own sense of history and seeking to move its constituents into a more prosperous world, their sharp differences defined the political dynamics of cotton, pitting organized agriculture against labor. With the national debate over the Brannan Plan, it became clear that the effective voices of agriculture would belong to the quickly growing business-oriented producers. Already small-plot farming was disappearing because of mechanization, particularly the development of the mechanical picker.

CHAPTER 7

THE SOUTH TRANSFORMED
Cotton's Mechanization, 1945-1970

"The Deep South is entering upon a process of change as dramatic, as rapid, and as profound as any of the major waves of the Industrial Revolution." That prophetic statement, written in 1946, expressed the impact of cotton mechanization over the next twenty years. "The backward and poverty-stricken agriculture of the old Cotton Belt," the author continued, "is likely to become efficient and moderately prosperous." This anticipated advancement would, however, drive "millions of people ... off the land, with results which will deeply affect the social structure of this nation, North as well as South."[1] Mechanization ended the remnants of the feudalistic rural order in the South and freed both landlord and sharecropper to become more viable participants in the new economy. Much has been written about the subject, with scholars and analysts examining, revising, and proposing varied interpretations but never questioning the significance. The tenants and sharecroppers uprooted by machines felt pain, while those who left voluntarily, sometimes with the machines at their heels, felt joy and relief. For the growers who successfully plodded through the transformation, however, it was the triumph over an inefficient and outdated farming culture. Mechanization came amid developments as dynamic as the emergence of the Cold War, the rise of television, and the new affluence and mobility of the populace. In this supercharged atmosphere of change and movement, mechanization pushed cotton farming swiftly into modernity and left preindustrial practices, both social and economic, behind.

Mechanization came late to the cotton kingdom. Southerners continued swinging hoes and picking bolls by hand long after Cyrus McCormick invented the reaper in 1832. Cotton farming desperately needed a similar breakthrough, but the technical barriers to building a machine to pluck the lint gently from bolls were insurmountable until the mid-twentieth century.

During World War II, sharecroppers had begun migrating northward, but many remained on the land. Landowners still needed them, so they lived in mutual dependence like lord and serf. In 1945 this old practice still thrived in the Cotton Belt, where plowmen walked the day behind mules or moved together in gangs chopping weeds, or hobbled along with bent backs to peel raw lint from stalks and stuff it into the long sacks they dragged behind them. Such scenes were the quintessential image of the South, causing it to be seen as a time warp of the nineteenth century. Cotton's antique methods of farming contrasted starkly with the modernizing American workforce, accounting in great part for the South's reputation as a backward region imprisoned in unchanging routines.

Mechanization had a revolutionary impact. It modernized the management of farming, ended the institution of sharecropping, and set off a fresh wave of black migration into the North. The end of hand labor and animal power penetrated deeply through layer after layer of the southern way of life, hastening the end of small-plot farming, jolting the unprogressive growers from their outworn practices, and forcing the business ethic on the old agrarian order. Mechanization alone did not drive the toiling masses off the land, for many were lured away by the call of better livelihoods in the expanding U.S. economy, but the total movement of people was so vast that it has been called the Second Great Emancipation.[2] Just as the locomotive had reshaped the West, so the new "steel mules" reconfigured the South, striking the last blow at the old social and economic structure of the cotton kingdom. The long overdue use of machinery enabled the United States to say good-bye to the antiquated cotton culture.

Mechanization has a celebrated status among the reasons for the migration of southern blacks and whites off the land. Some writers demonized technology as the "push" responsible for the displacement of America's poorest, the cotton laborers, who had to give up their livelihoods as growers replaced human labor with machines. Landowners were thought not to care or to misunderstand the impact of their profit-driven decisions on the cotton serfs. According to some analysts, a "pull" effect occurred, meaning workers left for better jobs in industrial areas. All agreed the status quo was overthrown. Stoop labor ended, and the landlord-tenant relationship on which the rural South had depended since the Civil War passed into history. In 1959 the Census Bureau collected statistics on sharecroppers for the last time. But mechanization had further dimensions: it forced out inefficient small-plot farmers and made the rest either incorporate the progressive and scientific advances of good farming or perish. Viewed in the largest perspective, the experience proved to be painful for those caught in the suffocating grip of the new iron masters, but it freed the producers of the nation's major crop from stifling backwardness.

It would be wrong to see mechanization solely as a gladiator fight between the effects of push and pull or to apply one interpretation, either to demonize the use of machines or to regard them only as an example of agricultural progress. The power of cotton was too great not to put mechanization in a global context, in which American growers had to compete worldwide or face ruin. Indeed, the adaptation to machinery contributed to the decline of small farmers producing only a few bales per year, but also spurred the rise of cotton farming in the western states, where "white gold" remained an appropriate term. With large tracts of land and machines, westerners produced cotton at the lowest cost and, along with the growers in the South who adapted to new conditions, kept the country's chief commercial crop viable in the world market. Small growers in the South who were unable to match the new efficiency fell victim to world competition. This meant the loss of small family farms so highly cherished in American culture, but the impersonal forces of the global economy lay behind their demise. Adding to the pressure to grow cotton at minimum cost was government policy, which reintroduced planting restrictions just as mechanization took hold.

COTTON CULTIVATION

Though mechanization began in earnest at the end of World War II, elements of it stretched back for a generation, generally appearing first among owners of large operations who experimented with tractors and various implements, though some refused to embrace new ideas. D. J. Pledger and his son, owners of Hardscramble, a 360-acre scrap of poor land adjacent to the Delta and Pine Land plantation in Bolivar County, Mississippi, insisted on mechanizing to the fullest extent. They acquired the property in 1930 and proceeded to improve the soil and make it among the first mechanized farms in the state. Hardscramble became known for outstanding yields. The Pledgers experimented with and practiced "cross-plowing," driving the tractor at a 90-degree angle to the rows with a small plow designed and built in their shop that eliminated the use of gang labor to hoe weeds. The Pledgers sought every means to reduce hand labor and animal power because, in their opinion, agriculture had "stood still," and they marveled that "backwardness and under development of agriculture can exist in the same community with a modern factory."[3] Until the late 1940s, growers acquired machinery sporadically, however, particularly in the older areas where small-plot agriculture was common, but also because many farmers, regardless of the size of their operations, were reluctant to acquire machines, since weeding and picking, the two most labor-intensive stages of cultivation, still required hand labor. Only when herbicides became available and the mechanical picker was available did full use of machinery become feasible. Using these engineering and scientific advances was the route to survival, however, so speed characterized the conversion to machines once the riddles of perfecting them had been overcome. "And it had to come in fast," the head of the Mississippi Experiment Station later remembered.[4]

A popular perception grew that mechanization so reduced the manpower required for farming that it created a surplus population in the cotton counties and forced the most unfortunate off the land. But the conditions responsible for overpopulation were evident before 1945. Improved sanitation, better maternal and infant care, the reduction of infant mortality, immunization from child-killing diseases, and improved general knowledge of medicine and public health made large families more common and accounted for the lingering of the elderly. Moreover, the old value of high fertility persisted in the agrarian areas, made easier by lack of birth control.[5] Temporary labor shortages induced by wartime migration, however, forced cutbacks in acreage planting before the onset of machine pickers. This rush to leave did not break stride after the war as black and white tenants, whether they had plowed cotton with mules or tractors, moved their families to towns and cities. America's burgeoning economy offered a richer and fuller life, a godsend for the impoverished.[6]

THE MECHANICAL COTTON PICKER

Once it became commercially available, the new picker was hailed for bringing about an agricultural revolution, for being the "missing link in the mechanical production of cotton," but it did not come easily.[7] Its inventor, John Rust, described as "a wonderfully eccentric genius with more than a touch of the older Henry Ford to him," belonged among the "uniquely American dreamers."[8] Born and raised on a Texas cotton farm, Rust had an insatiable curiosity for mechanical things; as a young lad, he built engines and gadgets from scraps. As a young adult he drifted from job to job, occasionally taking correspondence courses in mechanics. Rust saw the need for a machine to glean cotton from the fields and dedicated himself to solving the puzzle that had baffled inventors and tinkers as far back as the slave era: how to pluck the lint ripened in the bolls without damaging it. Since 1850 about eight hundred patents had been filed for mechanical pickers, and these efforts had developed the spindle, a spinning cylinder with protruding teeth that grabbed the lint, but the fiber clogged the machine. Despite many attempts, no one had overcome the barriers. Rust doggedly persisted, living with friends and relatives, working odd jobs and borrowing money to pay bills as he continued his quest. One night in 1927, he recalled, he remembered how raw fiber, when wet with the morning dew, had stuck to his fingers when he picked cotton on his father's farm. He rose from bed, wet a nail, and rolled it through a handful of cotton. It stuck to the nail. "I knew I had hold of something good."[9] With support from his brother, Mack, Rust built prototypes incorporating his discovery. In 1931 his machine picked a bale in one day at a field demonstration in Waco, Texas. In 1936, using an improved model, he picked five bales per hour in a demonstration at the USDA Experiment Station in Stoneville, Mississippi. Such speed equaled the work of forty to fifty hand pickers and caught the attention of the press, which led to pronouncements of the end of tenant farming in the South. It also brought cries of alarm that sharecroppers would flood roads and cities; Memphis political boss Ed Crump wanted a law to ban the mechanical picker's use, and Rust recalled the suggestions that the machines be dumped into the Mississippi River.[10] These prototypes broke down too often, however, to be used steadily.

Rust worried that his invention would displace sharecroppers once it became operational. *Time* magazine quoted him as saying that "75 percent of the labor population would be thrown out of employment."[11] He thought about leasing the machines only to growers who would promise not to use child labor, agree to pay minimum wages, and set a limit on working hours. In view of the practical difficulty of obtaining and enforcing such agreements, he abandoned this thought. He arranged for an FSA settlement in Mississippi, the Delta Cooperative Farm, to use his picker, but the farm faced such a multitude of difficulties, particularly financing, that it hardly used the machine. Commercial use evolved as the only feasible outlet for the pickers, but they broke down too often in field operations. Rust could not resolve the imperfections, and in 1942 he dejectedly put the machine under a shed and decided never to use it again.[12] His wife Thelma, however, believed her husband could overcome the technical barriers. She had taken an afternoon-evening job to assist him during the morning, and now she gently nudged him to try again. He began a thorough reevaluation only after Thelma persuaded him not to forsake his dream. "She urged me to go ahead and put my ideas on paper, pointing out that at least I still had my home and drafting board."[13]

Rust had competition. Farm implement manufacturers saw the opportunities in a mass-produced mechanical picker, but they had not been able to overcome the technical barriers any faster. In 1944 Rust agreed to let Allis-Chalmers manufacture the machines based on his years of work, but this effort failed to put more than a small number on the market. The John Deere Company developed a picker independently of Rust but could not lower the manufacturing cost.¹⁴ The real success came from International Harvester, which by 1940 made considerable progress when it mounted the picker on a tractor and drove it in reverse with the hopper and most of the picker placed behind the driver. This modification reduced the damage to the plants and gave the operator a better view of the machine as it swallowed the stalks and the spindles performed their magic.

The revolution exploded with a dramatic demonstration of eight mechanical pickers on the Hopson Plantation in 1944. Owned and operated by the Hopson brothers, Richard and Howell, the four-thousand-acre site just outside Clarksdale, Mississippi, had been a model of the traditional plantation with 130 families, a commissary, offices, and shops. It was near the intersection of Highways 61 and 49, where the blues singer Robert Johnson made his legendary pact with the devil to play the guitar. When the brothers inherited the place in 1933, they began turning it into a model of the new commercial plantation, a fully mechanized operation that had only forty employees by 1944. "The Hopson farm has none of the amiable lassitude of most cotton plantations," reported *Collier's* magazine. "It is run like a Detroit assembly line." Richard and Howell had arranged in 1937 for International Harvester to use their place as a testing ground for farm implements, so in 1944 when the company wished to display its first commercial picker, it conducted a demonstration there. The event was advertised, and twenty-five hundred people, some from Canada, came to see if the machines would work. Eight one-row pickers painted bright red moved across the fields, each leaving a swath of stalks clean of lint. Each machine picked sixty times faster than one worker. "Together," *Collier's* reported, "they were doing the work of 480 pairs of hands."¹⁵ The leaves of the stalks had been defoliated a week earlier, a necessary step for mechanical picking to keep them out of the fiber. It was the most successful demonstration of a mechanical picker thus far and put International Harvester in the forefront of development. The company announced it intended to build a manufacturing plant in Memphis and hoped to make the machines available in 1947 or 1948.

Collier's anticipated the displacement of tenant families, a consequence that the magazine called the "unfavorable side." It reported that estimates of displaced laborers went to 80 percent and that some observers predicted a mass migration of black tenants to northern cities. Such a movement would mean economic ruin for the South, at least some thought. Small towns were expected to dry up as small businesses went bankrupt. The magazine referred to Secretary of Agriculture Wickard's suggestion that subsidies should be withdrawn over five years, during which large landowners could mechanize.

During this shift, industry would hopefully expand and absorb displaced workers and small-plot owners. *Collier's* thus acknowledged the dilemma of forcing tenants off the land, but quoted one Hopson brother's response: "Is it better to take steps to improve the lot of people who remain on southern cotton farms, or to tolerate a stagnant arrangement which holds no hope?"[16]

THE CONFERENCE ON MECHANIZATION

The demonstration at the Hopson Plantation generated much excitement and anticipation. Full mechanization of cotton farming remained questionable, however, and research into all possible uses of machinery had to be undertaken. Research had proceeded haphazardly with no organization, which motivated the NCC to declare mechanization a primary objective at its annual meeting in 1947. It intended to rectify the lack of coordination.

In August 1947 the NCC, acting in conjunction with the USDA, held its first Beltwide Cotton Mechanization Conference at the Delta Experiment Station in Stoneville, Mississippi. About 250 attendees came from implement companies, land grant colleges, and farm organizations, along with many growers. Ransom Aldrich, head of the Mississippi Farm Bureau, presided and set the theme of the meeting when he explained that mechanization offered an opportunity for the South to improve its standard of living through more productive and efficient agriculture. Oscar Johnston spoke and urged small landowners not to fear machines but to see them as a means to better farming and higher incomes. He saw mechanization as the logical result of innovation and progress, not chaos, and thought mechanized farming would be essential to keep cotton viable. Johnston noted how the McCormick reapers and mechanical corn pickers provided benefits to farmers and consumers alike and foresaw similar advantages with the cotton picker. He noted, too, how the large flat fields in the West and Mississippi Delta made machines easier to use and acknowledged that no mechanical picker was currently available for small hilly farms, but hoped implement manufacturers would soon meet the need. J. T. McCaffrey, president of International Harvester, noted that his company had manufactured 175 pickers for 1947, but confidently announced that it would have 1,500 available for the next year. Other speakers discussed the need for improved ginning equipment to clean trash gathered by pickers and the need for agronomists to breed new varieties of cotton to accommodate the new machines. Frances L. Gerdes, head of research and testing at the Delta Station, stated that full mechanization required advances in defoliation of stalks, weed control, and the application of insecticides. Bell Helicopter conducted a demonstration of crop dusting.

Probably the most revealing comments came from C. R. "Jerry" Sayre, director of the Experiment Station. Sayre was young and belonged to the new agricultural generation that sought to gear farming and its related operations to the business ethic. He had earned a Ph.D. in agricultural economics from Harvard University, and his dissertation was titled "The Economics of Mechanization in Cotton Production." Sayre's perspective and career went beyond academics, for he spent his professional life in cotton affairs.[17] He later became director of D&PL, head of the Delta Council, and eventually president of the cotton marketing cooperative Staplcotn. At the Beltwide Conference in 1947, he stated that approximately 30 percent of cotton growers eagerly sought to mechanize and another 30 percent were amenable to it. But the remaining 40 percent would be a tough sell, presumably because they operated small farms or were wedded to old practices. His comments supported the claim that not all growers rushed to abandon mule farming, that conversion rested on considerations such as farm size, lack of capital, and the availability of cheap labor. Sayre pointed out that under old methods of farming, labor needs dropped from one hundred man hours per acre in the Delta to forty hours with the use of machinery, and that it went as low as fifteen on the Texas High Plains.

E. D. White praised the conference because the USDA had no program to foster mechanization, but he noted that the department had agronomists and other trained personnel to help with the needs of cotton farming. White acknowledged the potential to displace tenants but did not think it had reached "serious dimensions" so far. If a social problem arose, he thought job training programs should be inaugurated.[18] Other speakers reported that the conversion to mechanization varied from farmer to farmer, particularly east of Fort Worth, and that switching to mechanized equipment displaced tenants, but not in all cases. A panel discussion held on the second day further addressed the question "How can mechanization increase the income of the cotton worker?" Only one member of the panel, a spokesman for an implement company, did not avoid the sensitive question of the impact of mechanization on farm labor. He thought mechanization would be necessary to keep workers because it would bring an increase in wages. Better pay would hold them, he insisted, but he meant the newly trained equipment drivers and mechanics who would remain. He offered no thoughts for the unskilled hordes.[19]

Within the larger cotton community there was concern over the potential effects of mechanization on sharecroppers and workers. Frank J. Welch of Mississippi State College had recently studied the topic and concluded that approximately two million people would have to find new employment over the next twenty years; he believed that Congress should establish programs to help them find new jobs. Welch urged the establishment of federal farm insurance for small landowners. Thomas Linder, Georgia commissioner of agriculture, predicted a movement of five million off the land in the next five years. H. L. Mitchell, an activist with the old STFU now connected with the National Farm Labor Union, expected to see millions of tenants and workers affected. Some would remain as trained handlers of equipment, he predicted, but the mass of workers would have to seek their future in factories and workshops. "If industries can't employ them, what will it be—relief rolls?"[20]

Despite the recognition of these painful effects, a sense of relief and often euphoria was apparent when it became clear that mechanization was finally coming to cotton farming. The notion that the South could uplift its agriculture to the level of the rest of American farming generated excitement and hope that the preindustrial practices of the southern way of life were at last ending, but the concept that the government had a responsibility to furnish training for unemployed masses came too late to help displaced tenants. Growers complained of labor shortages and watched the most prized workers, the young and creative, abandon farming, which left only the old and physically challenged. Landowners, however, considered tractors and mechanical pickers a lifeline thrown to them in the swirling waters of change.[21]

FURTHER ADVANCES

A full analysis of mechanization appeared in one of the technical reports that accompanied the Pace Committee's special study released in 1947. Undertaken by the USDA Agricultural Research Division, it examined mechanization in a broad perspective by measuring the variations in cultivation across the Cotton Belt. Probably the study's most important observation was the need to recognize that "a new type of agriculture must be superimposed on an old and well established one."[22] Use of tractors came late in the South, but they were becoming more common, and farmers used them for many tasks not related to growing cotton. Besides general plowing and disking soil, tractors easily formed long rows or furrows with raised beds on each side in which seeds were planted. Delinted seeds were now available and could be placed in hills or rows at intervals of a few inches to eliminate "chopping." Fertilizer could now be applied in bands alongside the seeds, and thanks to tractors and attachments, these steps could be carried out in a single operation. Experiments at the Delta Station had shown that the deep placement of fertilizer, as much as six inches below the seed, slowed the growth of weeds and grasses because cotton had a taproot and could feed on the nutrients while weeds could not. Mule-drawn equipment could not place fertilizer deeply enough to gain this advantage. By placing seeds either in intervals along a row or in hills, the need for chopping, or thinning sprouts, disappeared. Cross-plowing had already eliminated or reduced thinning in large fields, but this practice did not work well on rough or hilly land. Planting in rows at three- to four-inch intervals with a seeding drill was expected to prevail and eliminate cross-plowing. A new and popular device was the flamethrower, the "weed-sizzer," for killing weeds. Mounted on the rear of a tractor with a fuel tank, this device shot a flame at the tender weeds and young cotton sprouts but, if correctly applied early in the season, damaged only the undesirable plant. It could be used simultaneously with a cultivator that cleared weeds in the easily accessible space between rows. Until the development of herbicides in the late 1950s, farmers relied on the weedsizzer or the hoe to clear weeds and grasses that grew too close to the cotton stalk for removal by the cultivator. Farmers had some skepticism toward weedsizzers because a misdirected flame would kill the cotton plant, a likelihood on cloddy or rough land. Bill Pearson, a grower in Mississippi, once lost acres of sprouts when a driver incorrectly set the flame.[23]

The use of mechanical equipment to spray or dust plants for insect control presented no difficult engineering barriers. Spraying rigs had been used with mule teams, and farmers adapted them to tractors. Crop dusting with aircraft became more common after World War II as young pilots returning from the war started crop-dusting businesses. Light aircraft could apply insecticide when muddy fields kept out tractors, and planes could cover large fields in a short time, an important consideration when trying to kill insects at the most vulnerable stage of their life cycle. The same methods could be used for spraying defoliants, which killed the leaves on the cotton plant and made them drop to the ground. Many new implements were available, but they required a tractor. Weed removal was only partially successful with flamethrowers and cultivators, however, so growers continued to rely on fieldworkers with hoes.

The authors of the 1947 report felt that mechanization offered promise, particularly since the Delta Experiment Station had grown cotton with no hand labor for the past three years. Savings in production costs were obvious because of the drop in man-hours, but the report cautiously pointed out that mechanical picking lowered the grade of lint and thus its final price. In the western states where cotton farming had just started expanding, mechanization provided the greatest benefit, but in the South the advantages remained less clear because of so many small plots. More engineering and scientific research must be deployed, the report continued, to improve ginning and to develop new varieties of cotton better suited to machine cultivation; but mechanization would improve efficiency, raise farm incomes, and replace the workers lost to the cities. For the South, the USDA researchers believed that "the attainment of an efficient agriculture would alter the profile of the region's agriculture significantly."[24]

In the aftermath of World War II, when some veterans returned to farms while others chose to avoid them, mechanization promised to keep the cotton kingdom from returning to the despair of the previous generation. It offered an opportunity to put cotton farming on an equal footing with wheat and corn. Few spoke out against mechanization, but there was concern about the ability of small-plot farmers to take advantage of the new opportunities. If farmers persisted with hand labor and mules and did not increase acreage, a wave of failures might come. Sharecroppers faced a frightening choice: they could either join the throngs choosing to abandon farming or take their chances by remaining on the land. For them the future was bleak.

Of all the advances made in cotton farming since Eli Whitney's cotton gin, the mechanical picker accounted for the "revolution." By 1947 its value and convenience had become clear, but the machines were in short supply, and picking by hand continued. Not until 1972 did machines harvest 98 percent of the crop.[25] Two types of pickers were developed: the spindle and the stripper. The stripper evolved from the use of a wooden "sled" with a V-shaped groove that ripped bolls from the stalk. These horse-drawn rigs originated on the Texas High Plains in the 1920s, with the mechanical version becoming available only after World War II. Stripper cotton, a short staple variety, dominated the Texas Plains because soil and climate conditions were not suitable for long staple varieties. The stripper took all bolls from the stalk, including unopened bolls, which brought a reduction in grade, but since the machine made greater acreage possible, it brought savings in labor costs. The advantages gained with the stripper nonetheless offset the side effects, so that by 1949 approximately 40 percent of the crop on the High Plains was harvested by strippers.[26]

Public awareness and the attention of the press focused on the South, where the spindle picker was the preferred machine. As the implement companies improved the design and engineering of the spindle, research on its use and impact got under way at the USDA Experiment Stations, particularly the Delta Station at Stoneville, Mississippi. Since spindles plucked the lint from the bolls, a gentler process than stripping, they were preferable for long staple varieties that grew in most areas outside the Texas Plains. Spindles matched strippers in advantages: lower costs of labor and faster harvest to avoid rain or frost. Whether spindle or stripper, machine picking gathered trash—leaves, bits of stalk, and burrs—that had to be removed at the gin.

This threw "all the burden of creating mechanical picker efficiency and of maintaining the natural qualities of cotton upon the ginner," reported the USDA.[27]

COTTON GIN ADVANCES

Machine picking indeed would not have succeeded if advances in ginning had not likewise occurred. Growers recognized their dependence on the ginners, so advances in ginning raw cotton, the first step in processing the lint after it leaves the field, pressed another task on research engineers. "The key to the success of the mechanical picker," asserted the 1947 study, "lies with the ginner."[28] This situation explained the move by the NCC to get funding from Congress for the appointment of two ginning engineers at the Delta Experiment Station during World War II. Cotton picked by machine presented two difficulties: it had more trash and carried higher moisture content than hand-picked lint. John Rust's breakthrough of moistening the spindle's teeth partially accounted for the higher content, but the additional trash also added moisture. Engineers at the Delta Station had by 1947 taken out twenty public patents on ginning improvements since 1930, which had been incorporated into new gins, but removing the trash and reducing moisture still required a solution. Ginning engineers at various Experiment Stations across the Cotton Belt developed the tower-drier to lower moisture and cleaning feeders to remove trash. For stripper cotton, the green boll trap removed the unopened boll.[29]

Gin capacity expanded to handle more bales, so along with the inclusion of new techniques, gins got larger and more expensive. Old-fashioned gins built for a small volume of hand-picked cotton, sometimes located on the plantation, dwindled in number. Just as the sight of mules pulling trailers to the gin became rarer as farmers switched to tractors and small trucks, the small gins that had previously dotted the landscape disappeared. From approximately 29,000 gins in 1900, the number had fallen dramatically to 8,632 by 1945, so the modernization of ginning had been under way along with other features of cotton mechanization before the advent of mechanical pickers. Most of this prewar change related to increased gin capacity, caused by the installation of better extractor-feeders and high-speed "saws" that grabbed the lint and pulled it from the seed. Higher volume made it possible to compress bales at the gin, so cotton compresses, once a fixture in many towns, began to disappear. Ginning could no longer remain an extension of farming or a small entrepreneurial activity; it now required considerable capital investment. As improved roads and small trucks made it easier to pull trailers greater distances to the fewer gins, advances in ginning demonstrated how modern cotton farming required more capital. Since the ginning season lasted three to four months, gins needed to process a large number of bales to be profitable. But machine picking introduced an unanticipated effect: growers brought cotton to the gin much faster and had longer to wait. The grower Bill Pearson saw trailers waiting in lines for miles.[30]

Converting from preindustrial practices inflicted pain on a large number engaged in growing cotton, for along with sharecroppers, many small farmers also had to eventually abandon home and soil. Owners of large tracts fared best, but they too faced risk of failure if they dawdled too long with mules and hand labor. Switching to new methods became mandatory and tolerated no comfort of delay. Mechanization proceeded in that context, reinforcing the claim that survival depended on mechanization.[31]

HERBICIDE DEVELOPMENT

Weed control remained the last barrier to full mechanization after the development of the picker. Until that hurdle could be overcome, hoeing by hand persisted, along with the use of flamethrowers. Weeds injured the cotton plant by stealing nutrients and moisture from the soil, robbing sunlight, and taking space. Fungi and diseases could also strike the roots; root rot ended cotton farming in pockets of blackland Texas. Mechanical harvesting made weed control more pressing, since the machine grabbed unwanted plants that stained the precious lint. Cultivators had been used with mules and tractors, but they only uprooted weeds between the rows of cotton. Only the sharp edge of a swinging hoe effectively removed the pathogenic plants growing in the drill row of cotton, and though a field could be chopped free of weeds, they would reappear after the next rain. Weeding outranked picking as the most labor-intensive step, and the combined pressure of fewer and fewer workers, along with the tendency of machine pickers to grab weeds, pressured scientists and engineers at USDA Experiment Stations to find a solution. They partnered with chemical companies to develop preemergent and postemergent herbicides.[32]

In 1947 the first use of a herbicide, dinoseb, occurred in Mississippi on a small scale, but in 1952 growers using dinoseb killed several hundred acres of cotton. This experience made farmers skeptical of herbicides, and while research continued, the commercial application of chemical killers lagged. Scientists learned "a valuable lesson that ultimately was helpful in the future development of herbicides for weed control in cotton," according to a researcher at the Coastal Plain Experiment Station in Georgia.[33] In 1950 chlorpropham became available, followed by dalapon and diuron in 1953 and 1954 respectively, all of which outperformed dinoseb. Development nonetheless proceeded slowly, partly due to the lack of trust by growers, so that by 1963 only chlorpropham, dalapon, and diuron were used much. Labor costs by the late 1950s began to make hand chopping prohibitive and softened resistance to herbicides. A new generation of growers, young and often educated in the land-grant schools, thanks partly to the GI Bill, looked more favorably on chemical weed control. Herbicide use in cotton entered its "golden decade" of the 1960s with the introduction of several effective preemergents.[34]

Postemergents, chemicals sprayed on grasses and weeds after the appearance of the cotton stalk, grew in use. Surveys made by the USDA Agricultural Research Service demonstrated a jump in treated acreage by 1962, with most usage occurring in the mid-South and Southeast. Mississippi growers who trusted chemical control used over fourteen chemicals to fight weeds on nine hundred thousand acres in 1969. In the West herbicide use never reached comparable levels because unwanted plants were not as abundant, but growers in all regions incorporated weed killers into their operations by the 1970s. Westerners used fewer chemicals but did not shun them.[35]

THE DECLINE OF SMALL FARMERS

By 1970 the greater part of mechanization had been achieved, though hand hoeing and picking might occur, generally on a small scale as a follow-up behind machines. Gone was the sight of hoers treading the fields, swinging their hoes to the chant of a song. Gone were the sharecropper shacks and the children playing around them, and gone were the mule teams. Town squares no longer hummed at picking season. The rural South grew quieter and seemed more subdued without the presence of thousands of workers and tenants of all categories. In the western states, the annual swing of migrants slipped into history.

There remained the question of small landowners, the segment of farmers concentrated in the South that produced only a few bales per year. Would they survive with hand labor and mule power? Could they grow cotton at a cost low enough to compete with mechanized farming? In 1948 a mechanical picker sold for $8,200, and to be able to purchase tractors and other equipment, a grower needed to raise between 100 and 150 bales per year.[36] Johnston had the most optimistic outlook for small farmers, asserting they could "share in the blessings of increased efficiency and productivity which the mechanized era certainly holds for the Cotton Belt."[37] He recommended they continue to handpick their crop so that it could command a higher price than the trashier machined cotton, and they should raise livestock and other crops to supplement their incomes. Johnston noted how small farms were concentrated in creek bottoms or hilly land, and thought only grasses for livestock should be planted in the hills, but the richer bottomland should remain in cotton. Small tractors and a few implements on these diversified farms would ease the drudgery. Johnston confidently predicted that under these circumstances, small-plot farmers could survive.[38] Johnston was, however, more optimistic than many. William D. Anderson of Bibb Manufacturing in Macon, Georgia, told Congressman Pace that "a large part of the Eastern Belt is going to go out of the business of growing cotton."[39]

Small farmers seemed certain to lose with the advent of mechanization. They tended to abandon cotton and grow other crops as soybeans, grains, and forage as livestock farming grew rapidly in the South. Oats and alfalfa were popular replacements on small farms. In the Carolinas and Georgia, where cotton farming first began to decline, poultry began to rise in its place shortly before World War II. Thanks to the Rural Electrification Administration (REA), electricity had started to become available in the late 1930s and continued to spread after the war. Using equipment powered by this source of energy, a lone farmer could raise fifty thousand broilers per year. Southeastern farmers "learned that it is far easier to let a machine pick a chicken than for them to pick 'bumble bee' cotton for a living."[40] Owing to the capital investment required to mechanize cotton farming and its comparative disadvantage in the Southeast, poultry production so expanded in Georgia that it prompted one resident there to ask, "Who'd ever thought a dad-blamed chicken would scratch cotton off the land!"[41] In Gainesville, Georgia, stands a monument to poultry symbolizing the appreciation for new farming. This retrenchment of cotton farming in the Southeast relieved the cotton kingdom of one of its poorest areas, the one that had most engendered the image of stagnation.

Attributing the decline of small farms and the displacement of tenants solely to mechanization overlooks the impact of another factor, the reintroduction of planting restrictions by the federal government. At the end of Korean War, Congress had to impose restrictions, and farmers throughout the Cotton Belt felt the effect. Commercial operators trimmed acreage just as small farmers reconfigured their plantings. The small farmers could least afford reductions in their principal cash crop or undergo the transition to livestock and other crops. With the introduction of the Soil Bank program in 1956, acreage fell even more. Nothing so "changed the landscape of the South as the Soil Bank," wrote the historian Gilbert Fite.[42] Evidence of hardship began to mount. There were record increases in food stamp dispersals, a surge of rural residents unable to pay electric bills, and testimonies about soup lines, increased crime, and a decline in the sale of agricultural implements. "The small farmer has lost heart.... He's given up," reported the *Commercial Appeal*.[43]

Cotton growers of all sizes were hit by cutbacks in acreage and rising costs of production, but prices for raw cotton stagnated. To survive in this environment of staggering squeezes on profits, growers resorted to strict management efficiencies, seeking to reduce costs to the rock bottom. Labor costs crept upward, however, so mechanization was the route to lowering expenses. The conversion from mules to machines occurred when severe adjustments in production were made to avoid plunging cotton back into the conditions of the Depression. Congress had to make cutbacks in the national allotment, which was also responsible for the displacement of tenants and sharecroppers off the land, a painful experience demanding empathy for them. Remaining on the farm and relying on cotton was not an acceptable alternative for those at or near the bottom rung of the ladder. It became increasingly clear that only the financially secure could afford to raise the royal plant.

AN AMERICAN DIASPORA

The history of mechanization acquired extraordinary significance during the civil rights movement of the 1960s. Suddenly interest grew in the reasons behind the mass movement, known as the Great Migration, of African Americans into cities. There was a popular conviction that the advent of the mechanical picker had forced thousands upon thousands of sharecroppers and laborers off the land, much in the theme of John Steinbeck's *The Grapes of Wrath*. "The attractiveness of this hypothesis," wrote one analyst, "derives in part from the timing of the mechanical picker's adoption in relation to the African-American outmigration."[44] Because mechanization preceded and overlapped the civil rights movement, it received more attention than usually given a topic of agricultural development, and acquired a legacy not found elsewhere in agriculture.

Oscar Johnston witnessed the migration during World War II and saw it continue afterward. To get a reasonable grasp of the number of croppers and tenants leaving the Delta, he stayed in touch with the train conductor at the Greenville depot of the Illinois Central Railroad who handled the daily passenger receipts.[45] To his correspondents, Johnston reported the loss of sharecroppers on D&PL and how he had to cut acreage devoted to cotton. He observed the potential of mechanization at Hardscramble Plantation adjacent to his operations and followed the progress of the Hopson brothers. Among his closest friends and advisers was James Hand, who always insisted that full use of machinery would be necessary to save cotton from oblivion. From the perspective of the Delta planters, mechanization filled a void left by departing workers, the position taken by Johnston in his oft-quoted article in the *Saturday Evening Post*. For him, mechanization did not force people off the land; they left for a "new life more attractive than the old."[46] He acknowledged how mechanization had a revolutionary effect due principally to the picker, but stuck to his belief that no harmful effects would come to small farmers. He urged them to pool their resources to buy and share equipment, and by handpicking they would receive the premium price. He expected them to diversify so that cotton would come to be not their sole source of support but an "important spare time crop."[47]

In his classic study *The New Revolution in the Cotton Economy* (1957), James Street wrote as the mechanical revolution occurred. An agricultural economist trained in the Institutional School of Clarence Ayres, Street had worked in the Bureau of Agricultural Economics, which enabled him to travel through the Cotton Belt and "to study its problems at close range."[48] As an academic at Rutgers University, he brought his practical experiences to the study. The book's title aptly described the transformation occurring as the long-sought escape from the outmoded way of life in the cotton kingdom had begun. Street's study came too early to see the full development of herbicides and credited the picker for the revolution. He recognized that tractors, the new flamethrowers, improvements in ginning, and other advances also contributed. For him, a strong current of change was flowing through the Cotton Belt, and more should be expected. The book's frontispiece showed a row of machine pickers sweeping across a field in the Tulare Lake Basin of California, the site of the huge Boswell operation. The book's subtitle, *Mechanization and Its Consequences*, pinpointed the question of the day: what will be the social impact of mechanization?

Street saw mechanization as progress, the achievement of a goal stretching back to the antebellum era. As sharecroppers and tenants, white or black, departed, he detected a decrease in child labor and felt wages for farmwork had inched up. He saw attitudes changing as the sense of entrapment, the serflike dependence of a poor class, began to melt away. He predicted improved racial relations because "there is no doubt that the peculiar relation of southern Negroes to cotton as a cheap-labor crop has long fostered the more coercive forms of racial exploitation in the region." On the question of displacement, Street concluded that only in select areas had growers adopted machinery because of labor shortages: the Mississippi Delta, California, and the Texas High Plains. Elsewhere "mechanization will have to push its way in against relatively abundant cheap manpower resources."[49] Street landed on the side of "push" but recognized that "pull" significantly affected the conversion.

A sizable body of literature on the subject followed Street's study, driven by the need to understand the resettlement of the southern rural poor into urban areas. Historians, economists, and journalists sought to explain the role of mechanization in this modern-day diaspora. With impressive credentials and equally impressive research, they examined data, old and new, to get as close as possible to a scientific answer, with the "push-pull" question as the central issue. Some studies cast mechanization in a negative light, concluding that machine pickers could have been used in the 1930s, that growers switched to power equipment only when cheap labor started to disappear, and that once the conversion began, growers threw out their tenants and bulldozed their shacks. Or mechanization became a way to end dependence on black labor just as segregation began to crumble. To be sure, variations in the process of removal occurred, with the larger growers in some cases allowing tenants to remain as workers, but not make a share crop. One writer believed that "the smaller and rougher planters simply kicked out their sharecroppers and left them to fend for themselves."[50]

Other assessments put the planters in a different perspective, acknowledging, however, that the push factor accounted for a portion of the displacement. Generally these writers saw a combination of socioeconomic forces at work, with the attraction of cities pulling workers. One analyst offered a comparative percentage: 79 percent independently left for a better life, while the remaining 21 percent were forced out.[51] In 2001 another study concluded that both push and pull operated simultaneously.[52]

Some analysts took a microeconomic approach by comparing costs of hand labor with mechanical equipment, examining wage structures, and tracing the transition from tenant farmer to day laborer before migrating. Others took a macroeconomic perspective that saw the migration, including the painful experience of the downtrodden, in a broad context that involved government policy as far back as the 1930s and the rising popularity of other crops, particularly rice and soybeans. For them the reduction of tenants had begun earlier via the AAA and continued through World War II. "The war reconfigured the southern labor system," recounted one writer, "phasing out sharecropping and utilizing wage labor, and many people who did not fit into the new scheme fled the land—and often the South forever."[53] Livestock farming also grew rapidly in the once tenant-ridden South. However varied the factors behind the end of tenancy, no one denied the suffering of the unskilled field-workers, whether pushed or pulled or just caught in the sweep of change and modernization.

The voices of blacks captured the anguish and relief of the tenant class, and the writer Richard Wright expressed deep anger over the plight of his race on the eve of mechanization in 1941. The promise of sharecropping would never come true, he insisted, despite the attraction of living on rich land and the "possibility of happiness." He saw a land of lynchings, the Ku Klux Klan, and severe oppression, where there was no opportunity and no justice. From his perspective, life on the plantation meant only misery for blacks who tried to balance credit and debt, but only the grave, "the final and simple end," awaited them. Wright anticipated the widespread use of mechanical pickers and regarded them as a threat to black livelihoods, but like the white establishment he saw danger in the rise of synthetic fibers and worried that if "Queen Cotton" died, many southern blacks would "die with her." For Wright, industrial jobs offered the only hope, but he wrote of black despair in cities, too. He wrote on the eve of the "cotton revolution" when the harshness of the Depression remained.[54]

In 1961 a small tent city sprang up in Fayette County, Tennessee, occupied by African Americans who had tried to register to vote. When denied the vote, they were ineligible for jury duty, which meant their brethren had to face all-white juries. Newspapers reported the eviction of 135 sharecropper families by landlords as part of a general boycott of blacks attempting to register. Eight families lived in the tents, named Freedom Village, whose interests were represented by the U.S. Department of Justice.[55] Cotton acreage allotments had recently been cut in half, and sharecroppers had started to leave farming. The evictions had originated over issues of voting, but the incident had implications for the allegations of evictions wrought by machine farming.[56] The tent city gradually faded away.

The legendary bluesman B. B. King remembered his youth in the Mississippi Delta as a time of tempering, learning to persevere, and seeing good and evil. He saw beauty in the land and the annual blossoming of cotton; he admired the seasonal routine of plowing and picking. In cotton farming he saw a poetry that had given him a sense of belonging and mattering. But all this occurred, he made clear, where "plantation bosses were absolute rulers of their own kingdom" and where some landowners were "cold blooded racists."[57] When his plantation boss showed kindness and gave King a job as a tractor driver, it made him "feel like a superstar."[58] Like thousands of black Americans, however, cotton could not hold him, and King fled to Memphis, where he started down the road to international fame.

Janice Kearney, daughter of black sharecroppers, rose from the cotton fields of southeast Arkansas to become the personal diarist for President Bill Clinton. Her memoir *Cotton Field of Dreams* recollected life on a cotton farm as one of seventeen children. She took pride in learning to chop and pick cotton as a small girl and knowing that her labor contributed to her family's upkeep. In her world, blacks and whites remained segregated but depended on each other to grow the white fiber. Her family experienced the backbreaking drudgery of hoeing and picking and living with bare cupboards and ragged clothes. The arrival of machines and herbicides sent away laborers and small farmers but enabled the area economy to diversify, making progress "good and bad."[59] Like thousands of others, no longer did Kearney and her siblings have to depend on landowners, and therein lay a major consequence of mechanization and postwar industrialization.

In black culture there was an obvious and understandable resentment over the oppressive conditions and poverty that went with cotton farming. At the same time there was a fondness for rural life and an admiration for the beauty of land. But African Americans were tired of paying tribute and homage to King Cotton for a meager life. Their hope and trust in the unknown North was stronger than their knowledge of the known South, so by the multitudes they fled, to "leave for a better life," as Oscar Johnston had stated.

A NEW CULTURAL LANDSCAPE

Location on the agricultural ladder determined outlook for those caught in the transformation. For the midsize and commercial operators, anyone intending to farm cotton had no choice: mechanize and practice sound management. Rarely did the switch to mechanization occur within a year, for when spindle pickers first became available, growers liked to follow them with hand-pickers to glean the remnants. Until herbicides arrived, some hand hoeing was necessary. Power equipment encouraged large-scale operations, but fierce competition from artificial fibers and foreign competition drove the need for cotton to expand, which explained the NCC's aggressive push for mechanization. Use of machinery varied from farm to farm, but the organization encouraged landowners to convert, and it promoted research of mechanization at USDA Experiment Stations. The NCC had concluded that sharecroppers and day workers would understandably continue to leave for a better life, but the council knew that only mechanization could keep growers in business. "There was no way for cotton to continue as a major and viable farm enterprise with the hand methods that were being used," recalled Charles Sayre, "so there was a pull in the use of technology, not an effort to simply reduce the manpower use."[60]

Machines eradicated sharecropping and forced landowners to act like the urban managerial class. Mechanization made cotton farming capital intensive, and the survival of growers required the adoption of an agribusiness ethic. Success tolerated few bad decisions, or else the sound of the auctioneer's chant would be heard. Remaining workers had to develop skills as machine operators or mechanics; recognize the effects of herbicide poisoning of plants; be alert when placing seed and applying fertilizer; and note if land drained properly or required further leveling. Machines needed field conditions to meet their specifications and could not adjust like man and mule. To pick cotton with a machine required evenly spread rows, usually at thirty-eight inches, and plowing the "guess" row in preparation for planting took precise driving of a tractor. Everyone engaged in the production of cotton had little margin for error. Gone was the lassitude of the former days as the impersonal tractor replaced the mule, the weekly wage replaced extended credit from the plantation commissary, and the whirring teeth of spindles replaced laborers' callused hands.

The appearance and physical arrangement of the southern and western landscapes changed. Where once stood tenant shacks and mule barns now appeared uncluttered fields and sheds full of tractors, pickers, and shop equipment. Vanished were the sights of sharecroppers or day laborers swinging hoes or dragging long sacks through the fields. Families of sharecroppers and small farmers no longer crowded into small towns on Saturday afternoons. Faded into oblivion were the small grocers who extended credit to field hands or the plantation commissaries that stocked provisions for the tenants. Many small shopkeepers went out of business or offered new services to a different clientele. No longer did small gins dot the countryside, nor did trading barns selling mules and horses stand on the edge of towns. Small farms operated solely by family members dwindled as they were consolidated into larger holdings, perhaps converted into pasture for livestock or planted in timber. Among the most remarkable disappearances were the large plantations that once served as small communities. The Hopson Plantation was chopped into smaller parcels, and its central area eventually became a museum along with bed-and-breakfast accommodations. Even the D&PL operation ceased raising cotton and parceled out its land to other growers, but it retained its seed breeding operations and became a leading developer of new varieties.

With the end of the plantation came the end of America's last vestige of landed gentry whose lifestyles had set the social standard. They had stood atop the pyramid of a paternalistic order as they supervised their estates and oversaw the employment of others. A plantation "was a world unto itself," wrote B. B. King. "Plantation owners were absolute rulers of their own kingdom."[61] The power and recognition of governance once accorded them now evaporated, and farming became more impersonal. Owners of plantations had practiced the social graces and enjoyed the benefits of being vassals of King Cotton, but once mechanization became common, they survived only as businessmen and no longer were the caretakers of an agrarian order. The tractor, the mechanical picker, and the chemistry of herbicides shoved aside the preindustrial practices of farming and left the old order, the southern way of life, a painful memory or sweet nostalgia. Mechanization broke the status quo and brought upsetting changes. "The old pattern of exploitation and exportation is breaking up," wrote Frank Smith, "and the Delta is happy to see it go."[62]

CONCLUSION

The growing competition from foreign producers made full mechanization imperative. Growers could not afford to drag their heels and expect blind faith to keep them viable in the new world order after 1945. The South had too many people employed in an endeavor of low productivity; their transition from farm to city was painful, but for them to remain in the squalor and stagnation of cotton farming was unacceptable. It had been commonly anticipated that Rust's invention would displace tenants, and nearly everyone, including Rust, saw the painful consequences of mechanization. Discussion of programs to assist the displaced resulted in little action, but the United States fortunately had an expanding economy that offered them a better life. Regarded a few years earlier as peasants, tenants and workers began to swim in the mainstream. For these reasons, a USDA study in 1947 concluded that "in all the many discussions of the cotton problem, mechanization has been mentioned as the most promising of all suggested remedies."[63]

The migration of black Americans out of the South has been called "one of the largest and most rapid mass internal movements of people in history." It occurred simultaneously with the mechanization of cotton farming, and how much the two were linked remains a subject of debate, but the development of machine farming unquestionably had an impact. Blacks had gone hand in hand with cotton for nearly two centuries, whether as slaves, freedmen, or sharecroppers, but their separation, "the decoupling of race from cotton," set into motion still-unfolding social developments. This diaspora, as stated in one account, "made race a national issue in the second half of the century—an integral part of the politics, the social thought, and the organization of ordinary life in the United States."[64] The role of cotton in this panorama is visible, but too slippery to grab and hold, and therein lies part of the mystery of cotton.

CHAPTER 8

THE WHITE GOLD RUSH
Cotton Moves West

In the expansive climate that seized the United States at the end of World War II, the exuberance of the country shone in the rise of cotton farming in the West. From that point forward, cotton no longer confined itself to the land of magnolias and kudzu but reached to the mountain ranges of California's Pacific shore. Youth and speed of movement characterized this extension of the cotton kingdom, and a new breed of growers met the challenges of the West, where wind and sand were abundant but water was scarce. New names appeared in the language of an old culture, each bespeaking location but also designating a particular kind of farming: the Texas High Plains, Far West Texas, New Mexico's Mesilla Valley and Pecos River Valley, Arizona's Salt River, California's San Joaquin Valley, and the Imperial Valley. In some areas, the introduction of cotton farming was the first step in population growth. Westerners, unlike southerners, were not wedded to agrarianism as a way of life but saw themselves as entrepreneurs practicing business principles. Apart from these developments, this expansion broadened the political and economic base of King Cotton's domain.

Young men returning from war and seeking to earn their livelihood from the soil looked on the vast West as the land of opportunity. Not averse to risk, they were driven by a work ethic and a determination not to let the pitfalls of cotton farming overpower them. With their reckless optimism, they brought virgin areas under the plow, transforming raw desert into an oasis of green that turned white each growing season. From these fields dotted across arid lands sprang a fresh energy and hope for the cotton industry, bringing a new perspective and outlook.

Jack Stone, founder of Stone Land Company, exemplified the courageous young men of the Far West. Mustering out of the U.S. Army in 1945, he returned to the land of his childhood, the western edge of California's San Joaquin Valley near Stratford. Before the war he had leased land near Five Points, where he had lived alone in a homemade shack, but sold his equipment and enlisted in the army in 1941. In 1946 he leased 300 acres and planted corn, wheat, and barley. The next year he bought 320 acres still in its natural state, unleveled and never broken by a plow. Stone leveled the land and prepared seedbeds, making his first crop of California's Acala cotton, relying on Asian Hindu laborers to pick the crop. He expanded his holdings, leveling land and planting cotton, and managed to purchase two used mechanical pickers in 1949. His yield of Acala ran to two and one-half bales per acre on the valley's rich sandy loam.[1]

In similar manner, Morgan Nelson returned from military service in 1945 and farmed with his father on the outskirts of Roswell, New Mexico, in the Pecos Valley, where water for irrigation was available but limited. In 1946 Nelson planted 270 acres of Acala 1517, which performed well in the Pecos Valley. He rotated his cotton with alfalfa and barley, taking care not to exhaust the soil from repeated plantings of cotton. Nelson constructed irrigation ditches with hand labor and horses because the available tractors were not powerful enough for earthmoving. Nelson relied on migrant workers from the Rio Grande Valley in Texas to pick his crop, which ran two bales per acre in 1946. He did not significantly increase his acreage, owing mostly to limited water. The outbreak of verticillium wilt in the valley tended to retard expansion, though the introduction of Acala 1517 WR provided strong wilt resistance. Acreage also tended to be smaller in New Mexico because the growing season was short.[2]

After the war, Don Anderson, a former dive-bomber instructor, started farming near Crosbyton on the Texas High Plains. Not until 1949 did he plant his first cotton, using 320 acres of rented land. He started buying land and continued to lease. He installed the first irrigation well in the Crosbyton area and farmed 1,800 acres by 1951, selling his crop that year at 41 cents per pound. So rich was the virgin soil that he used no fertilizer in the first few years, commenting later that he needed only "to water and weed" the plants. Like other westerners growing cotton, Anderson benefited from low insect infestation, particularly the absence of boll weevils, which meant less use of chemicals. He relied on hand hoeing and flamethrowers for weed control. With another grower, Howard Farris, Anderson designed a bed planter that quickly became popular on the High Plains and won him an appearance on the cover of *Progressive Farmer*.[3]

Young men like this trio did not settle the West in the manner of pioneers traveling in covered wagons, but they acquired raw land untouched by plows and converted it into farms. They nearly always had backgrounds in farming as descendants of southerners or were second-generation farmers who had spent their childhood under the western skies. Their outlook was not traditional, for they did not feel obligated to follow previous customs and practices of cultivating crops. Small farms with exhausted soil did not litter the landscape. Cotton carried no cultural baggage in the West; landowners were known as growers, not planters, and there was no identity with slavery or sharecroppers. History had no binding effect as it did in the old Cotton Belt, for the young westerners carried no badge of poverty and no image as stagnant holdovers from the past. Just as the horizon was open and vast, the cultural perspectives of the West remained uncluttered in 1945.

The western states' share of the country's annual yield had been small. In 1930 California, Arizona, and New Mexico accounted for only 608,000 acres, or 1.3 percent of the U.S. harvested cotton acreage.[4] The region's portion changed little, hitting its highest pre–World War II point at 3.2 percent in 1937. A similar story happened in West Texas on the Llano Estacado, the area stretching from Amarillo southward to Midland-Odessa. In 1923 production on the Llano reached 200,000 bales.[5] Texas ranked, however, as a giant in production, but before the war cotton farming there remained concentrated in the eastern blackland strip and the Rio Grande Valley.

Although cotton farming was negligible in the West, a fascinating development occurred in Arizona during World War I, auguring things to come. By then automobiles and trucks were becoming increasingly popular and necessary in American life, and along with them the manufacture and sale of rubber tires. In that earlier era, tire cord, the ingredient that reinforced the rubber carcass, was made from cotton, specifically extra-long staple (ELS) because of its greater fiber strength. Tire manufacturers had relied chiefly on Egypt for their supplies of ELS, but Britain imposed an embargo on this highly prized cotton, which left U.S. tire manufacturers searching for another source. In 1916 the Goodyear Tire and Rubber Company sent Paul W. Litchfield to the Arizona Salt River Valley to explore the feasibility of establishing a large plantation to grow ELS. In 1917 the company planted 1,500 acres. Goodyear then bought 24,000 acres and created the Southwest Cotton Company as a subsidiary. It built eight cotton gins. A variety of cotton known as Pima quickly became popular and dominated fields in the Salt River Valley. From 1916 to 1920, the profits for Goodyear and farmers in the valley were astounding.

As described by one writer, the "long staple cotton craze" swept the area in 1920, and planted acreage increased to 243,000, triple that of the previous year. It was like "a gold rush on Nome's beach or Leadville's great silver rush."[6] But the boom ended by picking season of 1920, and the year ended as a disaster for Pima growers. Goodyear's operations had served as a stimulant to growers in the area, many of whom had started the war as alfalfa or dairy farmers but switched to cotton in view of the profits. By 1922, however, most of them had returned to their original crops or dairy farming, and Goodyear liquidated most of its holdings except for the operations at Litchfield Park. In 1923 tire cord manufacturing switched to ordinary long staple grown in the Mississippi Delta, and the market for Pima fell dramatically. Litchfield Park became a research station used by the University of Arizona Extension Service.[7] Gins, banks, and other supportive apparatuses fell in the collapse, so "the bust did more than bankrupt farmers," stated one account.[8]

Cotton did not catch on in the West before 1945 for the simplest of reasons—the price was too low to warrant the investment required to plant and nurture a crop. Sliding prices in the mid-1920s, followed by the Depression, had made the cotton farmer the symbol of poverty, so putting venture capital into the depressed cotton economy of the 1930s appealed to no one. Other factors came into consideration, such as inadequate irrigation and the greater costs of shipping cotton to textile plants in the Southeast. Irrigation had to become more available, and the cost of drilling wells, as well as having to level the fields for controlling the flow of water, added to the up-front cost of beginning operations. The Texas Plains needed a "storm-proof" staple that could withstand wind and sand. Until the end of World War II, the comparative advantages of growing cotton remained in the South.

THE RISE OF THE WEST

In 1945 the prospects for lucrative returns began to appear. Water, or the likelihood of it becoming available through public irrigation projects, was a factor. Construction on California's well-known Central Valley Project got under way in 1937 but had sporadic delays during World War II. The storage, flood control, and power-generation features were complete in 1945, with the irrigation facilities only beginning to reach fields. New growers in California, most of whom were located in the western half of the San Joaquin Valley, drilled wells and took water from the Central Valley Project as it became available. In Arizona's Salt River Valley, a similar pattern ensued. On the Texas Plains, farmers drilled wells into the large underground Ogallala aquifer. In New Mexico's Mesilla Valley and the Texas El Paso area, known as Far West Texas, the Elephant Butte Reservoir, built in 1916 on the Rio Grande River north of Las Cruces, provided water for agriculture, but cotton became important there only after the war. Competition for water between urban and rural interests in the West was not severe at the time, and cotton farming was perceived as a means to economic development. The 160-acre limitation for use of water provided through federal auspices did not hinder development, since growers used wells or found means to get around the limitation.[9]

The availability of land at low prices figured in the western expansion. In the mid-South and Southeast, as well as the traditional cotton areas in Texas, land had been trampled by man and mule for over a century. Large portions lay exhausted from lack of care and overuse, making the still-fertile land such as the Mississippi Delta expensive. Large tracts in the West, though undeveloped, held rich loam, and bargains could be found in the San Joaquin Valley, where the Southern Pacific Railroad and Standard Oil, holding vast acreage, welcomed buyers. In 1939 Southern Pacific owned over four million acres in California.[10] Jack Stone got started by purchasing land grants given to the railroad by the state of California. Much farmland in California and Arizona had also gone into foreclosure or out of production during the Depression and was now available at bargains, so young entrepreneurs snapped it up.[11]

In 1946 Will Clayton and Lamar Fleming had enough confidence in the West to purchase a large tract near Fresno, about 19,000 acres, to test their convictions that cotton farming could be profitable there. Fleming reported the availability of the land and recommended it to Clayton, who had always kept ACCO out of farming on the grounds that corporations should not compete with "a good farmer living on his land." Fleming predicted that the purchase and operation might be criticized in the older Cotton Belt, but thought westerners would not object. The two men wanted "to have a laboratory for mechanized, large-scale farming," Fleming wrote, "which I believe is the farming of the future."[12]

With the advances in mechanization in the late 1940s, western growers could now efficiently cultivate fields and keep only a few workers on a year-round basis, compared with southerners who continued to rely on sharecroppers, though sharecropping was declining in the older growing areas. During the height of sharecropping, Oscar Johnston had kept one family per twelve acres in cultivation at the Delta and Pine Land Plantation.[13]

For westerners, nothing equaled the absence of the boll weevil to justify making a new start. In 1927 Don Anderson's father had moved his family away from Johnson County, Texas, to the High Plains to escape the dastardly devil.[14] The weevil could not survive in arid climates or find sufficient wooded brush in which to hibernate over the winter. Its line of infestation stopped roughly along the hundredth meridian, arguably the major factor in the spread of cotton to the West. In the folklore of cotton, it is held that "when the boll weevil arrived in America, cotton moved west, cattle came east, poor folks went north, and the economy went south."[15]

Only with the rising prices of raw cotton, however, could the emerging advantages of the West come into play. In 1945 the average price was 22.52 cents with the low crop of 9,015,000 bales. For 1946 the crop dropped to 8,640,000 bales, and the price jumped to 32.64 cents. Each of the far western states increased production; Texas fell slightly, but its major area of fiber production remained in the old Cotton Belt, though the High Plains was growing fast.[16] No federally imposed restrictions existed on planting, and the support price had been set at 95 percent of parity. For westerners, the federal price support meant little in the expansive atmosphere when domestic and foreign markets were strong. The efforts of the NCC to move cotton into Europe and Japan were succeeding, so much that Barringer warned Truman in August 1946 that measures might be necessary to ensure an adequate supply for the coming year.[17] For the next three years, prices remained strong, and production rose quickly throughout the United States, but the most phenomenal increases occurred in the West as all the elements favorable to the region fell into place. The white gold rush had begun.

The coincidental convergence of events explaining the rush into the West manifested itself with the development of Acala 4-42, the new variety of cotton that overcame the drawbacks of farming in the San Joaquin Valley. The previous staple grown there, Acala P18C, had to take a discount in price because textile manufacturers in the Southeast complained that it had "neps," small knots in the raw fiber that gave it irregular dying characteristics, and it had less moisture than rain-grown cotton. Textile companies would not mix Acala with rain-grown varieties. Growers in California had accepted this practice because their high yields offset the discount. Most California cotton, about 90 percent, had gone to Japan and China, where there were no reservations over the spinning characteristics and because transportation charges to the southeastern mills put the West Coast at another disadvantage.[18]

The story of Acala 4-42 began in 1934 when George J. Harrison became the senior agronomist at the USDA Cotton Field Station at Shafter, California. A native Texan, highly disciplined and devoted to field laboratories, Harrison spent over a decade in the Arizona USDA station. When he arrived at Shafter, cotton in the valley was not a popular crop; growers preferred wheat, barley, alfalfa, fruits, and vegetables. The Acala varieties grown at the time, P7, P18, and P21, had been developed at Shafter in the 1930s.[19] In 1942 the Shafter station released a superior variety, P18C, but it still received the textile discount. Harrison continued to struggle until 1947 when he developed Acala 4-42, which had superior staple length, tensile strength, and fewer "neps" than the P18C variety.[20] Harrison's new strain originated from a single plant of New Mexico's Acala 1517 and produced bolls that matured more quickly and uniformly, advantageous for machine picking and the San Joaquin Valley's perilous early frost.[21] By 1949 the new variety covered nine hundred thousand acres in the valley, and various strains of Acala dominated Arizona,

New Mexico, and the Texas High Plains. Some growers, reaping wealth from Harrison's years of labor in seed breeding, offered him employment at enormous salaries, but he always refused, preferring to work in the scientific field laboratory. *Collier's* magazine reported that landowners referred to the San Joaquin Valley as "Harrison's Cotton Patch" because he had "literally turned California cotton from a sickly agricultural side line into the most valuable field crop in the state."[22]

The new commercial operators in the western San Joaquin Valley saw cotton as a business venture compared with the east side of the valley, which was settled with family farms that grew fruits or engaged in dairy farming with hay and silage crops. Cotton grew there, but as a side crop. Farm size might be as small as twenty-five acres; the average reported in one account was fifty-seven acres.[23] Farmers operated these units with a sense of permanency and the traditional perspective of family agriculture. Where cotton grew exclusively, the average size went to eighty acres. Land costs were higher on the eastern side, often reaching $1,000 per acre, owing to the presence of wells for irrigation, roads, electrical service, and further development. On the western side, much land was available for strikingly low prices, often at $50 per acre, but it required considerable capital investment to make suitable for cultivation. An irrigation well, enough to handle 640 acres, cost about $35,000. Heavy equipment such as caterpillar tractors had to be employed for leveling. Capital investment often surpassed $1,000,000.[24] High development costs required large farms to be profitable, a fundamental principle common to nearly all businesses. For these reasons, growers in the new lands had to operate in the context of a corporation.

A limited amount of cotton had grown in California's Imperial Valley before World War I, but landowners by the late 1920s generally abandoned the area owing to insect infestations and unreliable water supplies. By the early 1950s, however, favorable conditions returned with deep wells and canals to carry water, and farmers leveled land to ensure better water penetration of soil. Facing no government controls on planting and seeing the wealth generated northward in the San Joaquin Valley, growers in the Imperial Valley and Riverside County planted 34,560 acres in 1951, using Acala 4-42. This variety yielded two bales per acre but produced too much vegetation for reasonable handling. Not until 1957 did growers learn that Delta Pine varieties from the mid-South performed better in their soil and weather.

The white gold rush caught national attention, at least for the San Joaquin Valley. Popular magazines with nationwide circulation featured cotton's growth: *Collier's*, *Fortune*, *Life*, and the *Nation*. Portraying cotton as a benefit that brought new wealth to the valley, writers credited mechanization for the development of raw land. Thanks to seed breeding conducted at Shafter Station, the new Acala 4-42 had made California the source of a premium fiber that brought a few extra dollars per bale. The advantages of irrigation, meaning the application of water at the precisely desired stages of stalk growth, almost guaranteed a healthy crop. So assured that each planting would yield an extravagant two bales or more per acre, California growers now easily found credit and often sold their crops in the field. No comparable press coverage went to New Mexico and Arizona, likely because production was smaller with less-dramatic impact. West Texas stood as the giant, outproducing California but not overshadowing it. Expansion of cotton in the Lone Star State, though far from the old Cotton Belt, did not engender excitement. As cotton became the major crop in California, outranking fruits, grapes, and vegetables, it stirred curiosity because the state had been established in folklore as the land of opportunity for people from all points eastward. In *The Grapes of Wrath*, Steinbeck's Okies had been tractored out of the cotton fields and sought opportunity in the vineyards and orchards of the West. Ironically, the southern crop synonymous with sharecropping that Okies had wanted to escape now became the principal crop in a state renowned for diversified agriculture.

Yet this development came under criticism. Carey McWilliams, a well-known journalist, saw "murky intrigue" by large landowners trying to take advantage of the one-variety law and monopolize the distribution of Acala 4-42 seed through the California Planting Cotton Seed Distributors' Corporation. McWilliams deplored the use of underground water for growing cotton and predicted that the western valley would become a "future dust bowl." He noted the rising use of mechanical pickers but reported that the migrant camps housing southern blacks recruited to pick cotton were full of prostitution, racketeers, gambling, and liquor concessions. He predicted that California cotton growers would continue to recruit southern labor and set off "a new current of migration which will have great social importance." McWilliams regarded cotton in California as "less a legitimate branch of agriculture than a highly profitable racket" because farmers grew cotton to take advantage of government subsidy. For him, California growers had lamentably joined the Cotton Bloc and acted as a special interest to foster their newfound niche in agriculture. The South would not be able to compete with California, he concluded, and that would cause production to expand in the West. "California is slated to inherit the deplorable social conditions incident to cotton production."[25]

McWilliams's critiques had little impact. Rather, the rise of cotton in the West encouraged the hope that a revolution in efficiency would occur, propelling cotton farming into a new era by making America's chief crop more competitive on the world market. Such an accomplishment would bode well for the general welfare, since related industries would be impacted, too. Best of all, competitive farming might end the reliance on subsidy and enable farmers once again to stand independently. In this context, the extension of the cotton kingdom became an object of close observation because all the elements required for efficiency existed in the arid zones: large tracts of flat land, no boll weevils, sunny weather, water for irrigation, and easy adaptation of new advances in machinery. "The only sensible solution for the U.S. cotton problem is to produce it cheaply on the most suitable flat land," Fleming told Clayton in 1944, "with a maximum use of machinery and agricultural and engineering science."[26] Fleming believed there could no remedy for the small farmer on "poor and hilly land."

Indeed, the new conditions favored western farming. Between 1945 and 1949, Arizona production climbed steadily from 158,000 bales to 543,000; a similar increase occurred in New Mexico, where output jumped from 106,000 bales to 276,000, but New Mexico never became a producer of volume because water supplies were limited. A surge came in California, however, where the yield leaped from 353,000 to 1,268,000 bales.[27] And in Texas, where cotton farming continued in its traditional areas, the opening of new lands on the High Plains resulted in an ocean of fiber, growing from 1,794,000 bales to 6,040,000. In Lubbock money flowed, and farmers in counties on the plains, wrote one scholar, "received a quarter-billion dollars cash income and led all other sections of the state."[28]

GUNFIGHT AT THE OK CORRAL

Like any gold rush, the cotton stampede into the West had a short life span. Cotton prices began slipping in 1948, CCC stocks began climbing because domestic demand declined, and despite the special export programs and the Marshall Plan, it appeared that planting restrictions might be necessary. Congress imposed no planting restrictions for 1949 but specified that plantings for that year could not be included in the historical basis for computing future allotments. The United States produced 16.1 million bales in 1949, the largest crop since the historic bumper year of 1937. Texas and the Far West accounted for 50 percent of the crop, and in some areas where production costs were low, growers reaped a 250 percent gross profit. A surplus now appeared imminent, and because the 1938 Agricultural Act prevailed, which required acreage restriction if the expected carryover and export reserve exceeded 30 percent of annual consumption, restrictions for 1950 seemed unavoidable. The 1938 act further specified that allotments would be apportioned to growers based on their history of production for the past five years. Since westerners had a smaller historical base and would thereby receive large cutbacks in acreage, this method discriminated against them "in favor of farmers in the older areas," wrote one analyst, "who had by this time largely turned away from cotton production."[29] This possibility had been anticipated across the Cotton Belt, which prompted a search for a proposal that would satisfy both the western growers and the traditional producers based in the South. Westerners conducted a series of meetings, particularly among the Californians, to formulate a compromise.

Each region took a different perspective on the allocation of allotments. Simply put, westerners wanted to maximize their planted acreage, while growers to the east wished to protect their allotment rights. The issue became a contest between the old Cotton Belt and the new western sections. Much was at stake because it was expected that large-scale and highly mechanized farming would have to advance further for U.S. cotton to compete in the world market. Small and marginal plots in the Southeast, some with only five to seven acres of allotment, were not competitive, but farmers there kept a grip on their historical base because it held up land values.

Realizing they could incur a serious loss, producers in the West organized the Western Cotton Growers Association in 1949 to fight for an amendment to the 1938 legislation. California led the fight, but the organization had the support of all western interests expecting to be hit by the cutback. This new group was militant in their belief that growers in the Cotton Belt held them back.[30] To find a resolution least harmful to all, the new organization sponsored five meetings of western growers, which ended with the Beltwide Cotton Conference held in Memphis in April. Representatives of all cotton-producing regions attended, and C. R. Sayre, executive director of the Delta Council, chaired the two-day gathering. The congressional Cotton Bloc wanted to receive a proposal from the conference that could be enacted as a compromise, and when the Senate Subcommittee on Agriculture and Forestry conducted a four-day hearing in June, it received a proposal from the conference. At the hearing, Sayre acted as the spokesperson for the growers, but a variety of cotton interests from across the Cotton Belt were represented, including the Farmers Union, the National Grange, the Farm Bureau, and Rhea Blake of the NCC. In this dispute between South and West, the NCC played no role because the issue divided growers, and when no agreement could be reached, the organization could not by virtue of its bylaws take a position. Blake endorsed the compromise presented to the subcommittee, however, on the grounds that it constituted a consensus across the Cotton Belt. He kept his remarks brief and otherwise did not participate.[31]

The consensual proposal called for a national allotment of 22.5 million bales and stipulated that extra-long staple should be exempt from restrictions. It recommended that computations for planting restrictions be made by using the history of acreage planted in all states from 1945 to 1948, but no state should sustain a reduction over 5 percent of the 1947–48 planted acreage. To protect the small farmers, the proposal further recommended no change from the 1938 legislation that guaranteed a minimum five-acre allotment.[32] By coming to this agreement, cotton growers, at least those active in farm organizations, presented Congress with a relatively easy solution, and it passed legislation establishing a national quota of 21.7 million acres that was expected to yield 12 million bales. In December growers approved the imposition of restrictions in a referendum. Only 19 million acres went into cultivation in 1950, owing partly to unfavorable weather, but it was nonetheless an unusually small planting because of clumsy arrangements for reserving acreage minimums to each county.[33] The U.S. crop barely surpassed 10 million bales in 1950, over a 35 percent drop from 1949.

This episode engendered only minor suspicion and tension between regional interests, thanks to the preemptive efforts by growers to present a compromise to Congress. Protectionist sentiment came from both sides, but the widespread recognition of a looming surplus had maintained a sense of harmony and compromise, particularly since it prevented the severe cutbacks in the West.

In the midst of the growing season, in June 1950, the outbreak of the Korean War caught everyone by surprise. Now a shortage of cotton appeared likely, and the United States imposed emergency quotas on exports to guarantee sufficient supplies of raw fiber to textile mills. "In the long struggle to preserve foreign markets," wrote James Street, "no one had contemplated that export quotas for cotton would ever be necessary."[34] Prices for raw cotton rose, climbing to 41.4 cents for the year, well above the 30.3 cents for 1949. The Department of Agriculture suspended planting allotments for 1951; acreage planted went to 26,949,000, and the country produced over fifteen million bales. For that year, the price dropped to 38.4 cents.[35] For 1952 controls remained suspended, but exports fell because other countries took advantage of the strong world price and increased production. The unexpectedly bullish market wrought by an unexpected war masked the potential rift between the West and South, but only temporarily. One year's experience with allotments had demonstrated the significance of reduced planting to westerners, and they intended to protect themselves. When two large crops came in 1953 and 1954, acreage controls seemed inevitable. Tension between the West and South mounted.

Congress began new hearings in 1953 and faced the old dilemma of 1949: how to draft legislation instituting acreage controls without harming the efficient western growers or the historic sections of the South. The act of 1938 still prevailed and provided for allotments to be based on a five-year history, which in this case meant 1947–48 and 1950–52, since a farmer's acreage planted in 1949 could not apply in calculating future restrictions. To reach a national quota of ten million bales and abide by the total allotment of 17,500,000 acres as specified in the 1938 act, the West would endure severe cutbacks. This measure did not account for the expansion into the West that occurred when no restrictions were in effect after the war. Table 1 shows the severity of the anticipated cutback.

Table 1. Expected Cutbacks in Cotton Planted for 1954	
State	Percentage
Arizona	54
California	51
New Mexico	42
Texas	39
Source: Congress, Senate Committee on Agriculture and Forestry, *Hearings, Cotton Marketing Quotas and Acreage Allotments*, 83rd Cong., 1st sess., June 30, July 1, 10–11 (July 1953), 3.	

For westerners, cutbacks at these levels would be disastrous. Most of their expansion occurred with borrowed money, and they could find no alternative crop that generated enough revenue to overcome their indebtedness. In some cases, especially in Arizona, growers had raised less-profitable alfalfa while getting their land ready for cotton. If they took land out of cotton in the San Joaquin Valley, predicted an economist with the University of California Extension, producers of fruits and vegetables would face serious consequences, since their market would likely be glutted by the switch from fiber to food.[36] It had been expensive to level land, drill wells, and install irrigation equipment, as well as acquire powerful tractors and the new mechanical pickers, and if acreage fell to the anticipated levels, there would be a major setback.

Powerful political figures fought on behalf of the West. Speaking to the Senate Committee on Agriculture and Forestry that conducted hearings in July 1953, California senator Thomas H. Kuchel succinctly outlined the issue, putting the case of the West into perspective and drawing on a statement by the USDA solicitor that only legislative action could provide the relief needed for the West. Arizona senator Barry Goldwater reminded his colleagues that westerners grew cotton at lower cost and obtained higher per-acre yields than southerners, which enabled the cotton industry to stay competitive in the world market. He joined other witnesses in pointing out that a geographic expansion had occurred, that the Cotton Belt now extended to the Pacific, and that the growth had happened so fast it went largely unrecognized. Arizona governor Howard Pyle wanted to avoid a "catastrophic situation" and thought the notion of planting restrictions should not be raised when it was so important to his state's economy. "It seems to us that justice should not be without a voice, and that is what brought us here."[37] Westerners hoped to amend the 1938 act to guarantee that no state would have a reduction over 25 percent and that the USDA should compute allotments based on three years, 1950 to 1952. Goldwater stated that the three-year proposal "provides an acreage allotment in a fair and equitable manner."[38]

John H. Davis, president of the CCC, and Marion Rhodes, director of the USDA Cotton Branch, Production and Marketing, brought their perspectives to the debate. Davis pointed out that the CCC and USDA had conducted meetings over the allotment issue, including one with the National Advisory Committee of Sixteen appointed by President Eisenhower to advise the secretary of agriculture. From these meetings came the recognition that cotton farming had undergone a geographic shift, and that any legislation affecting allotments should acknowledge the legitimate interests of the West as well as the South. Davis emphasized that allotments must be divided fairly among counties and apportioned equitably to individual farms. Rhodes foresaw problems in obtaining data from farmers about acres planted in previous years because the USDA could not reconcile data from the Bureau of Agricultural Economics, which had gathered records on plantings. He predicted dissatisfaction with an allotment program. Rhodes thought the three-year historical base was not long enough to show the southern trend of acreage decline, but agreed the five-year plan did not recognize the geographic shift. Davis and Rhodes offered no alternative and wanted further study before Congress amended the 1938 act. Senator Kuchel reminded the committee that time was short.

While the forceful words of senators and governors set forth the western position, the more personal tone of farmers also resonated with Congress. W. G. Kirklin, a farmer from Reeves County, Texas, on the South Plains, testified that he could not "survive this cut in acreage and be able to pay my honest debts, nor do I want to go back to the oil fields as a day laborer."[39] Like many of the new farmers, Kirklin was a veteran of World War II who believed he had responded to the government's call for production during the Korean War and now risked ruin, a predicament that evoked sympathy.

The fight over allotments provoked strong reactions. "They wanted to keep it all," recalled Californian Jack Stone.[40] Raymond E. Blair, a grower in Buttonwillow, California, who also operated a farm machinery business, stated, "There was just a feeling on both sides, and we thought they were trying to hide everything so that we couldn't get any more."[41] To put their case in a broad perspective and ease resistance, westerners described the shift of cattle ranching from its traditional home in the arid West to the greener South, a consequence of the new comparative regional advantages in agriculture that had been under way since 1945. Senator Goldwater recalled his surprise at seeing numerous Herefords in Virginia.[42] His anecdotal observation conformed with the general awareness of cattle moving east. *Collier's* magazine had recently reported on the growth of livestock farming in the Southeast, demonstrating how, in the words of one account, "the center of the new cattle industry became the old cotton South."[43] In this context, spokesmen for the West hoped to use the adage of "cotton moving west and cattle moving east" to justify their claim for the three-year plan.

If the West presented a compelling argument to minimize their acreage reductions, an equally powerful reply came from the older Cotton Belt. Southerners acknowledged the need for cutbacks but felt the western proposal discriminated against them, since their plantings had either dropped or held steady in the last three years. They preferred the five-year program written into the 1938 act. The South agreed that the national allotment of 17.5 million acres set by the secretary of agriculture was too severe and suggested a 22.5 million allotment. Southerners again agreed that cutbacks should not be carried out in one year but be stretched over several years, and they insisted the three-year plan gave preferential treatment to westerners because of the West's expansion, while in the South plantings had been scaling back for a generation.[44] Senator Allan J. Ellender of Louisiana expected his state to take a reduction of 18.21 percent with the western proposal, but he saw a smaller drop of 12 percent with the existing law. "I do not suppose it would make the cotton farmers in my state very happy, particularly when cotton has been the backbone of the southern economy for so long and these [western] states have just entered the picture in more recent years."[45]

This last point, effectively set forth in detail by Harvey Adams, general manager of the Agricultural Council of Arkansas, expressed a deeply embedded frustration in the Old Belt. Adams demonstrated how cotton planting had fallen since the inauguration of the New Deal programs, which Adams considered justified in view of the crushing surplus. He blamed the renewed surplus on the expansion into the West when the South had been making cutbacks since 1933 that averaged 45 percent.[46] Southerners had diversified into other crops and incurred debt, all the while watching thousands of small farmers and tenants seek new livelihoods. Now, Adams contended, the West's three-year plan would force more off the land. "You can readily see that thousands of individual farm families would be affected through losses of cotton acreages," stated G. C. Cortright Jr. of the Delta Council, "while relatively a few large producers would profit greatly."[47] Reductions based on three years would have a painful effect on southerners, Cortright continued, just as the five-year formula would harm westerners, but the proportion of stricken southerners would be much greater owing to the larger percentage of small farmers. Congressman Thomas G. Abernathy of Mississippi made a passionate plea on behalf of small farmers who had only five to twenty acres of cotton. The western proposal would make "40,000 beggars" in his District: "We would then have another Tobacco Road sure enough."[48]

Southerners saw themselves facing a speeding train. As early as January 1953, Hilton L. Bracey, executive vice president of the Missouri Cotton Producers Association, urged growers not to fail to report their 1951–52 plantings to their county USDA Production Management Administration (PMA) because he expected 1954 allotments to be based on the 1951–53 plantings.[49] Secretary of Agriculture Ezra Taft Benson urged growers in all states to voluntarily cut back in 1953 to avoid controls for 1954, but only farmers in North Carolina and Florida took his advice, because farmers in the other southern states thought such a move would penalize those with a smaller history while the "acreage hog" would benefit. Yet the acreage planted or harvested in 1953 fell below the expectation. Farmers did not plant to the fencerows, and the infamous drought of the 1950s, which hit Texas and Oklahoma particularly hard, had begun. The *Cotton Trade Journal* reported that the drought affected 1.5 million acres.[50]

Southerners in Congress expressed a sense of betrayal. They had supported the West in the past over water development; they had agreed to let western cotton gins and oil mills receive a preferential tax amortization schedule with the Internal Revenue Service; and they had voted for the Bracero Program for western growers.[51] Now there was a question of fairness for southerners, who saw the West as "trying to raid their cotton acreage." Westerners, stated Abernathy, shed "crocodile tears" in the interest of veterans, but he claimed to "represent more veterans than the people who support this bill."[52]

This contest over a farm bill amounted to the first public forum on cotton since expansion in the West began. New legislation could not impede the geographic shift to the region, where mechanization and large-scale farming offered the chance to compete in global markets. Westerners were suspicious, too. "This particular progress met an organized resistance," ACCO's Fleming told the Seventh Annual Cotton Research Congress in Dallas, "which found expression in the AAA."[53] It now fell to Congress to produce a coherent program. Each side had a logical and justifiable position, and even nonfarm interests got involved. From the Los Angeles Chamber of Commerce came a statement that "cotton is now the most important crop [in California]," and "cotton production has assumed major economic importance in recent years."[54] In the South, cotton remained the principal crop, and though cotton as a way of life had dwindled by 1953, the region's dependence on the fiber continued to be substantial. Southern yields had begun to rise in some locales and match those the West. What started as an economic question became a political matter, and Congress could not injure the general welfare, nor could it arbitrarily put people out of work. It looked for a compromise.

A solution would not come easily. One nonetheless had to be found, owing to the huge 16.5-million-bale crop in 1953 that stuffed CCC warehouses and fed at the trough of the U.S. Treasury as the Eisenhower administration attempted to balance the federal budget. Several bills representing the positions of the West and South, with variations, were introduced in both houses of Congress, but the leadership for finding a consensus fell to Senator Clinton P. Anderson, Truman's former secretary of agriculture, who commanded the trust of growers in both regions. As a member of the Senate Committee on Agriculture and Forestry, Anderson could cajole and persuade various interests, and when the committee's chair, George Aiken of Vermont, appointed a three-man committee to resolve the dilemma, he selected Anderson and senators James Eastland of Mississippi and Edward Ihye of Minnesota. A momentum to set the national quota over twenty-two million acres got under way, an idea presented by the Farm Bureau and supported by USDA and the Delta Council.[55] No agreement came before Congress went into recess in August 1953, but when it returned in the fall and the USDA projected a large crop for the year, an air of compromise emerged. Seeing the large crop and low prices glaring from the fields, westerners softened, and Anderson, who naturally leaned toward the West, reached an agreement with Eastland.[56] In December cotton farmers approved the proposal in a beltwide referendum.

With the approval in his pocket, Anderson submitted a compromise bill in January 1954, explaining that renewing allotments based on the 1938 act would be too severe if carried out in one year. He wished to avoid "the unnecessary hardship which would result from making the adjustment in a single year."[57] The hearings of the Senate committee, he acknowledged, had not been able to "work out adjustments," but he now seemed confident, particularly in view of the referendum, that an agreement could be reached. Congress passed his bill, Public Law 290, which provided a national quota of 21.4 million acres to be allotted to the states using the five-year plan. To further ease the impact on the West, the measure added another 315,000 acres, half of them designated for Arizona, California, and New Mexico. With the larger quota and the additional acreage, the West would take a softer blow with only slightly larger cuts than predicted in the three-year plan.[58]

In the South the extra acreage for the three states aroused some resentment. At its annual convention in Chicago, the Farm Bureau had floated the provision and even proposed another 59,000 acres each for Arizona and California. Congress did not accept the proposal, but the extra 315,000 acres, which the USDA had endorsed, prompted a swift response from Mississippi congressman James L. Whitten. "An equal effort should be made," he wrote Secretary Benson, "to relieve the drastic effect on real cotton farms in the historical cotton belt due to changes since 1950." Whitten urged Benson to fight for extra acreage to accommodate small farmers.[59]

This battle to protect acreage was the pinnacle of the differences between the two regions because the struggle to remain in farming was particularly severe in the older Cotton Belt. Medium-size farmers there were finding it tough to generate enough income to sustain a family. Cotton continued to lose market share to synthetic fibers, and the price of cotton locked at 30 cents for three years until 1959, when it dropped to 28 cents.[60] Price supports kept the most desperate farmers in the market, and it became customary for them to farm "for the loan." Stocks in CCC storage increased, owing partly to increasing per-acre yields. The 1955 crop, for example, was larger than expected at 14.7 million bales when the average lint yield hit 416 pounds per acre, a new Cotton Belt record.[61] Cotton farming was caught in the pit of surplus and low prices and having to be sustained by government assistance. Large operators fared better because they practiced the economies of scale, but hard times fell on all, enough to further erode the preeminence of King Cotton. Farmers in the South understandably continued switching to cattle, forage, soybeans, and other crops that presented fewer troubles than cotton with its boll weevils, bollworms, wilt disease, and the expensive machinery required for cultivating and picking. In Texas and Oklahoma the severe drought of the 1950s led to so-called bumblebee cotton, nickel-size bolls that had to be plowed under. In the Texas blacklands, families looked for a better life in the growing urban centers of the state.

In a decade known for the new age of baby boomers and the blossoming of television, farming continued to lose appeal. Sometimes known as the "fabulous fifties," this era witnessed the last breath of sharecropping and self-sufficient farming. Sharecroppers went directly to cities, but small landowners switched to cattle and other crops before they gave up farming, demonstrating that the demise of the family farm sprang from the difficulty of depending on small acreage. Former tenants expressed a wistful melancholy for the seasonal routine of planting seeds and picking the bolls of lint on the stalks. "If I was able to farm," recalled a Texas blackland renter-farmer, "that was the happiest life that I ever lived in my life, even if I didn't get nothing out of it."[62] But vacant storefronts, boarded-up gins, and empty shacks spread silently over a kingdom that had once vibrated with life and expectation. As stated by one astute observer, "They had been dispossessed . . . by the quiet revolution in the old Cotton Southeast."[63]

Some were sad and others gleeful not to be in the fields, but the end of small plots in the South related to the rise of the West. The adaptation of large-scale mechanized farming based on a corporate model drove out the yeoman farmer who followed the old ways, however idealistic and self-fulfilling they might be. Agriculture now competed in a global market, especially cotton, and all growers lived in a competitive atmosphere. Farming had to be conducted on a cost-benefit basis, a harsh change from the old mannered ways, but in the ruthless complex of international trade and competition from artificial fibers, no place remained for those unable to keep pace. This wrenching development, bringing real pain to those caught in its grip, affected farmers in all regions, but the experience was most pronounced in the South, where small-scale operations and tenants were numerous.

WESTERN FARM LABOR

The grinding poverty synonymous with cotton farming in the West often escaped notice. Conditions among migrant workers could be deplorable, since they lived on low wages and carried little with them. They lived in crude shelters or tents with makeshift sanitation facilities; a nearby irrigation ditch or small well might supply water. Their possessions consisted of an old automobile, clothing, and cooking utensils. Landowners varied in their management of labor, with some providing food as part of the earnings, but they nearly all furnished shelters in the camps with wells and pumps. Labor advocates deplored the conditions for workers, however, charging that growers provided only crude accommodations.[64] Growers and migrants alike complained of chicanery: landowners guarded against rocks or other items put in cotton sacks during picking to increase the weight of the cotton, while pickers insisted the weigh masters set their scales to cheat a few pounds. Migrants also worked in vineyards and on vegetable farms, perhaps in better or worse conditions, so the treatment they received from cotton producers usually followed the general pattern.

Western labor came from a variety of ethnic groups—Japanese, Chinese, Asians, African Americans, and white Okies—but Mexicans and Mexican Americans outnumbered all of them. Southern blacks picked cotton in California immediately after the war but were no longer used toward the end of the 1940s. Alice Mack of Arkansas starting going there in 1946, earning money each year until her earnings fell in 1948. "Within two years," wrote Richard Street, "virtually the entire California crop would be harvested by machine."[65] McWilliams's prediction of southern field hands descending on the state and bringing social strife did not occur. Most of the Okies found better employment or acquired plots of their own. On the Texas Plains, migrants swept across the area each picking season, gleaning the fields in forty-five to sixty days. Growers throughout the West kept a few local workers permanently for maintenance and preparation for the next year.[66]

Braceros had been the major source of farm labor in the West. The Bracero Program began in 1942 when the United States and Mexico made a commercial treaty that allowed Mexican citizens to enter the country primarily for harvesting crops. This arrangement was not made solely on behalf of cotton growers, but they were included and strongly supported it. Mexican nationals entering through the program concentrated in Texas, California, Arizona, New Mexico, and Arkansas and worked in crops such as cotton, sugar, beets, fruits, and vegetables.[67] "Bracero" means a laborer using his arms, an apt phrase for cotton pickers. They returned to their homeland at the end of the season.

Latinos were treated as second-class citizens, but the treatment they faced occurred throughout the Anglo culture. "No dogs and no Mexicans" might be written on signs at the entrances of cafés and grocery stores. Growers disliked such practices by retailers because the migrants would sometimes leave early to travel to the next county.[68] Cotton growers were often involved in disputes with Mexican labor, and when in 1950 Rio Grande growers were paying low wages, according to the President's Commission on Migratory Labor, Mexico rescinded the Bracero program for Texas. Agricultural interests pressured the U.S. Immigration Service to allow illegals into the state and transport them to cotton and other farming areas. The federal study and outcry against treatment of illegal workers led to Operation Wetback in 1954, a federal effort undertaken by the Immigration Service to return illegals to Mexico. But undocumented workers continued to enter the Southwest and, along with workers employed in the Bracero program, furnished the labor to get the cotton out of the fields.

Cotton interests defended themselves from charges of harsh treatment and strongly lobbied on behalf of the program, joining with other agricultural organizations to maintain congressional support. Labor reformers saw growers as a special interest intent on protecting a supply of cheap and docile labor, while the growers insisted that the program helped them overcome shortages of labor while offering Mexico's rural poor an opportunity to earn dollars. Growers also believed it reduced the number of illegal migrants. In the late 1950s, the use of Braceros came under increasing fire, and Congress, along with the White House, ended the program in 1965. A new era in America's use of Latino labor began that year with the new Immigration Act.

In the West, mechanization stimulated large-scale production and prevented the development of the landlord-tenant relationship synonymous with the older Cotton Belt. The body of literature about western labor, at least in cotton farming, was not comparable with the novels and social analyses that had cast the South in a negative light. No sense of permanence developed among western growers and migrant pickers, who traveled from county to county and returned to Mexico or South Texas after the crop was harvested. Indeed, the increasing use of machines undercut the Bracero program. "As machines replaced the bracero in the cotton fields," one analyst reported, "the supporters of Mexican labor were deprived of one of their principal arguments in behalf of bracero importation."[69] Western growers quickly adapted full-scale mechanization to practice the economies of scale essential to their survival and to resolve labor issues.[70]

EXTRA-LONG STAPLE

A special consideration of the expansion into the West featured extra-long staple (ELS) cotton, known as Pima, the premium fiber used for the finest cotton apparels. Cultivation centered in Arizona, where conditions such as weather and soil favored it, but ELS also grew in the Trans-Pecos of Texas near El Paso and New Mexico's Mesilla Valley. It was not a viable crop in the rest of the Cotton Belt. It had a bumpy history during the interwar years but had a renewed market when the War Mobilization Board wanted the fiber, which caused production to average over 64,000 bales per year during World War II. But the output of ELS fell again during the gold rush years of 1945 to 1949, dropping to a yearly average of 4,585 bales. "Pima cotton as an industry," reported one historian, "was close to extinction in 1947."[71] Upland varieties of Acala gave growers a better return. In 1950, however, production jumped because of defense needs and climbed to 93,467 bales in 1952. For the rest of the decade, production was lower, though consistent, because of a new variety, Pima S-1, developed by Dr. Walker E. Bryan at the University of Arizona. Arizona growers first planted the new strain commercially in 1954 and used the total allotment of 44,000 acres in 1955. This new variety matched Acala for high yield and hit slightly over 650 pounds of lint per acre between 1954 and 1957.[72] Pima always constituted a small portion of the U.S. crop, however, amounting to about 2 percent of the total each year between 1945 and 1960, but it sold at 50 cents to a $1.00 pound compared with Upland's 30 cents during the 1950s.

To further their market opportunities, ELS growers created the Supima Association of America in 1954, an organization similar, though smaller, to the NCC. Pima farmers sought to increase consumption of their staple because ELS grown in Egypt sold to U.S. textile mills at a lower price. Arizona growers had been farming "for the loan," putting their crop into the CCC to obtain the guaranteed price, when Egyptians began underselling the U.S. crop. In an unprecedented move, Supima persuaded Congress to lower the support price for ELS, which allowed Arizona growers to regain much of the domestic market in 1956.[73] The new organization also lobbied fashion designers to incorporate Pima into their creations, much like Ed Lipscomb's efforts on behalf of the NCC. Advertisements for Pima appeared in *Harper's Bazaar*, *Vogue*, the *New Yorker*, and *Town and Country*.

Producers of Pima got caught in the geopolitics of the Cold War. Egypt grew this highly prized cotton, which was its only means of raising currencies for foreign exchange. In 1952 Gamal Nasser seized control of the Egyptian government; he was a Pan-Arab who wanted to assert the sovereignty of his country as well as other Arab states by driving Britain and the European powers out of the Middle East. The Department of State sought to maintain harmonious relations with Egypt and allowed it to sell cotton to U.S. textile mills, which was cheaper until the CCC lowered the support price for American Pima. Because of pressure from the State Department, the USDA had placed acreage restrictions on domestic growers. At one point the State Department declared ELS an "expendable" industry and thought growers should produce vegetables. Pima farmers were helpless and had to compete with Egypt in their own home market. Pima growers complained in 1954 that they were allowed to produce less than one-third of the U.S. consumption of extra-long staple and appealed to President Eisenhower, Secretary of Agriculture Benson, and members of Congress to readjust the import quota for cotton from Egypt, but got no relief.[74]

World events rescued them. Egypt had asked the United States and Britain to furnish funds for building the Aswan Dam on the Nile, from which water would be used for irrigating cotton. Cotton interests in the western states lobbied against U.S. funding along with other interests opposing the structure. In 1955 Nasser made an arms agreement with communist Czechoslovakia, and the United States and Britain withdrew offers to finance the dam the next year. Angered by this withdrawal, Nasser seized the British-controlled Suez Canal. His action precipitated the Suez Crisis that ended with Egypt controlling the waterway, but the Soviet Union became Nasser's creditor. This development had ramifications throughout the Middle East, but U.S. Pima growers benefited because Nasser used cotton from the Nile Valley to pay for communist weapons. American growers increased acreage, ending the 1950s with growing production. Pima farmers gave much of the credit to the promotional efforts of the Supima Association.

CONCLUSION

The year 1960 closed the Eisenhower era and marked the establishment of the West as part of the Cotton Belt. By that point, the expansion of tillable land had halted as a new generation of entrepreneurs settled into their communities and strove to protect the niche they had made. The lucrative profits of the white gold rush had furnished the incentive and capital to transform arid lands into lush fields, and westerners quickly proved to be capable keepers of the soil. They proved that large-scale farming in conjunction with full mechanization lowered the cost of production. "To say the least," wrote one observer of regional development, "the Old Cotton Kingdom faces prospects of continued decline in cotton production and further outmigration."[75] Growers in the western valleys now stood as compatriots of the planters in the South and made clear they could organize and mobilize for their own well-being. They still felt underrepresented, however, and believed their southern brethren wielded power and influence disproportionate to their effectiveness in the cotton economy. "Almost all cotton growers east of Fort Worth have allowed their dog-in-the-manger attitude toward the irrigated states and the Plains to make them incapable of holding even the Russians in as intense animosity as they hold the Californians," Fleming wrote Clayton in 1950. "So, there is a really powerful resistance of uninformed blindness and prejudice."[76] A new term popped up in western circles, the "Mississippi Mafia," which referred to the disproportionate number of Mississippians who held leadership positions in the NCC and the powerful organizations located in the Mississippi Delta. By 1960 the tension had subsided, but the resentment would resurface.

Chopping cotton. Courtesy Memphis University.

Cotton pickers arrive by bus from Memphis. Courtesy Memphis University.

Pickers weighing cotton sacks in field.
Courtesy Memphis University.

Oscar Johnston's call for unity, Peabody Hotel, November 1938. Courtesy NCC.

Oscar Johnston and George Marshall. Courtesy NCC.

Two giants of cotton: Oscar Johnston and Will Clayton. Courtesy NCC.

Lew Barringer speaking at Peabody Hotel. Courtesy NCC.

Cotton at war with butter. Courtesy NCC.

A close-up of the original one-row mechanical picker. Courtesy Memphis University.

A row of ten mechanical pickers working in tandem. Courtesy Memphis University.

A cotton field in the Far West. Courtesy NCC.

Large-scale farming in the Far West. Twenty-five early model mechanical pickers operating in tandem. Courtesy NCC.

A field near Stratford, California, showing the vastness of large-scale farming in the Far West. Courtesy Kay Brown.

Boll weevils attacking a cotton boll. Courtesy NCC.

CHAPTER 9

BOLL WEEVILS, WORMS, AND MOTHS
A Hundred-Year War

In 1892 the Mexican boll weevil first appeared in the United States near Beeville, Texas, about one hundred miles above the Rio Grande border. This infamous beetle soon made its presence known by wiping out field after field of cotton in south Texas as farmers watched helplessly. It swept through east Texas and spread to the eastern seaboard, leaving ruin and devastation in its path. Named *A. grandis*, the weevil became the most dreaded insect pest of the South, riding the wind like one of the Four Horsemen of the Apocalypse. By 1921 it reached the Atlantic and threatened the lifeblood of the Southeast. Land values fell, banks refused to make loans to farmers, and merchants suffered as gins and cottonseed oil mills went out of business. Financial losses rose into the billions. In some cases, tenants and farmers gave up and abandoned their land. "A period of great poverty and distress [fell] among all classes of agricultural people," stated a report by the South Carolina Boll Weevil Commission.[1] In the worst-hit areas, the boll weevil pushed people into towns and cities and energized farmers to move to the West, where the dry climate proved less hospitable to the snouted devil. In 1923 President Warren G. Harding urged the cotton-growing states to "unite with the national government in the war on the boll weevil. It is no longer a local question," he continued, as "the greatest industry of the U.S. and world's supply of cotton are threatened by these little creatures."[2] This single insect was capable of destroying southern agriculture, a threat peculiar to cotton because wheat and corn in the grain belt, though faced with insect damage, had no comparable pest. The boll weevil exacerbated the poverty and uncertainty of cotton farming, which prompted William Skaggs, author of *The Southern Oligarchy*, to write that it "presents the most serious economic problem with which this country had been called upon to deal for many years."[3]

THE HUNDRED-YEAR WAR BEGINS

The struggle against boll weevils ranks among the extraordinary tales in U.S. agriculture. It was greatly responsible for the decline of the Southeast as a major cotton-producing region. For over one hundred years, the industry waged war against the weevils, gaining the upper hand only toward the end of the twentieth century. Other insects feasted on cotton, particularly the pink bollworm in the West, so insect control included more than boll weevils, but the weevil posed the greatest threat.

The weevils' appearance in south Texas shocked farmers. Insect devastation had never been such a serious threat, and no one knew how to cope with the invader. Federal and state governments had little support apparatus with which to fight back. Farmers scrambled to find methods of control and even resorted to handpicking the beetles off the stalk, but such efforts were futile. Estimates put losses at 300,000 bales in 1903 alone; another source put the figure at 1,350,920 bales from 1901 to 1904. Losses in bales and dollars mounted as the weevils spread to other states; the Census Bureau believed the country had lost $500 million by 1914. Average loss of the South's crop was about 17 percent, but it hit 30 percent in 1921. By 1950 the boll weevil was called "the $10 billion bug." By 1999 the total figure for losses to the insect, when adjusted for inflation, was reported at $102 billion. The invader had the ability to spread rapidly and quickly adapt to new environmental conditions. With its small wings, the weevil can buzz short distances, but strong winds can carry it over great distances. The Galveston hurricane of 1900, still the deadliest natural disaster in U.S. history, blew weevils as far north as the Texas-Oklahoma border.[4]

A major feature of the boll weevil, and the history associated with it, was the unknown. No mention of the weevil appeared in botanical literature until 1880, when USDA entomologists discovered the beetle in southern Mexico. They speculated that it depended on a shrublike tree there similar to cotton for a host and switched to cotton about the time of the American Civil War. This meant the weevil could switch hosts, not a common feature of insects. No winter occurs in Veracruz, Mexico, but when the insect reached north Texas, it survived cold weather. Its ability to hibernate in winter made it a formidable foe. Pressure for research to understand the weevil came as soon as it infected cotton, but a full knowledge for eliminating the pest would be long in coming.[5]

Inventors came up with devices to combat the weevils, including a quirky machine pulled by a mule with revolving brooms that knocked the insects off the plants. Farmers tried spreading lime and ashes between the rows or sprayed Paris green on the stalks. Nothing stopped the weevils until experiments with calcium arsenate during World War I at the Delta Boll Weevil Laboratory in Tallulah, Louisiana, showed promise. Farmers quickly accepted calcium arsenate and used it predominately until after World War II. Dusting machines pulled by mules or double-wing aircraft applied the arsenic mixture. Individual farmers commonly resorted to their own ingenuity to apply the chemical; one method was a mixture of calcium arsenate with molasses and water mopped onto the plants. In some locales, boll weevils forced a permanent change, and the most-cited example occurred in southeastern Alabama, where farmers switched to peanuts and prospered. In an ironic sense of gratitude for the pest's role in forcing farmers to abandon cotton, in 1919 the town of Enterprise erected a monument to the boll weevil. But the overall effect of the insect was hardship and despair, and by the mid-1920s, the cotton farmer had to stand guard against the new invader that could destroy his cash crop in short time.

Calcium arsenate presented drawbacks. In some cases, livestock died from eating grasses and hay contaminated by overspray, and beekeepers might lose hives. Some farmers believed the poison retarded the growth of other crops planted on soil that had been treated previously, and there were reports that the chemical left residue in cloth. Sharecroppers objected to exposure to the dust. Anyone who spent a day mopping a watery slush of molasses and arsenic on plants went home encrusted with toxic syrup. Impure supplies of the chemical from manufacturers caused farmers to mistrust it, and the arsenic tended to kill beneficial insects that held down other harmful bugs. Herein lay a fundamental difficulty with relying on insecticides: application of a chemical often led to outbreaks of secondary pests since the "beneficials," the insects that preyed on predators other than boll weevils, might be killed. No hue and cry arose over the use of calcium arsenate, but like all methods used in farming, the chemical had an economic threshold, meaning that its cost did not always justify the applications.[6] The chemical nonetheless continued to be widely used until the end of World War II and demonstrated that the boll weevil was not immune to poisoning.

To fight the scourge, researchers at USDA Experiment Stations quickly learned the value of "cultural control," which involved cultivation practices apart from the use of insecticides. This method helped reduce infestations but never eliminated boll weevils. Cultural control consisted of planting earlier and adding fertilizer to mature the plants before the weevils arrived; cutting and burning the stalks, or "plow and smoke"; destroying weevil-hibernating areas such as brushy or wooded areas next to fields; and rotating crops. Some states passed laws to make landowners destroy stalks.[7] Infestation generally varied from year to year, causing a heavy or a minor loss, and weevils might disappear for a while, but there was never an assurance that they would not reappear. This menace gave a common purpose to all classes of farmers and fostered a sense of unity in the search to eradicate or gain control over the weevil. "Farmers realized that for the first time since cotton was grown in the New World," reported the Cotton Foundation, "the industry faced a pest that could literally destroy their entire livelihood."[8]

By the mid-1920s the boll weevil had become part of folklore. Poets, novelists, and folksingers expressed the farmers' frustration over the insect and their resignation that it could not be driven out. In her novel about life on a cotton farm, Dorothy Scarborough wrote of the destruction of the devil bugs and the acceptance of them. Scarborough's character Jake, a black field hand, recites the poetry of the weevil:

> Oh, have you heard the lates'
> De lates' all yore own?
> All about the Boll Weevil
> What caused me to leave my home?
>
> First time I saw de Boll Weevil
> He was settin' on de squar,
> Nex' time I saw dat Weevil
> He was settin' everywhar,
> Jes' a-lookin' for a home—lookin'
> for a home.[9]

The poet and Lincoln biographer Carl Sandburg popularized the "Boll Weevil Song," which had the same theme, the hardship wrought by the pestilent bug. In 1946 the Disney movie *Song of the South* featured the hardship induced by boll weevils in the lyrics of a song. "When the weevil get the cotton everybody feels low, look out, / There'll be nothin' on the table when the dinner horn blow."[10] All classes of owners and tenants felt the curse of the weevil, which added to the misery of cotton farming.

AFTER 1945

Starting in 1945, a new era began in boll weevil history with the introduction of DDT, benzene hexachloride (BHC), and toxaphene. These new organochlorides gave growers better control because they had high toxicity to cotton pests and enough residual life to kill marauding insects from nearby fields. DDT did not harm weevils, but it killed the secondary pests such as bollworms and tobacco budworms that arose with the use of BHC and toxaphene for weevils. When DDT was mixed with these other substances, insect control became easier and more affordable, and aerial spraying became more common.[11] In 1951 tests at the Georgia Experiment Station led to the recommendation of applying insecticides every seven days, a "washday" or "womb-to-tomb" program that became common in the wetter regions of the Cotton Belt. Growers unquestioningly shot chemicals at cotton plants, a method that proved effective and affordable. Insecticide use went unabated; cotton accounted for one-third of the organochlorine insecticides applied between 1945 and 1960. Production of DDT grew, reaching 164,180,000 pounds in 1960. It appeared the long-awaited victory over the boll weevil had come. The cotton industry, along with most entomologists, embraced chemical poisoning, causing a general loss of interest in biological studies of insects and cultivation practices. "A lot of basic biological and cultural control . . . was sort of forgotten right after World War II and up into the early 1960s," remembered the famed entomologist Perry Adkisson, "because there was such heavy emphasis on the use and development of insecticide."[12] The application of chemicals was associated with progressive agriculture and came to be regarded as part of modern and scientific farming.[13]

Boll weevils began, however, to develop resistance to insecticides. As early as 1954, growers in Mississippi and Louisiana reported that weevils no longer seemed affected by chemical poisoning. In 1955 farmers in the Arkansas Delta experienced resistance in boll weevils near Lonoke, Pine Bluff, and the St. Francis and White River basins, where timber was plentiful and provided rich winter cover. Texas growers noticed resistance in 1956 and even wider resistance by 1958. Entomologists quickly responded with organophosphates such as malathion and parathion. These new chemicals killed the weevils but had little effect on secondary pests. To rid their crop of all pests, farmers mixed the old and new chemicals and threw in DDT.

This witches' brew worked, and a window of comfort from insect predators opened, leaving an uncluttered path to maximize and increase yields for the first time in a half century. Growers added more fertilizer, developed or expanded irrigation, and experimented with late-maturing varieties of cotton yielding more lint that had not been feasible when boll weevils were not under control. Per-acre yields climbed. In view of the acreage restrictions imposed by the federal government and the need to meet competition from synthetics, per-acre increases became mandatory.[14]

The triumph over insects quickly ended because bollworms and tobacco budworms became resistant to the organophosphates by the early 1960s. Growers tended to respond with more spraying, but the budworm resistance worsened to the point that some growers were forced out of business.[15] Control of budworms became more and more difficult, and there were incidents of human poisoning and damage to other crops caused by toxic drift. But the real culprit remained the boll weevil, and it had to be controlled without opening the door for attacks by secondary pests.

Reservations over the wisdom of the unfettered application of chemicals had never fully disappeared, and the "insecticide treadmill" led nowhere. Entomologists in experiment stations and land-grant colleges, on whose shoulders rested the responsibility of solving the riddle of boll weevil resistance and the confounding reappearance of secondary pests, had begun to rethink the problem.

In the 1956 entomologist E. F. Knipling spoke before the NCC's annual Beltwide Cotton Production Conference and pointed out how the fight against insects had focused on chemical poisoning. He urged more "basic research on many aspects of cotton insect problems."[16] A similar call had come from B. T. Shaw of the USDA Agricultural Research Service (ARS) that same year. Speaking later to the National Agricultural Chemicals Association, Knipling again urged that "more attention must be given to methods of insect control that do not require chemicals or require minimum use of chemicals."[17] To defeat the most devastating insect in American agriculture, entomologists would need help, and a major step occurred at the NCC's annual meeting in 1958 when the council formally committed to eradicating boll weevils. The cotton producer Robert Coker led the organization in taking this stand, but the delegates needed little prodding to declare war on the insect. Coker wanted to remove "from around the necks of southern cotton growers the millstone of needless insect losses and excessive production costs."[18] Within the same year, Coker testified before the House Agricultural Appropriations Subcommittee chaired by James Whitten of Mississippi, and explained the need for a concerted crusade against the boll weevil. The House then asked the USDA to study the status of research on cotton insects and make recommendations to Congress.

The USDA created a study group that hosted seventeen conferences of scientists and chemical engineers to gain input for its report. From this inquiry, the USDA learned that federal funding for boll weevil research had been grossly neglected, averaging only $75,000 per year from 1949 to 1958, when losses to the predator reached $350 million per year.[19] Funding had instead gone to chemical development and alternative controls. The study group recommended new approaches and more funding for basic research, including the construction of a laboratory devoted solely to boll weevils. Congress responded in 1960 with $1.1 million solely for the construction of a boll weevil laboratory. Mississippi State University at Starkville was selected as the site, and the "boll weevil lab" was dedicated in 1961. The entomologist Ted Davich of Texas A&M University became the director and belonged among those who wanted less emphasis on developing more chemicals. The goal of Davich and his staff, supported by the NCC and much of the USDA establishment, was eradication of the weevil.

In 1959 another important step in understanding boll weevils occurred when J. R. Brazzel outlined the weevil's wintering cycle, known as diapause. He demonstrated that weevils hibernated during the winter in off-field sites near the host cotton plants such as brushy fence lines, woods, or a line of trees. His research showed that late-fall applications of methyl parathion, along with destruction of the cotton stalks, reduced the wintering population. The next year Brazzel provided information about the reproductive biology of boll weevils and recommended making only four applications of insecticide per season. He claimed that up to 90 percent of the weevils could be eliminated in this manner, thereby reducing their population for the next spring. Knipling thought control could reach 95 percent, however, by increasing the insecticide treatments to seven. He also believed weevils returning from hibernation could be caught in pheromone traps, thus reducing their population still more. Between 1964 to 1967, his ideas were field tested in an experimental Texas High Plains suppression program that covered over one million acres where weevils had begun to infiltrate. The experimental program reduced weevil populations by 99 percent. Entomologists were encouraged. To suppress boll weevils and monitor their presence, however, entomologists needed a trap to catch them in the field. Research on finding a lure became a major objective, and in 1968 the responsibility went to the new laboratory in Starkville.[20]

When the special laboratory began operations, a lure to attract weevils was not an objective. But the development of the synthetic pheromone hormone grandlure proved to be a key factor in the struggle to eradicate the pest. One day researchers at the lab saw a large number of weevils on a shrub near the window of the basement where the insects were grown for study. They thought that food prepared for the rearing colony had attracted the weevils from nearby fields to the shrub, especially since their diet consisted of pieces of cotton plants. But the weevils were attracted by pheromones emitted by the hundreds of male weevils inside the building. Doctoral students and laboratory personnel tediously worked to isolate the hormone and find a synthetic compound as attractive to female weevils as the pheromone. Jim Tumlinson captained the team, and by 1971 they succeeded.[21] The NCC had meanwhile appointed a committee in 1969 to determine if an eradication test program was feasible. Now armed with pheromone-baited traps placed in the fields, USDA entomologists conducted a three-year field test in southern Mississippi, Alabama, and Louisiana beginning in 1971. This experiment proved encouraging, but not conclusive.[22] With these successes, however, the gate opened for an eradication program across the United States.

INTEGRATED PEST MANAGEMENT

An approach with minimal use of insecticides had already been under way in Texas. It began in 1958 when J. C. Gaines, head of Texas A&M's Entomology Department, brought Jim Brazzel and Perry Adkisson to Texas. Gaines went along with the "washday" program but had some reservations, while his two new professors opposed the heavy use of chemicals. Brazzel departed for a new assignment with the USDA Animal and Plant Health Inspection Service (APHIS), which left Adkisson to head the insect control studies. He saw danger in the ability of insects to develop resistance to insecticides and thought it wise to minimize, though not abandon, the use of chemicals. As he developed his concept, known as integrated pest management (IPM), he emphasized the importance of beneficial insects that preyed on the secondary offenders such as bollworms, tobacco budworms, aphids, leaf hoppers, and spider mites. IPM also employed cultural control for crops besides cotton, so that a minimum use of pesticides would be employed. Adkisson received the World Food Prize in 1997, but he is best known for his contributions to controlling cotton insects.

Born on a cotton farm near Blytheville, Arkansas, where cotton reigned supreme, Adkisson earned a Ph.D. in entomology from Kansas State University. After a few years as a project leader for cotton insect research at the University of Missouri, he went to Texas A&M in 1958 as an associate professor of entomology to lead a project on the pink bollworm. "Opportunities for work in cotton entomology were so much greater in Texas," he recalled.[23] Brazzel uncovered the diapause of boll weevils in 1959, and Adkisson and his team discovered the diapause of pink bollworms, which led to the Pink Bollworm Law in Texas, which regulates the time for planting, picking, and destroying stalks. By hitting the insect with chemicals only a few days before its diapause (similar to Brazzel's practice with weevils), the pink bollworm population can also be greatly curtailed. This breakthrough reduced much of the threat of the "pinkie" as a pest in cotton. Armed with this knowledge, Adkisson switched his attention to boll weevils, but they hibernate in the adult stage, not as larvae, and leave the cotton field for wintering. These characteristics accounted for the difficulty in combating them.

In 1966 tobacco budworms in the lower Rio Grande Valley became resistant to organophosphates, and cotton production fell there from three hundred thousand bales to seventy thousand. Within two years, some growers in the valley left farming, and the same development occurred in northern Mexico. Boll weevils remained susceptible to organophosphates, but this group of chemicals also killed the natural enemies of tobacco budworms. Adkisson advised Rio Grande growers to stop the washday program and incorporate his principles of IPM: a mix of cultural practices and limited spraying at certain times. As entomologists sought solutions, however, a new dimension began to impact their war against insects.

In 1962 Rachel Carson published her famous book *Silent Spring*, which described the use of chemicals, particularly pesticides, as a threat to the environment. Her favorite target was DDT, one of the most effective and popular chemical compounds for controlling insects. She pointed out that DDT survived in the tissue of birds and mammals that fed on insects killed by the poison, and they in turn had started developing birth defects and other harmful effects. Her book became a best seller and has come to be regarded as the milestone document that launched the environmental movement. Opposition to the use of pesticides in agriculture suddenly mushroomed, and cotton farming drew attention because it was a large user of chemical poisons. In 1969 Congress created the Environmental Protection Agency (EPA) and banned the use of DDT in 1972.

The crusade to overcome the boll weevil, while not allowing outbreaks of secondary pests that could be equally damaging, now faced the new public concern over soil and water pollution. Environmentalists had political clout and could not be ignored or shunned, but cotton entomologists were not upset because they had seen resistance among cotton pests for a decade and had already begun searching for methods and tactics to fight them without extreme use of poisons. Nor did the loss of DDT threaten the effectiveness of insect control. In 1969 the NCC had appointed Adkisson to its committee on boll weevil eradication. This committee arranged a ten-thousand-acre trial eradication program in Mississippi in 1971 that lasted two years and had encouraging results, but disagreement over the result erupted, so entomologists agreed to conduct another three-year trial in North Carolina and Virginia starting in 1978.[24] To test the IPM approach, another three-year trial was simultaneously conducted in Mississippi in 1978. Both trials brought the boll weevil and secondary pests under control while reducing costs up to 90 percent. The principles of both systems were incorporated into a large eradication program in the Carolinas in 1983, which proved so successful that cotton farming there began to expand for the first time in forty years. These advances led to the Boll Weevil Eradication Program (BWEP).

THE BOLL WEEVIL ERADICATION PROGRAM

The program in the Carolinas commenced only after the NCC lobbied Secretary of Agriculture John Block and Congress to ensure that the federal government would cover 30 percent of the costs.[25] Growers had responsibility for 70 percent, so only after a farmers' referendum of two-thirds approval did the program begin. As Congress made funds available and arrangements were made at the state and local levels, additional programs to eradicate the weevils got under way. The BWEP jumped next to California and Arizona in 1986 and then back to Georgia, Alabama, and Florida the next year. By 1987 six states in the Southeast had an eradication program in progress. Savings of up to 80 percent on insecticides encouraged further participation. In 1992 the BWEP took in Alabama and began incorporating states in a westward fashion. In 1995 growers in portions of Texas joined the program, but some bad luck occurred when the lower Rio Grande region nonetheless lost about 350,000 acres, and producers in the San Angelo area lost half their crop. Landowners in both areas blamed the BWEP, claiming it killed the beneficial insects and left nothing to block an onslaught of beet armyworms. When Mexican growers who were not participating had a much lower loss, about 15 percent compared with the astronomical Texas losses, emotions became bitter north of the border, and the program was terminated in south Texas in 1996. In 2004, however, the Texans passed a new referendum to rejoin the BWEP, and a new effort started in 2005 to eliminate the weevils in south Texas and the blacklands.[26]

The BWEP required treatment of a zone for three years, beginning usually with seven applications of malathion in late summer and early fall. Cultural control methods were employed, too, and pheromone traps were placed along the fields. Insecticide treatments followed only if monitoring showed an increase in weevils. Farmers reported about an average 10 percent increase in yield once the weevils had been eradicated and an overall reduction in use of pesticides. By 2000 it appeared that boll weevils no longer threatened cotton, though some pockets of farmers still had not gone through the BWEP.

NCC entomologist Frank Carter would not write off the devil bug but remained optimistic that with close guarding and ready treatments, the weevil could be kept under control.[27]

THE PINK BOLLWORM

While cotton growers in the western states had minimal worry with boll weevils, at least compared with the South, the pink bollworm gave them trouble. The history of this pest resembled that of the boll weevil, though the "pinkie" never wrought devastation on an equal scale and generally stayed, though not entirely, in the West. A worm that feeds in the larva stage inside the cotton boll, the pink bollworm was first discovered in 1917 near Hearne, Texas, in the Brazos River Valley. Entomologists and farmers had only rudimentary knowledge of the pest as it spread, but infestations were localized and not widespread. They learned that clearing fields of stalks and cotton trash removed the worms' food supply, which weakened them during the winter hibernation. Clean fields meant fewer pinkies, so cultural control became the standard means of fighting them. But cotton interests had learned a lesson from the boll weevil, that vigilance must be maintained, so whenever an infestation occurred, the area was quarantined by federal mandate or not allowed to grow cotton at all. A regulated buffer zone went around infested fields, and they were closely monitored. Before 1945 no insecticide had effectively worked, but cultural control worked well. Outbreaks occurred in south Texas near Mexico, and the Arizona Salt River Valley experienced pink bollworms in 1929, but the area had been cleared by 1934. Another infestation hit the same area in 1946, but the control method of quarantine and buffer zone enabled it to be cleared by 1950. Only in northern Florida and southern Georgia, as well as portions of southern Louisiana, did the insect invade the older South. Eradication seemed impossible, so "the wisest course," stated one writer, "was to learn to live with and control it."[28]

In 1945, with the introduction of organic chlorides, but mostly DDT, insecticide control of pink bollworms became possible, but only as a supplemental weapon to cultural methods. Applications of DDT at timed intervals would provide up to 65 or 70 percent control. Since the pinkie, like the boll weevil, thrived in Mexico, the USDA Bureau of Entomology and Plant Quarantine had a cooperative program there to hold down infestations on both sides of the border. In these early years, the USDA provided technical expertise to the Mexican Department of Agriculture.[29] The two departments would set schedules for stalk destruction and outlined programs for "sanitizing" gins of worm eggs. For the United States, cooperation made sense, since cotton acreage had increased in the Matamoros area next to the Rio Grande. As one entomologist stated, "The pink bollworm does not know what an international boundary is."[30]

Spraying and dusting DDT and organic chloride gave growers a temporary respite from pink bollworms until 1949, when insects destroyed about 2.8 million bales, about one-seventh of the U.S. crop. These losses represented the combined impact of all cotton pests, but the 1949 loss doubled from the previous year.[31] A cauldron of trouble had started bubbling. In 1952, Avery S. Hoyt, chief of the USDA Bureau of Entomology, explained that insects of all sorts—houseflies, roaches, and mosquitoes—were becoming resistant to insecticides. "We don't know just how long present insecticide controls . . . will continue to help farmers give this nation the bumper crops of food, feed, and fiber so necessary to the present high standard of living."[32] Only a few months after Hoyt's warning, the worst outbreak of pink bollworms in thirty-five years occurred. The USDA described the pink bollworm as the "most threatening insect pest of American cotton," and news of the damage caused a jump in the price of cotton, about $2.50 per bale on the New York Cotton Exchange. Compared with the total loss of the U.S. crop, the pink bollworm's share was small, but it nonetheless demonstrated an ability to remain a threat.

As acreage expanded in the West, the threat of the pink bollworm grew. DDT proved effective, but western farmers grew alfalfa for dairy cattle, and the drift of DDT from a nearby field of cotton might contaminate the alfalfa, and no traces of DDT were allowed in feed for dairy cows. Resistance in pinkies and other insects intensified too, and "farmers and scientists both stood aghast," recorded one observer, "as field after field of cotton, drenched with spray or powder from buzzing crop duster planes, promptly came alive again with pests that sucked and crunched and bored at the plants, relentlessly reducing yield."[33] Pink bollworms remained inside the cotton boll, destroying both lint and seed, which made a large portion of insecticides ineffective against them. Westerners now experienced the same hurdle as southerners who had crusaded against the boll weevil. They drowned with chemicals the pinkies and lygus bugs that harbored in alfalfa, but soon other pests appeared that had not been a problem before: spider mites, the cabbage looper, the leaf roller, and the salt marsh caterpillar. "Kill one bug and it came back in even greater numbers—or another struck," wrote an observer. The "insecticide treadmill" had come to the West, which caused an entomologist at the University of California to comment in 1963 that "things are becoming a mess."[34]

As is typical of insect pests, the pink bollworm would disappear and reappear; it would inflict serious damage in some locales and then go out of sight. By combining cultural control with insecticides, western growers tried to back pinkies into a corner, but the worms were masters of hide-and-seek. In 1966 the pink bollworm appeared near Blythe, California, and the USDA attacked the infestation with chemicals. Poisoning generally proved effective, but the portion of a crop saved with pesticide might not offset the costs involved. Insecticides often killed honey bees that pollinated the fruits and vegetables in western valleys. Relying on chemicals had the same drawback for combating pink bollworms as other insects; they became resistant. Where the pinkie appeared, growers attacked with heavy spraying, which never eradicated the pests. According to one source, there was some thought of abandoning cotton in view of the likelihood that pinkies would become too resistant.[35] But research continued, including experiments to disperse sterilized male moths through cotton fields, expecting them to mate with females and thus produce no offspring. Entomologists had employed the same practice against the boll weevil, but it never proved successful in either case. Growers settled on the methods employed through IPM for fighting pink bollworms.

In 1996 western growers planted Bt varieties (genetically modified cotton) and followed the same procedure as southerners fighting cotton bollworms. They established refuges of traditional cotton where pink bollworms would not develop resistance, but were expected to mate with moths surviving Bt cotton. Growers reported no cases of resistance, and by 2004 approximately half of Arizona's acreage had been planted in Bt varieties. Losses attributed to pinkies fell there by over 50 percent, according to the University of Arizona. That year field entomologists at the university found some resistant caterpillars but reported them as rare.[36] The NCC reported that in far west Texas and New Mexico a similar program reduced populations between 87 and 96 percent. The Mexican state of Chihuahua participated and reported similar drops. With these encouraging results, the NCC launched a proposal in 2004 to initiate an eradication program across the West. Chihuahua would participate, since the adult moths were wind-borne like boll weevils. The proposal called for USDA funding of $7.8 million, but the program would be financed principally by growers, whose share would reach 80 percent. Eradication of the pink bollworm had been considered impossible, but incorporating pheromone, Bt cotton, sterile moths, cultural control, and regulated applications of insecticide, the likelihood of eradication seemed possible, particularly since the boll weevil, the scourge of the South, had been brought under control. The program began in 2005 in stages across the West, and in Arizona, where the eradication started last, growers reported progressive results in 2008.[37]

BT COTTON

In 1996 the ongoing war between farmer and insect stepped in a new direction with the first planting of a genetically engineered cotton spliced with a bacillus thuringiensis (Bt) gene. Commonly known as Bt cotton, this new genetically altered variety appeared to be a breakthrough in controlling pink bollworm in the West and the cotton bollworm and tobacco budworm in the older Cotton Belt. Bt varieties did not control boll weevils. Monsanto had developed the genetically modified (GM) cotton and registered it under the trade name Bollgard and made licensing arrangements with D&PL, which produced the seed at the Maricopa Agricultural Center in Arizona. By 1996 D&PL had enough seeds to make them commercially available, and Arizona growers planted about seventy thousand acres in the Bt variety that year as a preventive measure against pink bollworms. In some areas of the South, growers planted large acreages of Bt cotton, up to 77 percent of acreage in Alabama, where a serious outbreak of cotton bollworms had occurred. About 1.8 million acres were planted in this promising new variety over the Cotton Belt the first year. Farmers had high expectations. Monsanto and the licensed seed companies, which had expanded to include Stoneville and Hartz, praised the merits of an environmentally safe method of insect control because it held out the hope of eliminating or reducing the use of pesticides.[38]

The new wonder in the cotton patch had mixed results. To begin with, growers experimenting with Bt cotton reported fewer outbreaks of bollworms and tobacco budworms. Jay Roberts of Georgia reported that his Bt plots produced smaller bolls but nonetheless yielded over twelve hundred pounds of lint per acre. He noticed, too, that beet armyworms avoided the Bt cotton and migrated to the conventional varieties. In his county and surrounding areas, insect pressure had been light in 1996, however, so the advantage of Bt cotton was limited. Roberts thought the new technology did not improve his profitability, since his licensing fees for Bt cotton, $32 per acre, were no less than his usual four applications of pyrethroid at $8 per acre. He saw, however, the promise of the new GM varieties and intended to plant a portion of his fields with Bt cotton in 1997; "But I would like to see," he stated, "the price come down so I can make some money."[39]

In other areas of the Cotton Belt there was dissatisfaction. Outbreaks of cotton bollworms occurred in Bt fields in portions of Texas and eastward through the Southeast, which prompted about 40 percent of the first-time users to apply insecticides. These growers were unhappy, since they had believed spraying would not be necessary and they now had to endure double costs—both licensing fees and insecticides—while sustaining yield losses due to bollworms. In Texas a group of planters initiated legal action against Monsanto and D&PL. The Bt variety also proved disappointing with lower-than-expected yields, which were reported in northeastern Arkansas and portions of Arizona. Complaints included erratic growth of bolls, more difficulty in scouting for insects than with traditional varieties, and misjudging the amounts of fertilizer to apply owing to the tendency of Bt cotton to set bolls. But the most dissatisfaction came from the costs incurred in spraying for bollworms that use of the Bt variety was meant to avoid.

Growers had to agree not to save seeds from their Bt crop. If caught, they faced a fine of $120 per acre because Monsanto wanted to protect its intellectual property rights. To prevent insect resistance to Bt genes, which entomologists feared, farmers had to plant "refuges," a few acres of traditional cotton amid the fields of GM plants to feed the damaging insects. For each one-hundred-acre plot, a refuge of four acres went into the field, a 96:4 ratio. These worm preserves, a concept strongly endorsed by entomologists, would allow bollworms exposed to Bt to mate with unexposed worms in the refuges and produce offspring with no resistance. This practice was expected to prevent the development of worms resistant to transgenic cotton. Growers complained that refuges should be reduced in size in order not to consume too many acres and pressure their narrow profit margin.[40]

Despite the drawbacks associated with Bt cotton, farmers accepted the adjustments necessary to use it effectively. Frank Mitchener, a grower in the Mississippi Delta, sustained losses with his first use of Bt in 1996, but he numbered among the growers who "swallowed their pride, made a few mental calculations, and signed up to plant the new crop."[41] Mitchener planted the Bt variety on more than 90 percent of his fields in subsequent years. Advocates of Bt acknowledged that it was "not a silver bullet" because insect resistance could occur any time, and since Bt reduced applications of pyrethroids, planters had to be vigilant for outbreaks of secondary pests. Farmers found themselves facing tricky management decisions as they had to guard against infestations of damaging worms while watching out for other predators. Danger was imminent in spraying too early or too late, and it remained necessary to use chemicals to control outbreaks of boll weevils. To help maneuver through this maze, growers started relying on professional scouts.[42]

COTTON SCOUTS

Insect control thus became increasingly complicated with the development of IPM and monitoring for boll weevils and other threatening insects. Growers needed to hold down costs by minimizing their applications of poisons or relying on beneficial insects to rid the fields of pests. Quick discovery of infestation became critical so as to hit invading pests at the optimum time. Other developments such as the new growth regulators that could be applied to prevent undesirable stalk growth, and the numerous seed varieties that allow customized planting for particular soils, added to the complexity of cotton farming. More and more, growers needed precise, accurate information to make management decisions. As farming operations increased in size, furthermore, growers could not personally examine all their acres. To provide such critical information to them, a new profession arose in the 1970s: the cotton scout, trained in insect recognition and control. Barry Aycock of Parma, Missouri, exemplified this new player in cotton farming.

With a Ph.D. in plant science from Southern Illinois University in 1994, Aycock opened a consulting business in his hometown, located in the heart of the Missouri Bootheel. Named *Cotton Farming*'s 1995 Consultant of the Year, he operated the Aycock Agricultural Services in Parma, where high-yield agriculture prevailed. Most of his work dealt with insect control, and working on a per-acre fee basis, he furnished an insect profile with a recommendation whether to apply insecticides. Aycock hired assistants during the summer to scout the fields. They walked diagonally across a field once a week to make an examination. Unlike many other scouts, he refused to use motorbikes, preferring a diagonal investigation and the closer observation provided by walking. His methods enabled him to gather precise information and, for instance, recommend a single application on only a hundred acres in a thousand-acre field. Today, cotton producers value Aycock's recommendations because he is not an employee of a chemical company. During the off-season, he attends the NCC Beltwide Conference and engages in research experiments with chemical and seed companies.[43]

The importance of growers having accurate information to make management decisions became evident when in 2000 public concern arose over the discovery of Bt genes in taco shells. Bt corn, under the trade name StarLink, had been approved by the EPA for animal feed, but not for human consumption. The implanted gene kept corn borers from destroying the stalks, but approval had been withheld for human use because the Bt protein Cry9C had qualities similar to an allegoric protein and could conceivably cause allergic reactions in humans. Opponents of genetically modified crops had discovered the presence of StarLink by testing corn products taken from grocery shelves in Washington, D.C. This discovery, accompanied with much publicity, caused a temporary setback for Bt technology in corn products but did not affect cotton. Corn farmers who had grown StarLink that year had no market because buyers worried about liability or their own customers avoiding StarLink. Some Asian countries refused to import StarLink corn.[44]

Aventis, the French company that marketed the Bt seed, made an arrangement to buy back the corn, offering farmers a 25 percent premium over the market price. Estimates for the buyback program involved about 80 million bushels, ranging in costs from $60 to $100 million. Public alarm over transgenic crops focused on corn, which preempted environmental objections to the use of Bt cotton, but the danger of incorporating new technologies into farming practices, regardless of the crop, was evident.

THE RISE OF PUBLIC HEALTH CONCERNS

In the struggle against damaging insects, growers faced rising concern over public safety. Although opposition to Bt cotton did not match the alarm over StarLink corn, the new gene-altered cotton did not escape suspicion. Growers of organic cotton worried that cross-pollination from genetically modified varieties threatened their nearby fields, and there was suspicion that Bt cotton would contaminate traditional varieties and eliminate pure seed for future breeding. The general public was concerned about risks to health that might arise in the future, since the long-term ramifications of Bt were unknown, particularly since cottonseed oil went into oleomargarine, salad dressings, and other table foods. There remained the additional fear that insects would develop resistance to the Bt gene and endanger not only cotton but other crops exposed to pollination. The counterargument that use of pesticides fell dramatically with Bt cotton failed to quell objections. Critics continued to point out that the total volume of pesticide use remained nonetheless high, since farmers continue to use organophosphates against secondary pests. This dispute pitted growers against environmentalists and proponents of organic farming.

After the mid-1990s, the use of insecticides was caught in the tornado of public discontent owing to population growth and the housing boom. Chemical drift from crop dusting alarmed homeowners, who complained of odors and toxic clouds. Chemicals used to defoliate cotton for picking brought complaints of eye irritation, sore throats, and nausea, known as "cotton flu," according to the North Carolina Agricultural Resources Center.[45] Californians for Pesticide Reform sued the state agency in California charged with protecting the public from exposure, insisting it had been too lax in enforcing regulations. Urban sprawl created by new retirement developments in the Sun Belt forced growers to defend their use of chemicals occasionally from legal action taken by environmentalists or activist groups.[46] Opposition to Bt cotton ran deep among environmentalists and amounted to more than a prickly thorn of irritation. In 2001 the Earth Liberation Front (ELF) set fire to a D&PL storage building containing Bt planting seed. The organization proudly took responsibility for the fire and felt justified "because it contained massive quantities of transgenic cotton seed in storage."[47] Described by the Federal Bureau of Investigation (FBI) as "domestic terrorists," the ELF has also claimed credit for firebombing a Hummer dealership and structures of the U.S. Forest Service. More activity against GM crops should be expected, the group warned.[48]

In 1996 the use of pesticides became more complex with passage of the Food Quality Protection Act (FQPA). This measure had strong public support and passed both chambers of Congress unanimously, and President William Clinton quickly signed the bill. It put stiffer safety requirements on agricultural insecticides that had long-lasting residues in water and foodstuffs. The law specifically targeted the safety of infants and children based on a 1993 study by the National Academy of Sciences that showed young people's greater susceptibility to toxic exposures. The new law mandated that a safety factor of ten be applied when measuring the sensitivity of children to pesticide toxicity. For cotton growers, the FQPA meant the reduction of the use of broad-spectrum chemicals such as pyrethroids and phosphates that had been successful against cotton pests. Farmers would now have to use insecticides designed to target specific insects rather than the broad-spectrum chemicals. Toxicity levels were expected to be lower overall, but growers would have to monitor their fields closely and apply an insecticide aimed only at a single insect. They would "have to know what the problem is in the field and target their spraying to that insect," stated Ron Smith, an entomologist with the Alabama Extension Service.[49] He expected the new system to offer more adaptability of crop treatment and reduce the likelihood of insect resistance, but growers would have to become more knowledgeable about insecticides. Preventive spraying under the old system would have to give way to applying a specific chemical to a specific insect only when it was discovered in the field. Scouting would become more necessary.

No hue and cry against the stiffer regulations surfaced across the Cotton Belt, although growers were not happy whenever they faced restrictions. In 1997 the EPA disallowed the use of carbofuran to combat aphids, mites, and nematodes. Before the passage of the FQPA, the EPA had approved carbofuran's use, but now the EPA felt that the stronger safety levels for children prevented it. The EPA based its decision on monitoring data that showed levels of the chemical in fields beyond the acceptable target. There was some expectation of aphid pressure that year, and the NCC stated that other insecticides for controlling aphids were more expensive, and estimates of cotton damaged by aphids ranged from one to five million acres.[50] Growers complained that the EPA applied the statute rules inconsistently, which forced them to frequently alter their risk management. And they cautioned that stringent cutbacks in pesticides would jeopardize the Boll Weevil Eradication Program. "We simply cannot lose the ability to eradicate the boll weevil," stated Texas producer and NCC president William Lovelady before the House Subcommittee on Department Operations, Nutrition, and Foreign Agriculture in 1998.[51]

To represent the interests of its constituents, the NCC teamed up with the National Food Processors, the Minor Crop Farmer Alliance, and the American Protective Association to forward recommendations to the EPA and USDA about implementing the FQPA. This coalition of agriculture, food processors, and pesticide organizations created a roadmap for regulatory agencies to consider when implementing the new measure.[52] Phillip C. Burnett, NCC executive director, expressed the general attitude of growers when he stated that standards to protect child health made sense, but the EPA had caused frustration among growers as it sought to establish rules and guidelines for implementing the legislation. As the EPA sought to make decisions, it was "often guilty of changing its mind."[53] Burnett preferred to work cooperatively with the agency to maintain the interests of agriculture without jeopardizing public health. The proposed roadmap or blueprint would, in his opinion, be a reasonable approach to carrying out the new standards, since it relied on a conjunctive effort by cotton interests and the EPA. In other words, there was no fundamental objection to the FQPA, but frustration over the baffling lack of consistency in its implementation.

Environmentalists and consumer activists nonetheless saw agriculture and pesticide manufacturers as a single special interest that, rather than cooperating with regulatory bodies, sought to thwart the intentions of the FQPA. The Consumers Union complained that too few representatives of public health, children's, and consumer organizations had been selected to testify before the House Subcommittee in 1998. Too much attention, continued the accusation, went to the concerns of agriculture and too little for the dangers of pesticides. Jeannie Kenny, policy analyst for Consumers Union, wanted Congress to arrange a more thorough study of "both the science and policy questions raised by FQPA by considering the wide range of views surrounding those questions." She complained that the EPA favored agriculture interests by reviewing "the 10 fold safety factor for surprisingly few (roughly 10 percent) of the pesticides reviewed."[54] To delay implementing the FQPA on behalf of agriculture would jeopardize child health.

As the charges and accusations continued, the terrorist attacks of September 11, 2001, directly affected segments of the cotton industry. In 2002 Congress passed the Public Health Security and Bioterrorism Preparedness and Response Act, which directed the secretary of agriculture to protect the public from a terrorist attack on the U.S. food supply. In October 2003, the EPA published regulations setting forth compliance requirements for domestic and foreign companies that manufacture, process, pack, or hold food for human or animal consumption. Such organizations were required to register with the EPA no later than December 2003 but were required to take no further action. In the event of an attack, the agency would be able to locate food processors and alert them. Foreign companies had to designate an agent, a broker or importer, who should live in the United States and be in the country for registration. This law affected cottonseed oil mills, cottonseed warehouses, and gins. Companies were allowed to register online and did not have to pay a fee. Failure to comply, however, could be considered a civil or criminal offense. Farms were exempt from the legislation.

New technologies to control insects had made the use of pesticides more than a matter of farm management because of the ramifications for public health. Young growers in the new millennium could not emulate their fathers, who only had to identify the culprit in their fields and apply a chemical poison. In the past, it was only necessary to apply chemical brews at regular intervals, routinely following the "washday" program. But insects, with their uncanny ability to develop resistance, forced the methods and practices of control to shift periodically, driving entomologists to keep one step ahead. Improved knowledge of insects reversed the original notion that heavy dosage of poison was the answer, a development already under way when the new environmentalism first appeared.

CONCLUSION

With the persistence of agricultural entomologists, dependable government support, and the NCC's efforts led by Bob Coker, the triumph over the boll weevil finally came. The struggle lasted about a century, with the real progress beginning around 1945, when the new era of insect control commenced. Much research and experimentation went into the struggle, which no one has claimed to be over, but hopefully the boll weevil, the greatest menace of all, could be kept subdued through unrelenting vigilance. Thanks to this accomplishment, cotton farming reappeared in the Southeast. Progress against the pink bollworm in the West also demonstrated the value of teamwork.

Despite the triumphs, the war against insects had to conform to the new environmentalism. To protect public health and preserve an uncontaminated environment, the U.S. government imposed stiff regulations on the use of cotton insecticides, which compelled growers to exercise restraint and caution in their chemical attacks on worms, weevils, and moths. The new technologies of genetic engineering became a valuable asset that reduced the use of insecticides, but chemicals nonetheless remained a part of farming. Because cotton had ramifications for protecting the environment, growers had acquired a part, perhaps unwittingly, in the greater goal of balancing economic welfare with public health and were now embroiled in the general issues of the day.

CHAPTER 10

MEMPHIS

The Epicenter of the Cotton Belt

"Memphis is the cotton center of the world!" That statement, made in 1959 by Gerald Dearing, a columnist for the Memphis *Commercial Appeal*, expressed how the city on the Mississippi River served as the throne room of the cotton kingdom. No city depended more on the white gold that flowed into its warehouses and onto its wharves, and no city embraced cotton as did Memphis. The sweet aroma of cottonseed oil flowed through the air, and downtown lunch counters were packed with cotton classers whose trousers carried clinging puffs of lint. The city's elite came from the cotton merchandising houses and presented their daughters to society in the ballrooms of plush hotels. One downtown hotel was even named King Cotton. When in 1931 the city began hosting its Cotton Carnival, it paid tribute to the royal plant. But splendor and elegance did not fully explain the city's preeminence in the cotton culture, because the sweat of black labor fused with the white pursuit of wealth and created a synergy in which paths led to and away from cotton, where the white establishment and black subculture toiled all day and enjoyed the fun of the city at night. In the heart of downtown, the Cotton Exchange, Cotton Row, and the famous blues capital, Beale Street, sent waves of energy in all directions that made Memphis the epicenter of the Cotton Belt.[1]

Location made Memphis the nation's cotton center. The city sits near the northern tip of the Mississippi Delta, the "most southern place on earth" and the area most commonly associated with cotton. This alluvial plain hosts rivers flowing southward: Deer Creek, Sunflower River, the Yazoo, and the great Mississippi. For eons these rivers deposited a fertile topsoil as rich as any in the world. Cleared of timber and drained of water after the Civil War, the Delta became known for its large-scale agriculture relying on black laborers and sharecroppers, where all manner of people conformed to the life cycle of the cotton plant. No place in the South so enthusiastically staked its well-being on *Gossypium*, which generated wealth and poverty alike. Until cotton farming mechanized, the feudal practices of plantation agriculture prevailed and drew the attention of social analysts and commentators. Writers such as William Faulkner, William Percy, Eudora Welty, and John Grisham used the Delta as a backdrop for their novels. From deep in this Nile, the Illinois Central Railroad ran straight into Memphis and helped make it the urban center of the Delta, drawing people, money, and business, all looking for opportunity to make or enjoy wealth or just a little relief from the drudgery of cotton farming. No city in the mid-South matched Memphis's urban attractions and cosmopolitan atmosphere, and for good reason David Cohn wrote that "Memphis draws sustenance from its immense surrounding territory, and the Delta is one of its richest tributary provinces."[2]

Across the Mississippi River and within sight of Memphis's Cotton Row lies the Arkansas Delta, another major cotton-producing area. This Delta, another alluvial plain within the lower Mississippi River Valley, occupies the eastern half of Arkansas, stretching westward from Memphis to Little Rock, and bound north and south by Missouri and Louisiana. It too has rich soil deposited by the Arkansas River, White River, Black River, St. Francis River, Cache River, and L'Anguille River, as well as the Mississippi. Running north from Helena for 150 miles into Missouri like the raised backbone of a dinosaur is Crowley's Ridge, a geographic oddity in the heart of the flat Delta, varying in width from one-half to twelve miles. No cotton grew there, and the way of life resembled Appalachian culture. But the Arkansas Delta had all the features of the cotton South: plantation agriculture, staggering poverty, a large population of black sharecroppers, rigid segregation, and one-crop farming. The town of Cotton Plant sits in the middle of this Delta, and Blytheville hosted the National Cotton Picking Contest until mechanization made it obsolete. Eastern Arkansas never produced literary giants like Mississippi, and it received less attention from social analysts, though the Southern Tenant Farmers' Union strike of 1935 caught the nation's attention. The poverty of cotton farming seemed worse there, however, owing partly to the infamous flood of 1927 that covered half the state's land in water and forced landowners to rebuild homes and farms. Arkansans sold their cotton in Memphis, and like Mississippians they enjoyed the conveniences and pleasures of the city.

The third section of the lower Mississippi River Valley with which Memphis has a mutually beneficial relationship is the Missouri Bootheel, the state's distinctively shaped southeastern corner, whose boundary origins are shrouded in the mystery of the early territorial history of 1820. Bordered by the Mississippi River on its east side and the St. Francis on the west, the Bootheel stretches northeast to Sikeston and northwest to Popular Bluff.

Much of the Bootheel had to be drained and ditches installed before it could be cultivated, and not until the early twentieth century had this been accomplished. It had the characteristics suitable for cotton farming: rich sandy loam, abundant rainfall, and flat land, but the seven counties there are the northernmost point of the cotton-growing Mississippi River Valley. Cool weather and the Missouri Ozark Mountains furnish the northern cultivation limit. Plantations were never numerous, but sharecroppers were common. Segregation thrived, and poor whites were numerous. Kennet, Malden, and Caruthersville were the principal towns, but Memphis was the big city that attracted Bootheel residents. For the more fortunate, major shopping sprees nearly always meant a trip to Memphis. "We always looked forward to Memphis," stated Hortense Russom, a longtime resident. "There were so many choices for clothes."[3]

THE COTTON CAPITAL

With natural ease, Memphis developed an infrastructure to make it a major cotton market. In 1950 the city handled 40 percent of the cotton transactions in the United States. Barges plied the Mississippi River, and the city built loading docks and wharves below the bluffs where stevedores loaded and reloaded bales. When railroads replaced barges, Front Street remained the trading center. A row of buildings on Front Street housed buyers where they bought and sold cotton. Each buyer had "classers" who graded each bale of cotton in transaction. Classing rooms had long tables holding samples cut from bales, and men skilled in classifying cotton assigned a grade to each bale. Classing required practice for determining the staple length and the various shades of white. With enough practice, classers made their work into an art, and it was their word that set the grade on which the price rested. Oddly enough, farmers generally knew little about grading and accepted the decision of the classers. When disagreements occurred, they were resolved through a second grading known as arbitration. Brokers were qualified classers, too, and since each bale going through the Memphis market had to be sampled and graded by hand and eye, a sizable number of men worked in the buying houses, which accounted for their appearance on the streets with cotton lint clinging to their trousers. This odd sight was a regular feature of the Memphis business district. The sidewalks of Front Street also had "snakes," large jute bags stuffed with the leftover samples. These puffy bags of cotton littered the street and curb until they were rebaled and sold. Businesses known as pickeries handled these secondary bales.[4]

Memphis buzzed during the months following cotton picking. Warehousing flourished as bales went into storage. Large buildings along South Main housed the cotton, and the sight of black workers handling bales on two-wheeled trucks and rolling them down ramps was common. Inside the warehouses, workers stood the bales on end and later loaded them into rail cars for shipment to textile mills or a saltwater port. The work was strenuous in the Memphis heat and humidity, so only the strong could hold the job. Lighter work involved the sample cutters who gathered and delivered samples to the classing rooms. Growers often visited the offices. The hustle and bustle of daily trading from September to December had a sense of urgency, since cotton prices fluctuated by the hour and buyers hedged their transactions with futures trading. Like all commodity traders, Memphis brokers sought to buy low and sell high, and each transaction had risk. People rushed from place to place, and what might appear frantic to the outsider amounted to business as usual. Front Street was the center of this activity and became known as Cotton Row; it even had its own poet laureate, William Johnston Britton. In 1948 he published *Front Street, A Book of Poems*, in which he paid tribute to white gold: "Here's to our native Southland and here's to cotton, our King."[5]

The Memphis Cotton Exchange, built in 1924, had been purposely located at the corner of Front and Union Streets to be in the midst of the activity because the previous location on Madison and Second, four blocks away, was considered too distant to be convenient. In the upper stories were offices, but at ground level was a high-ceilinged trading floor where cotton prices from exchanges in Chicago, New York, and New Orleans were posted on a large blackboard. The Liverpool Exchange was included, too. Posting included pertinent information such as acreage planting, seed sales, and estimates of world supply and demand. A ticker tape accessible to traders on the floor provided market quotes. Here the mainstay of the mid-South, cotton, traded, and here the earnings of farmers and planters were subjected to the capricious nature of the open market. The high windows, marble floor, chairs, spittoons, and ceiling fans gave the trading room a southern ambience. Brokers drifted on and off the trading floor, staying keen to price fluctuations, filling orders, catching up on news, and waiting. On slow days, a game of dominoes or checkers would likely be in progress. For all the complexity and risk of commodity trading, the brokers used an honor system, giving only their oral agreement on a sale. "The word of the cotton man," wrote William Bearden, "is all that is needed for a transaction running into the millions of dollars."[6] In 1945, when cotton prices had recovered from the Depression and future increases appeared certain, this description fit the Memphis Cotton Exchange as it hummed with activity.

Nothing better illustrated how Memphis was the hub of the Cotton Belt than its mule trading. Mules were part of southern folklore, kindred spirits in the cotton field where man and animal strained to make the soil yield its white gold. In the southern cultural landscape of moonlight and magnolias, the mule burdened with harness stood in front of the plow or wagon. Not uncommonly, mules were regarded as pets, and a favorite mule would be proudly shown like a new Ford or Chevrolet. Mule races, pulling contests, and judging shows were popular attractions at state and county fairs, where farmer and city cousin enjoyed the spectacle of well-bred mules. In 1920 the United States had twenty-six million mules, a large percentage of them in the cotton South. In 1937 D&PL had one thousand mules on its 38,000 acres, but tractors were beginning to appear throughout the Delta. For a few years after World War II, however, the mule remained in use, particularly in the southern cotton areas, and Memphis served as the largest center for buying and selling mules in the United States.[7]

"There's something sort of pretty about a mule plowing up good, black delta dirt," said one Memphis trader in 1947.[8] Cotton and mules went together like food and drink, and "when the mule was king the royal palace of the mule era was the Stockyard Hotel at 150 E. McLemore."[9] Customers at the hotel bar sat on saddles refashioned as bar stools, and poker and dice games took place. In 1941 police raided and wrecked the bar, but the liquor flowed and dice rolled, and the saddle stools remained until 1957, when the hotel was razed. Mule farming and trading had not long to live, but there was a flurry of activity and excitement in its last years when "farmers from Crittenden County in coveralls rubbed elbows with sleek, scarf-wearing agents of means from Missouri or Oklahoma."[10]

In 1946 twenty-two mule barns and yards were located in South Memphis and near downtown. In South Memphis, Owens Brothers at 1157 Stockyard Place had the largest operation, auctioning forty-five thousand mules in 1946 alone. Other trading barns operated on East McLemore Avenue, at Third and Monroe, and at 1121 Kansas. Another barn stood near the Peabody Hotel on the present-day site of the Memphis Redbirds' baseball stadium. Trading ran eleven months of the year, from New Year's Day to Thanksgiving. Sellers brought mules by train from as far away as Idaho, Montana, and the Carolinas. For 1946 the *Commercial Appeal* reported that 763 carloads of mules came into Memphis, accounting for the city's recognition as the "world's largest mule market."[11] This was a wholesale market, however, where buyers from smaller towns obtained mules in lots and resold them. The U.S. Army purchased mules there, and public bodies such as the Mississippi State Penitentiary in Parchman relied on Memphis for its supply of the long-eared beasts of burden. Foreign governments also used the Memphis market.[12]

M. R. Meals, known as "the Colonel" and regarded as the dean of mule auctioneers, worked at Owens Bros. and proudly let people know that during an auction he sold a mule every two minutes. A portly man with a large frame, Meals bellowed the auctioneer's chant, watching the buyers' subtle bid signals, which might be as small as a pull on a coat lapel or the lifting of a few fingers. Mules considered unworthy for fieldwork were sold at lower prices to be slaughtered for dog food.

For all the feverish activity of mule trading, mechanization had begun to push the beasts off the land. In 1951 Owens Bros. sold about twenty-five thousand mules and acknowledged that business was not what it used to be. Across the city, sales reached about a third of the peak years from 1946 to 1949, when cotton prices were high and no planting allotments were in place. Trading was fading into memory as some barns sold only twelve to fifteen mules per week. Barns and corrals collapsed under the wrecking ball, and in 1962 the last mule barn, located at 1121 Kansas, was razed and replaced as a trucking yard. Colonel Meals had died, and the era of the mule passed into history.

Mule traders complained that the shortage of labor had driven out mules and had forced growers to buy tractors for the lack of mule drivers. Their comments reinforced the claim made by landowners that sharecroppers and field-workers had moved away for better jobs and compelled farmers to modernize. Specifically, tractors and trucks drove out the mules, but not mechanical pickers, since only a few animals were used to pull wagons when hand-picking prevailed. For the chores of plowing and towing numerous pieces of equipment, tractors proved to be superior in power and convenience. Tractors did not require acreage and labor for growing hay and feed, nor did they demand daily care, but tractors required gasoline and replacement parts, which hastened the farmer's switch to a cash economy.

Memphis had no monopoly on cotton. Exchanges in cities besides New York, Chicago, and New Orleans figured prominently: Dallas, Los Angeles, Galveston, and, starting in 1947, Lubbock. Brokers and cotton merchants were scattered across the Cotton Belt, often in small towns in the same manner as gins and compresses. The cotton kingdom was too vast and diverse, too powerful and independent, to fall under the hegemonic influence of any single entity. Cooperatives were also growing, channeling more and more bales through their farmer-owned marketing operations. Names like Calcot, Staplcotn, Southwest Irrigated Cotton Growers, and the Plains Cotton Cooperative Association that started in 1953 became more and more powerful in the market. Other cotton-trading cities had more diverse economies such as Los Angeles, Dallas, and Chicago, where manufacturing, finance, defense, and transportation had strength. But Memphis sat in the midst of the American Nile, where there were so many advantages for farming, and it logically served as the country's cotton capital.

After 1945 cotton began losing relative importance in several cities such as Dallas, New Orleans, and Galveston, but Memphis continued to rely on it. Cotton-related businesses rose and fell and younger leaders replaced the old, but cotton remained essential to the city. Manufacturers and suppliers of ginning equipment liked the central location in the mid-South just as farm equipment and tractor dealers did. One dramatic example occurred when in 1948 the International Harvester Company opened a factory for manufacturing cotton pickers, the one-row model known as Old Red that it had demonstrated at the Hopson Plantation in 1944. Mechanization boosted the farm supply business with parts dealers, mechanics, and fuel suppliers as mules and harness equipment disappeared. Cottonseed oil mills thrived, and seed companies kept offices in Memphis. Once chemical farming took hold, herbicide suppliers maintained offices and equipment yards in Memphis. Chemical fertilizer companies grew, too, as the use of fertilizers increased in cotton farming. The USDA Marketing Service operated a cotton-classing branch and a fiber technology section in Memphis, and the Extension Service kept a cotton branch in addition to its regular Shelby County office. The Agricenter International near Shelby Farms opened in 1985 and began field experiments with cotton and hosted conferences with farming organizations. It dealt with livestock and other crops, but cotton, for obvious reasons, received much attention.

Numerous cotton-related organizations headquartered in the city, including the National Cotton Council, the National Cottonseed Products Association, the National Cotton Compress, the Cotton Warehouse Association, and the National Cotton Batting Institute. One of the most powerful organizations, the American Cotton Shippers Association, kept its headquarters in Memphis, from whence it directed the operations of six affiliated associations, including the Southern Cotton Association. In 1959 Memphis handled nearly 4.5 million bales compared with Fresno, California, which had the second-largest volume of traffic, 1.5 million bales.[13] Most of the bales arrived in Memphis by truck and departed by rail. Up to eighteen thousand railroad cars were required to transport the cotton each year. The heavy traffic made the city a bustling truck and rail center.

THE COTTON MEDIA

Memphis had a strong agricultural press, particularly publications bearing directly on the cotton industry. The most widely read was Hickman's *Cotton*

Trade Journal. Francis Hickman and his newspaper were players in the affairs of cotton, which his tour and report of textile-manufacturing conditions in Europe after World War II demonstrated. His regular coverage included political developments, economic trends, weather, advances in technology and mechanization, insect damage, marketing, and feature stories. Hickman published an annual international edition with reports of developments in foreign production and markets. His editorials, though not limited to cotton issues, spoke on questions of the day. Guest articles by cotton leaders appeared. Hickman was devoted to the affairs of cotton, and his newspaper provided the industry with updated news. The weekly paper ceased publication in the mid-1960s, and no comparable news journal replaced it.

In 1957 a new Memphis publication appeared in magazine format, *Cotton Farming*. Founded by Walter F. Little, it originally focused on growers and did not reach out to their related interest groups. By the 1970s, however, the magazine broadened its reach and arguably became the most comprehensive organ covering cotton affairs after the cessation of the *Cotton Trade Journal*. Still in publication in 2009, *Cotton Farming* reports mechanical advances in farming and ginning, improvements in seed breeding and herbicides, developments in international trade, and farm legislation pending in Congress. In 1993 it started a column by the executive director of the NCC, who typically explains the position and activities of the organization on current issues. These comments provide the public stance of the organization on a variety of topics such as world trade and congressional action. *Cotton Farming* devotes considerable coverage to technology, which reflects the importance for growers of the advances and breakthroughs accomplished through research. It carries advertising from seed companies, herbicide manufacturers, biotech agricultural firms, and implement companies.

The two newspapers in Memphis, the *Commercial Appeal* and the *Press Scimitar*, both owned by Scripps-Howard since 1936, covered cotton extensively. They reported on conferences held by cotton organizations, business developments within the industry, congressional action pertaining to cotton, visits by dignitaries, and other newsworthy events. They included daily prices of cotton by grade in the Memphis Cotton Exchange and prices on spot markets in other exchanges through the Cotton Belt. Weather received attention, particularly rainfall and frost. Frank Algren, editor of the *Commercial Appeal*, toured Europe in 1946 as part of an official party organized by the U.S. military, and longtime columnist Gerald Dearing won wide acclaim with his "Cotton Commentator." Dearing started writing news about cotton in 1926 and continued until his retirement in 1970 from the *Commercial Appeal*; he went on to furnish columns for *Cotton Farming* until his death in 1976.[14] Along with the *Commercial Appeal*, the *Press Scimitar* covered cotton-related news such as the activities of the Delta Council and Farm Bureau. Occasional human interest stories appeared about farmers or sharecroppers, and questions of economics or politics received much attention. The newspapers were part of the establishment but saw cotton as the livelihood of renters and croppers as much as of well-heeled planters and wanted a viable industry for the sake of all classes. In 1983 the *Press Scimitar* went out of business, but the *Commercial Appeal*, combined with publications devoted solely cotton, showed how the media reinforced the importance of cotton in the life of Memphis.

THE PEABODY HOTEL

Two blocks from the Cotton Exchange sat the Peabody Hotel, the downtown centerpiece of the cotton capital of the South. Renowned for its elegance and lavish ambience, the Peabody served as the foremost hotel in the mid-South and the popular site for charity balls, debutante presentations, wedding receptions, and proms. It hosted the city's elite, plantation owners, and all manner of business meetings. According to legend, plantations changed ownership over a roll of the dice or a card game there. Cotton was the business of the city, the Peabody was the gathering place of businessmen, and one writer described the hotel as the "semi-exclusive bailiwick of cotton men."[15] The founding meeting of the NCC in 1938, where Oscar Johnston made his plea for cotton's leaders to unite or watch their industry become extinct as the dodo bird, appropriately took place at the Peabody.

It became regular practice for the NCC to book meetings and hold conferences at the hotel. Read Dunn customarily requested Blake's secretary, Lucille Boswell, to make reservations for rooms and meetings there for the NCC Foreign Trade Committee. The soothing ambience of the Peabody encouraged goodwill and camaraderie when differences among the NCC segment delegations had to be resolved, and whether they met in the lobby for a relaxing drink or attended an upstairs cocktail party, the effect was the same. For dinner meetings or banquets, the plush surroundings added to the formality that enhanced the identity and confirmation of events, whether they were recognition ceremonies or annual celebrations of the like-minded. The hotel sat on the edge of the cotton district amid restaurants and shops, which made it convenient for businessmen visiting Cotton Row and for their spouses who enjoyed shopping. Today the Peabody still features a prominent tourist attraction, a handful of mallard ducks that spend the day in the hotel lobby fountain. Kept overnight in a pen on the rooftop, they walk to and fro from an elevator to the fountain as crowds watch them. A grand hotel adds much to a city, and the Peabody contributed to the aura of Memphis as the focal point of the cotton kingdom.

BEALE STREET

The city was segregated until the 1960s. Whites worked in office buildings, traded in the Cotton Exchange, ate at downtown lunch counters, and danced in the Skyway Ballroom of the Peabody. But there was an area of the city distinctly black, where the food and drink, the music and dance, the aura and atmosphere, were African American. It was Beale Street, which ran east–west in south downtown, a short distance from the Peabody. Here the toiling stevedores, the gin hands, the sharecroppers, and day workers from across the river or from Mississippi plantations took refuge from the white world; here they followed their own preferences and did not defer to white tastes or judgment. Described as the "Harlem of the South," where blacks frequented legitimate businesses and services offered by fellow blacks or enjoyed gambling and moonshine whiskey, "Beale Street," stated the blues musician Muddy Waters, "was the black man's street."[16] White or black, however, all the people of Memphis were subjects of King Cotton, and his kingdom included Beale Street.

If Cohn commented that the Mississippi Delta began in the lobby of the Peabody Hotel, a black perspective would name Beale Street as the site of origin. By sheer numbers, cotton's black working class stamped its culture on this section of Memphis, which ran from the river to Church's Park auditorium past Turley Street. Sharecroppers and field-workers went there to escape the plantations and backwoods, if only temporarily. Some remained in Memphis or used it as a stepping-stone to live elsewhere, generally Chicago, St. Louis, Cincinnati, or other northern cities. Beale Street thrived on account of cafés, dental offices, clothing stores, record shops, and various small businesses that catered to blacks, and it had a raucous night life. Juke joints accounted for the fame of the street, and the music was the blues.

Beale Street began to develop as the home of the blues in the early twentieth century. The oft-repeated history of the blues includes W. C. Handy, known as the Father of the Blues, who early in his career catered to white audiences, but when he saw whites respond enthusiastically to small ragtime African American bands playing blues, he quickly adopted the sound. He composed and arranged famous songs such as "In the Land Where Cotton Is King," "Memphis Blues," and "The St. Louis Blues." Throughout his career, Handy played Broadway melodies and big band hits, but his fame rested on the blues. A statute of Handy stands in a small city park on Beale Street near the site of the old Palace Theatre.[17]

In the 1920s and 1930s, Beale Street drew black musicians who came mostly from the Delta and whose sounds and lyrics reflected the raw vitality of the cotton fields. Many traveled from plantations where they worked as sharecroppers or mule drivers. Jug bands played on the streets for tips or entertained at parties and dances for whites. After World War II, the blues and Beale Street blended the old with the new. Audiences wanted less of the old music that emphasized pain and struggle and more of an upbeat message. Electronic amplification gave a new sound. Chester Arthur Burnett, better known as Howlin' Wolf, represented the old with his forceful style of playing the harmonica. He was born in hilly eastern Mississippi at West Point but moved at the age of thirteen to a Delta plantation near Ruleville. After serving in the military in World War II, he returned to farming in Arkansas, but Wolf formed a small band in 1948 and went to Memphis, where he worked as a disc jockey on radio station KWEM. In 1951 he recorded two hit songs with the legendary Sam Phillips, who first recorded Elvis Presley. Wolf stayed in Memphis till 1953, when he moved to Chicago.

COTTON AND THE BLUES

No blues figure better demonstrated the connection of cotton with Memphis than B. B. King. He left the Johnson plantation near Indianola, Mississippi, in 1948. Johnson was a kind and considerate landowner, but King wanted more from life than farmwork. When he accidentally damaged a tractor, he fled to Memphis, or more specifically to Beale Street. "I didn't think of Memphis as Memphis," he recalled. "I thought of Beale Street as Memphis."[18] Music was calling him. He found the city almost overwhelming with its factories, trolley cars, and people shopping all day in so many stores. Beale Street held excitement and promise: "There was three movie palaces, cafes, hotels, pawnshops—I'd never seen a pawnshop before—variety stores, and musicians everywhere."[19] He landed a job as a disc jockey on the new all-black-format radio station WDIA on Union Street, singing a few songs and commercial jingles while spinning records. He played with various bands in Beale Street nightspots, but he also worked at a manufacturing plant. He returned to Indianola for a short time, but Memphis called him back. Chicago never appealed to him as it did to Howlin' Wolf or Muddy Waters because King saw Memphis as the "blues heaven," and it remained his musical and cultural home. In the ensuing years, King rose to fame, traveling far beyond Beale Street, but never cutting his roots there.

Cotton never left his mind, either for what he remembered fondly or what he wished to escape. In his childhood, cotton had been "a force of nature." It was everywhere. "I saw it, felt it, dealt with it every day in a thousand ways." He remembered the cyclical life of cotton farming—plowing, planting, chopping, and picking—year after year. "There's a poetry to it. . . . It's a study in patience and perseverance." The Mississippi fields molded King, and when he first worked in Memphis and got low on money, he would go across the river to Arkansas "and pick a few hundred pounds" of cotton.[20]

Blues reflected the life of a black sharecropper or plantation worker. Loneliness, love, and escape are the common themes of the blues, but endurance runs through these songs because survival in the world of cotton demanded the strength to persevere under extremely demanding circumstances. Lyrics alone did not convey the full message of a song. Stretching and bending musical notes created a moaning, a sound of loneliness and distress. Language could capture only part of the subtleties of living as a sharecropper; it could not fully express the feelings from living with the resignation that the future offered little. Blues artists relied on the effect of dragging or sliding notes on a guitar, often with a broken bottleneck inserted on a thumb, to create a pleading or crying sound to convey pain in a world where wealth, scorn, and color divided. King recalled his loneliness after the death of his mother and how he found comfort in the music of Lonnie Johnson and Blind Lemon Jefferson. "The blues was bleeding the same blood as me," he said, and he understood the 'mystery of pain'" in the music, "in the cries of their guitars."[21] Loneliness on a remote cotton farm was universal, not just a black experience. Speaker of the House Sam Rayburn often spoke of sitting on a fence as a child on his father's remote Texas farm, wishing someone would come down the road.

Analyses and descriptions of sharecropping have understandably focused on housing, diets, child rearing, health, debt, and landlords—the outward or visible features. The subtler dimensions, particularly the private feelings of people who lived on a low rung of the social ladder, known as peasants, affected their music. Child and parent alike knew their lowly status and knew they were valued only for their strong backs. Many were tucked away in the backwoods or scattered along remote fields, only occasionally getting to town, which made loneliness and boredom part of their day. Here novelists like Faulkner wrote with empathy, but there were too few African American writers with their own perspective. Painters such as Remington and Russell captured the western cowboy, but the cotton cropper had no comparable artists. Only the cameras of the FSA photographers such as Dorothea Lange and a few others captured images of these people. With its poetry of lyrics and sound, blues survived as the best expression of the cotton peasants.

Since sharecroppers lived apart and had no electricity for radios, joy and fun had to be manufactured in their remote environment. That is why music, simple and emotional, became important to them. Impromptu gatherings of neighbors would become an outdoor dance party between shacks, a place to hear and play music or to watch and listen. Musical instruments had to come from available sources: a jug, washboard, or barrel. A tautly pulled single strand of wire nailed to a cabin wall or post for strumming notes was common. "The need was to evoke rhythms and timbres from whatever lay close at hand," reported Frederic Ramsey in his study of rural blacks in the 1950s.[22] Players would drift in and out of the band, while others might dance. There was no sheet music; the players improvised on repeated twelve-bar stanzas. Conversation on the sidelines blended with the singing. Not all blues singers worked in fields or came from the Delta; some lived in towns or other states, but the blues originated in the cotton South. The music was self-generated because the artists had no musical schooling; W. C. Handy was an exception. No form of music with such stature emerged from the wheat fields on the American plains or the cornfields of the Midwest. Only the cowboy ballad or the bluegrass of the Appalachians was a parallel.

So the blues had many dimensions, and whether sung alone in a field or before audiences, the music carried the heartfelt emotions of black southerners. A song might be sad and regretful, but a boogie, an upbeat, danceable rhythm, meant joy and enthusiasm. Blues were many things, but the field holler was the root from whence more complex and sophisticated music grew.

The agrarian order of cotton, southern and biracial, explained much about the blues. Cotton farming was labor intensive, and generated only enough revenue to amply reward the large landowner. Since the mass of croppers tended to live on small plots scattered among the fields, often hidden from sight, they dwelled quietly among themselves. White overseers might be their only contact for periods of time. Sharecroppers moved frequently from plantation to plantation. There were exceptions, but the recurring change of residence and entering into a "share contract" with a new landowner characterized this semifeudal practice. Moving generally grew from disputes with landlords or a desire to find more productive land. Indebtedness might also prompt a cropper to move, or the refusal to live under the supervision of a bad-tempered manager or owner. The harsh social atmosphere of the cotton culture carried over into the blues through the themes of broken love, constant drifting, and loneliness.

Like much of the United States, Beale Street entered a new era at the end of World War II. Memphis political boss Ed Crump cracked down on vice and kept up the pressure until he lost power in 1948. White entrepreneurs opened retail and entertainment spots, and activity blossomed, but the all-black character of Beale Street began to wane. During Truman's presidency, the civil rights crusade began to stir and gained momentum through the 1960s. To be sure, Memphis had a color line, but the city's population of blacks grew owing to the mechanization of cotton farming, and they increasingly gained economic power. When white downtown businesses added lunch counters for African Americans and the *Commercial Appeal* began to capitalize the word "Negro," the color line softened.[23] As Memphis opened itself to blacks, Beale Street began losing its appeal as a refuge. White tourists began flocking there until the famous street became a commercialized tourist attraction.

Segregation continued to crumble as a new social atmosphere apparent throughout the United States challenged privilege and exclusiveness. In downtown Memphis, new businesses began to overshadow the old. The Peabody lost guests to motels and fast-food chains. It went through foreclosure in 1953 but remained open. In 1955 the hotel closed its popular coffee shop and shut down the upscale Venetian Room in 1957. Formal dancing in the Skyway Ballroom seemed outdated with the rise of rock and roll and teenage culture. When in 1961 the hotel accepted black guests, many whites shied away. "Basically the hotel was abandoned by the Memphis upper crust," reported one account, "as a result of its inclusive policy toward African-Americans."[24] A new owner acquired the Peabody in 1974 but went bankrupt the same year. For a while the hotel was part of the Sheraton chain, which sold the hotel to the Belze Investment Company, which restored the famous structure and reopened it in 1981.

These two icons, Beale Street and the Peabody, represented the changing order of Memphis when cotton began to share economic prominence with new industries. In 1973 Federal Express chose Memphis as its central hub for routing air freight around the country. It grew rapidly and made the city a world leader in overnight delivery services. Auto Zone, a Fortune 500 nationwide chain of auto parts stores founded in 1979, located its headquarters and distribution center there. In 1987 International Paper moved its center for corporate operations to Memphis, erecting a modern office campus on Poplar Avenue that employed four thousand. Population growth and economic diversification did not extinguish cotton in the city, but the changing times brought the industry down from its lofty peak.

THE COTTON CARNIVAL

Before cotton slipped from the top of the Memphis economy, it enjoyed the city's adulation. No testimony of cotton's importance surpassed the Cotton

Carnival, which ranked behind only the New Orleans Mardi Gras in size and splendor. For the brokers on Front Street and other business leaders, the fabric of life was truly white gold, and they organized a weeklong celebration that paid homage to King Cotton. All sorts of citizens, white and black, participated in the parties and parades or eagerly watched from the sidelines. For a city dependent on—nay, addicted to—cotton, the annual festival was appropriate because all manner of businesses and "ordinary citizens not directly employed in the cotton world," wrote the historian Perre Magness, "were keenly interested in the fluctuations of the cotton market because it affected them all."[25]

In 1931 several business leaders, including Everett R. Cooke, president of the Memphis Cotton Exchange, determined to organize a festival as a means of stimulating activity and business in the depths of the Depression. They quickly settled on the theme of cotton and named the celebration the Cotton Carnival. To garner support, they recruited the directors of the Cotton Exchange. They drew inspiration from the former Memphis Mardi Gras celebrations of the 1870s, which had enjoyed a short-lived success until the yellow fever epidemic of 1878. The first Cotton Carnival, in 1931, featured a downtown parade of eighty-six floats that passed under a specially built arch of cotton bales at Main and Monroe. A fashion show arranged by the Junior League, held on a stage built of cotton bales, featured cotton designs for evening and afternoon wear. A king and queen reigned over the festivities, and despite unexpected late-spring snow flurries, the event "was judged a big success."[26] From this adventuresome beginning, the Cotton Carnival became both a social event for the elite and a public festival dedicated to advertising and promoting cotton. Exclusive fashions for women were almost nonexistent in cotton, so Memphis retail merchants, prompted by the Junior League, began ordering fashionable dresses of cotton from their apparel suppliers, who had to place orders with textile mills. In 1938 the carnival originated the Maid of Cotton contest and tour as a means of promoting fashions made of cotton. The carnival grew and expanded each year, except during the war years of 1942 to 1945, when the citywide party was suspended.

In 1946 the Cotton Carnival swung back into action and became larger and more elaborate. That year it featured an air show by pilots from the Millington Naval Air Station. Day and night parades had various themes such as a tribute to the history of Tennessee or to former President Franklin D. Roosevelt. Secret societies known as krewes held private parties in which the city's social set danced and enjoyed the merriment. Memphis did not allow the sale of liquor, but it could be served at private functions, a fact that added to the popularity of the krewes. A children's parade and a social ball for the carnival queen were part of the jamboree. But the theme of cotton predominated at dances, home tours, and other events. The carnival queen preferably came from a family prominent in the cotton business, while the king, wrote one author, "must come of a family with lint on its clothes for several generations."[27]

The Cotton Carnival was not a private celebration. Memphians and townspeople from surrounding communities enjoyed the parades and special attractions. For the 1947 carnival, a U.S. Navy submarine, the *Conger*, docked at Memphis and allowed sightseers to come aboard. Certainly the most elaborate event was the arrival of the royal barge bearing the king and queen. This elaborately decorated barge made an impressive sight, featuring the royal pair on thrones and surrounded by their court. Crowds would gather at the foot of Union and Riverside Streets to welcome them as bands played and fireworks exploded in the air. Their arrival, wrote William Bearden, "was generally the most spectacular part of the entire carnival week."[28] For the general public, the carnival, which occurred in May, offered a chance to see fancy dresses and costumes, pretty girls posing on cotton bales, marching bands, and, by the mid-1950s, perhaps the chance to spot Elvis Presley, who performed for an overflow crowd in 1956 at the Ellis Auditorium. For anyone who could get to downtown, there was a chance to enjoy the "South's greatest party."[29] Attendance over the week would run as high as 250,000.

The Cotton Carnival continued to arrange more attractions. In 1973 it moved most of its activities from downtown to the fairground and operated a midway. Attractions included hot air balloon races, demonstrations by military precision teams, a horse show, antique cars, and similar exhibits. The Memphis Arts Council also hosted concerts. Popular singers including Neil Diamond, Roberta Flack, and James Brown were a regular feature. The Maid of Cotton modeled in fashion shows and made public appearances. Each krewe had its own queen, and along with the carnival queen and her royal attendees, attractive young women bedecked in beautiful gowns with escorts were plentiful. The carnival bragged that the queen of 1975 had "a waist so tiny it would make Scarlett O'Hara jealous."[30] Along with the pageantry and splendor, the "nation's party in the land of cotton" remained a family-oriented spectacle with an emphasis on charitable causes. To offset the regality of the festival, the Boll Weevils unexpectedly appeared in the 1965 parade, dressed in green costumes concealing their identity. They randomly attacked the royalty with brooms. They were so popular that the next year the Boll Weevils were founded as a secret order of the Cotton Carnival "to introduce some levity and mischief into the regal atmosphere of the court."[31] Keeping their identities secret, except for the unmasking of the group's leader each year, his

Evil Eminence, they visited children's hospitals, schools, and promoted food drives. Like all krewes, the Boll Weevils devoted much effort to charitable causes and described themselves as the "Merry Mutants of Mirth."[32]

For all its admirable features, the Cotton Carnival was originally segregated. In 1936 a prominent African American dentist, R. Q. Venson, took his nephew to watch the parade. The young boy did not enjoy the parade because, as Venson's wife recalled, he said that "all the Negroes were horses."[33] The boy's remark referred to the employment of black men wearing long white coats to pull the floats. This practice had disappeared by 1946, when motorized floats were used, but the image of blacks pulling carts like horses resonated through the black community. In 1936 Venson wanted to bring African Americans into the carnival, but he was rebuked. This refusal motivated him and several black leaders, including the powerful Robert B. Church, to organize the Cotton Makers Jubilee, which resembled the white festival but held its parades on Beale Street. W. C. Handy served as the first grand marshal in 1936. In the post-1945 era, the jubilee also grew and offered new attractions. A royal cortege with a king and queen reigned over the festivities each year. Onlookers could enjoy several parades. Teas, fashion shows, and a grand ball were part of the jubilee. It also kept the theme of cotton. In 1956 it sponsored an essay contest for black schoolchildren based on the theme "King Cotton in the Atomic Age." No outbreaks of violence or fisticuffs occurred between blacks and whites; the jubilee had the support of the white establishment, though it remained separate from the carnival. When Martin Luther King Jr. died by an assassin's bullet in 1968, the Cotton Carnival suspended its celebration for the year.

Criticism of the jubilee came from some blacks who argued that cotton had brought misery to African Americans. Critics wanted no homage paid to the crop that put blacks in slavery and then sharecropping. Defenders replied that the jubilee was an opportunity to show the black contribution to the South, to keep alive the history of black labor in the cotton kingdom. No other means was available, they continued, to make this point. Whatever the disagreements, the Cotton Makers Jubilee thrived, and in 1981 the two organizations began to integrate when the Cotton Carnival invited the jubilee to furnish one of the grand krewes for the white organization. In 1984 the king and queen of both groups formally reviewed the parades. Cooperation and cross-participation have since prevailed, but each organization has retained its own identity. In 1999 the Cotton Makers Jubilee adopted a new name, the Memphis Kemet Jubilee, as a tribute to black Egyptian culture. Its new krewes had names like Pharaoh and Nile. In 2004 the Kemet hosted golf tournaments, barbeques, and balls.[34]

For all the frivolity and gaiety, the Cotton Carnival remained a "party with a purpose." It began in 1931 to "further the uses of cotton and make people in all sections of this great land of ours more cotton conscious" and devoted one division of the carnival strictly to cotton promotion. The parades and secret balls were meant to improve cotton consumption. In the 1950s, the carnival received national television and press coverage and even appeared in Tennessee Williams's play *Cat on a Hot Tin Roof* (1955). The emphasis on cotton fashions, as well as the generic advertising by the fashion shows and appearances by the Maid of Cotton, caused sales to jump for finer cotton clothes in the local market. More important, however, was the connection of cotton clothing with high society, the creation of a new image of fashion consciousness among women by wearing cotton evening gowns and stylish day outfits. Such promotional activity gave cotton an identity with something other than work or street clothes and began to change the association of cotton with southern poverty. However imperceptible this improvement might be, the Cotton Carnival demonstrated how Memphis not only celebrated but sought to improve the image of America's fiber.

With the passage of time, however, the singularity of cotton in the Memphis economy dwindled. The city had been steadily diversifying and reducing its reliance on the staple. There has always been a feeling that the carnival was a function of the social elite, and when in 1984 the crowd booed the king and queen at their barge landing, reassessment was due. The next year the organization changed its name to the Great River Carnival, and in 1987 the carnival adopted a policy of recognizing a different segment of Memphis business each year. "Cotton remains an important industry to our community," stated carnival president J. Stuart Collier Jr., "but there are other industries which contribute to the local economy."[35] That year the carnival recognized the agribusinesses in Memphis and in subsequent years concentrated on finance, wholesalers, real estate and construction, tourism, and other interests. In 1993 the annual Maid of Cotton pageant stopped owing to lack of funds from its principal sponsor, the NCC. That year the *Commercial Appeal* described the carnival as "a minor social event."[36] The newspaper complained the event no longer attracted national television and generated no publicity. Should the city continue to police and clean behind the parades, it asked?

Cotton declined relative to its earlier importance, and the Great River Carnival had a new name in 1987, Carnival. That changed still again in 1995 to Carnival Memphis. No longer did the carnival king use the title "King Cotton." Like many cities, downtown Memphis declined starting in the 1960s, sparking the growth of suburbs in eastern Shelby County. The parades and royal barges disappeared, the midway and fireworks stopped, and the float factory sold. The original concept that had emphasized royalty, plantation homes, fashion shows, indeed, the empire of cotton, was gone. New events such as the Memphis in May International Festival and the Beale Street Music Festival drew large crowds and captured the spirit of globalization by highlighting the culture of another country each year. In the carnival's seventy-fifth anniversary in 2006, Carnival Memphis executive director Ed Galfsky still saw, however, a future for the festival: to salute business, to raise funds for children's charities, and to "provide a forum for social interaction."[37]

COTTON IN THE NEW MEMPHIS

In the 1960s, the world of cotton began to change, and Memphis changed with it. Toward the end of the decade, "King Cotton" no longer seemed an appropriate term, and no longer did the "cotton kingdom" describe southern agriculture. The royal plant had lost its crown, and everywhere realignments and adjustments were in order. In 1961 the famed house of Anderson, Clayton and Company began to question the wisdom of remaining in the cotton business.[38] The Memphis Cotton Exchange had been largely, though not totally, responsible for the importance of cotton to the city, but trading began to decline after the 1950s with the growth of cooperatives that transacted growers' sales. Buyers and sellers began to bypass the Cotton Exchange by using computer-related technologies. In 1964 the New Orleans Cotton Exchange permanently closed, and by 1970 the Memphis Exchange had lost much of its power and glory. Activity on Cotton Row had dwindled, and the historic street conveyed more nostalgia than trading. Gone were most of the traders cubbyholed into offices on the curb, and gone were the "snakes" of leftover cotton on the sidewalk. "Inside the Exchange you see the inevitable game of dominos and cotton men watching the statistic boards," wrote an observer. "The Exchange still provides many of the services it had always given its members but to fewer of them."[39] Technology continued to change the marketing of cotton until in 1978 the trading floor of the Memphis Cotton Exchange was closed in favor of computer trading conducted within buyers' offices.[40]

Long-established cotton merchandising houses and the families associated with them began dropping out of the Memphis business scene. In 1961 Elkan Hohenberg, founder of the Hohenberg Bros. Company, died, and Cargill, a huge trader of grain in the Midwest, took over the firm and combined it with Ralli Brothers of Liverpool in 1981. This combination created an extremely large merchandising company, which used computer technology to conduct business.[41] The large merchandisers dwindled in number as the founders died and their heirs had no interest in maintaining the business. In 1973 Anderson, Clayton and Company formally withdrew from all aspects of the cotton business. Though ACCO was based in Houston rather than Memphis, this decision by the grandfather of the traders indicated that the days of the giant cotton houses had passed.[42] Cotton trading remained risky and required a highly individualistic entrepreneur to succeed, and the profits gained from the size of investment were generally not attractive compared with the returns from other opportunities such as technology, pharmaceuticals, or energy. By 2000 a handful of large firms handled the bulk of the trade, while most of the midsized operators had disappeared. Small niche traders remained plentiful. "If you look through all the cotton merchandising fraternity," stated Rudi Scheidt, a retired broker, "there is a departure . . . of all the post World War II babies that went to work in the industry in the '50s or late '40s.[43]

An exception to the trend occurred in Memphis: the emergence of Dunavant Enterprises. It traced its origins to 1929, when William Dunavant Sr. opened an office on Front Street with T. J. White. Dunavant's son, known as Billy, started working there in 1952 and took over operations when his father died in 1961 and White retired. At twenty-nine, Billy established Dunavant Enterprises and approached the business more aggressively. He demonstrated a knack for timing in the market, being one of the first to engage in forward contracting. Dunavant also concentrated on the Chinese and Indian markets, which accounted for much of his success. He made the first U.S. sale to China in 1972. His volume of trade quickly went from one hundred thousand bales per year to three million, which enabled him to withstand smaller profit margins and operate very competitively. Dunavant opened offices in Hong Kong, Singapore, and Osaka, and when he purchased the old-line McFadden firm, he acquired trading offices in Australia and Latin America. Dunavant Enterprises expanded into cotton warehouses, ginning, and real estate and developed a truck brokerage company. This new player became the largest privately owned cotton trading merchandiser by 2000 and was listed in the 2004 Forbes 400 list of privately owned companies, handling over four million bales of cotton that year.[44]

The Memphis Cotton Exchange had shut down operations because traders now operated solely within their offices. This development led to decentralization away from downtown as merchandisers relocated just outside the city in Cordova and other suburban areas. Downtown Memphis had already begun to languish with old buildings in need of modernization. Movie houses, department stores, and restaurants—retailers in general—began departing for the more lucrative outlying areas of eastern Shelby County, where new housing and shopping malls sprang up. Young adults of the baby boomer generation cared not for the structures and business climate of the old center core but wanted the freshness of new surroundings. Across the United States, cities repeated this process, but for Memphis it meant the end of Front Street as the concentrated center of the cotton trade. With this change departed the aura of Cotton Row as the epicenter of the cotton culture, for no longer did men walk the streets with cotton clinging to their trousers, and no longer did growers traipse through offices to conduct their business. In earlier years, it had not been uncommon at board meetings of the exchange for directors to discuss city affairs rather than the business of the day, but now the elite cotton families had dwindled. A few cotton families, led by the Dunavants, remained active philanthropists and supported social causes, but they took their place alongside the business families of the city's Fortune 500 corporations.

An example of computer technology that replaced the trading floor of the Memphis Cotton Exchange came in Seam, which was launched in December 2000. Headquartered in Memphis, Seam was an online trading exchange that described itself as a "totally neutral Internet-based global marketplace for the buying and selling of cotton and related products."[45] A coalition of merchandisers founded the organization: Allenburg of Memphis; Dunavant Enterprises; Hohenberg Brothers in Memphis; and the Plains Cotton Cooperative in Lubbock, Texas. They recruited NCC executive director Phil Burnett to head the new outfit. It based its operating system on EELCOT, an online service that originated in Lubbock in 1972. When Seam opened, it provided an online exchange service between growers and buyers only for Texas and Oklahoma but expanded to a nationwide basis in a few months, known as G2B (grower to buyer). Seam guaranteed payment to the seller and delivery to the buyer. A grower lists his crop on the Seam Web site, including specifications such as grade and staple length. Buyers navigate the postings looking for the type of cotton needed to fill a particular order. Growers and buyers anonymously negotiate price strictly online. Within ten months of startup, Seam reached the one million mark in bales traded.[46]

Seam also began offering buyer-to-buyer service, B2B, in 2001, which meant that merchandisers could trade among themselves. Growers prefer to sell their crop on the spot market and thus want a maximum of buyers to review their cotton. Through Seam, buyers on a nationwide basis examine the farmer's crop. Speed of sales and efficiency are attractive features of the electronic exchange. In 2001 Seam announced that cotton put into the CCC program by growers would be eligible for posting. New partners joined Seam, including the powerful Calcot cotton cooperative in Bakersfield, California, and Staplcotn in Greenwood, Mississippi. A few textile mills climbed on board. In 2001 Seam expanded its client services by featuring editorials from *Cotton Farming* on its Web site. Burnett envisioned Seam becoming the global marketplace for cotton, and by 2006 it had transacted sales for over eleven million bales.[47] This new Internet-based operation demonstrated how cotton trading continued in Memphis, but out of sight.

If King Cotton had to relinquish his throne in Memphis, there remains a testament to his glory. In 2006 the Memphis Exchange Cotton Museum opened, the result of a five-year, $1.1 million project to convert the old trading floor of the exchange into an entertaining and educational experience. The museum recalls the history of the exchange as it operated in 1939, showing how traders worked on the floor and Cotton Row. The old telephone booths where buyers made calls, the telegraph machine, and the huge chalkboard with its catwalk for posting prices all give a sense of times past. But the displays incorporate the full range of the cotton culture: the poverty, the advent of mechanization, blues music, and the Cotton Carnival extravaganza. With audio and video exhibits, touch-screen displays, historic films, murals, and artifacts that include an old hand-turned gin, the museum encapsulates the pathos and drama, the pain and glory, of the cotton kingdom. In the hall of fame section is a display of industry leaders such as Eli Whitney, Will Clayton, and Oscar Johnston. Artifacts and displays cannot demonstrate, however, how in former days the members of the exchange were the elite of Memphis, nor can the exhibits convey the pathos that rested on the white gold of the South. So the history of cotton is shown with its far-reaching social and economic ramifications, acknowledging the unpleasant but nonetheless giving cotton a positive image.[48]

CONCLUSION

By 2000 the visible imprint of cotton had disappeared from Memphis. Mechanization had dried up the stream of sharecroppers and workers flowing into town; Beale Street had become a tourist attraction. In some cases, new recreational developments had changed the flow of traffic. In place of the King Cotton Hotel stood the Keegan office building. Now Memphians traveled into the Mississippi Delta for the gambling casinos, and surrounding towns offered shopping facilities. Some trading still occurred on Front Street, but the district had lost the excitement of the past. With cotton transactions handled through computers and the rise of direct sales to mills, bales of cotton no longer flowed through the rail yards and warehouses. Gone, too, was the infrastructure of eateries that supported the labor force, whether trained classers or stevedores. Downtown buildings were being converted into lofts, boutiques, and restaurants catering to young professionals and tourists. Only remnants of the empire of King Cotton were visible. Memphis traders had been the social elite, but they now shared that distinction with powerful corporate leaders in paper, logistics, finance, and travel.

Cotton nonetheless retains a strong presence in the city. In 2007 Memphis still handled 40 percent of the cotton transactions in the United States, and there remains a concentration of related businesses and organizations. In 2008 the NCC moved from its historic building on North Parkway to a new site in the suburb of Cordova. Periodicals such as *Cotton Farming, Cotton Grower,* and *Delta Farm Press* in Clarksdale, Mississippi, still regard Memphis as the logical point for watching and reporting on developments within the industry. The USDA operates several installations there. Around town the legacy of cotton stands out, whether through the sale of miniature bales in gift shops, cotton bolls made into Christmas tree ornaments, or books of photographs of farming. Tourists snap up such items as they begin to grasp the mystique of cotton and its heritage in the city on the Mississippi.

CHAPTER 11

"THE FABRIC OF OUR LIVES"
Cotton Incorporated

"By the 1960s, the age of the miracle fibers, polyester and synthetics had arrived in full force." This statement by Morgan Nelson, a grower from New Mexico, explained the threat facing growers and why they organized Cotton Incorporated in 1970 and gave it a clear purpose: to raise cotton consumption. In 1970 Cotton Incorporated (CI) went into operation and quickly became the organization the public associated with cotton. It became the advertising arm of cotton and produced catchy commercials for national television, of which the "Fabric of Our Lives" campaign is the best known. The NCC had promoted consumption since its inception in 1938 but could not raise the funds to carry out advertising on a level necessary to combat the promotional and research efforts by the manufacturers of synthetics. Cotton Incorporated originated owing to this threat, which by the late 1950s had reached such proportions that it prompted NCC's Clifton Kirkpatrick to comment, "We have got a devil, and a real devil ... in synthetics."[1]

The roots of the new organization lay in the NCC. Indeed, in the 1930s Oscar Johnston and his colleagues had sought to increase consumption to prevent the surpluses that depressed the price of cotton. In the early stages of forming the NCC, growers saw this objective as a benefit of unifying the various segments of the cotton kingdom. Johnston considered synthetics the worst danger, and though artificial fibers had not yet grabbed a large chunk of the consumer dollar, he anticipated further losses to them. At the end of World War II, synthetics began expanding in the market, but the strong price of cotton overshadowed the threat and slackened the anxiety.

Troubles soon mounted. After the Korean War, cotton glutted the market, and the price declined. Small operators and tenants were disappearing in the South as a consequence of the "revolution in cotton." Others operated on the edge of collapse. Members of Congress and the USDA received pleas for relief as farmers struggled to keep their land and possessions. For them it was a question of survival, while the commercial growers saw a gradual slippage in their ability to prosper. It was commonly accepted that the unrelenting growth of synthetics—rayon, nylon, Dacron, and polyester—into the fiber market worsened the cotton farmers' predicament.

SYNTHETICS' CHOKE HOLD

The NCC had not failed to challenge artificial fibers. It fought back through radio and television advertising, magazines, and promotional campaigns. In 1957 it arranged for Maid of Cotton Helen Longdon to appear on *The Ed Sullivan Show*. In 1938, when the NCC established its fund-raising practices, television was not a factor, but the new medium now commanded consumers' attention. Industries dependent on the mass market could not ignore television, but this new form of advertising was costly. Giant chemical corporations such as DuPont could afford television advertising, but the meager budget of the NCC allowed only limited spending for it. This inability to compete in advertising occurred when "women found that there was life outside the home," recalled Morgan Nelson.[2] Housework had come to be regarded as an impediment to a fuller life, and the drudgery of ironing 100 percent cotton clothes became anathema. Textile mills had started blending polyester with cotton at a 65/35 ratio. Synthetic manufacturers became so bold that they forced textiles to weave blended sheeting or lose their supply of polyester. It became obvious that research and promotion had to be undertaken on a much larger scale for cotton to remain viable in U.S. households. "We were just badly outgunned," commented the NCC's Macon Edwards.[3]

The NCC continued to rely on the financing program it designed in 1938 and raised membership dues periodically, which worked satisfactorily until the injury from artificial fibers became severe. That point arrived in the mid-1950s, when the organization again raised the dues, with growers paying twenty cents per bale beginning in 1957.[4] Growers now paid twice as much as other segments and felt they carried the burden of financing while the market for cotton shrank steadily.

ORIGINS OF THE CPI

The 1950s were a decade of frustration for other reasons. Boll weevils began to show resistance to organophosphates and would pop up like a killer disease and then go into hiding. The West experienced sporadic outbreaks of the pink bollworm. Mother Nature played her part by foisting a severe drought on the Southwest, comparable to the years of the Dust Bowl. This prolonged dry spell ended in 1957 with severe flooding that sent lobbyists to Congress, pleading for an adjustment in allotments to overcome a shortage of premium cotton wrought by heavy rains and high waters. The number of small-plot farms fell drastically. Frustration over synthetics was just part of the general angst in the cotton kingdom, and when the newest assessment hike put the burden of financing on growers, they became restless and dissatisfied. The traditional system used by the NCC for raising revenue for research and promotion appeared to fail, and further increases in dues were unacceptable. More money had to be found. "Oh my Lord," Rhea Blake remembered thinking, "what can we do?"[5]

Growers in the West grew restive as they saw the NCC almost paralyzed by the dilemma of increasing funding for research and promotion. They knew that textile mills were not likely to help because the synthetic manufacturers gave them incentive payments to use artificial fibers. For the textile mills to enter into an agreement with growers for promoting cotton might also violate antitrust laws.[6] If the NCC hiked assessments for gins and compresses, the fees would likely pass back to the growers, so the conviction grew among growers that if they had to pay the new cost in hidden charges, it would be better for them to absorb the costs and control the expenditures. With their large investments, western growers felt that they had more at stake and that if an answer came forth they would have to provide it. Hence the momentum to erect a system for raising more money began in the West among large-scale operators and cooperatives like Calcot, Ranchers Cotton Oil in California, and the Plains Cotton Growers Association in Lubbock. Their objective was to raise "some real folding money," Kirkpatrick recalled.[7]

This bone pile of short budgets, loss of market share, falling prices, abandoned farms, and the new immunity of boll weevils generated a feeling of helplessness. Growers wanted answers. Unrest was strongest in the West, perhaps owing to the youth and adventuresome spirit there, but unease stretched across the Cotton Belt. Mindful of this malaise, in 1955 the NCC created a special task group, the Industry Wide Committee on the Future of the Cotton Council, to examine the future prospects of the organization.[8] The committee had sixty-eight members from all the cotton growing states, who were expected to study all the issues of the day and provide advice and recommendations with no implications for making policy.

D. W. Brooks, general manager of the Cotton Producers Association in Atlanta, made one of the first calls to set forth the concept that led to Cotton Incorporated. Brooks was a thinker, a longtime correspondent with Oscar Johnston, known for offering proposals to deal with issues facing the cotton industry. In 1955, as a member of the Industry Wide Committee, Brooks made known his conviction that the NCC needed to employ more technical personnel to work with the textile mills to make cotton fabric more competitive. While he admired the NCC for a marvelous job in promotion, he saw research lagging. To fund more research, Brooks proposed an assessment of $1.00 per bale on growers and urged the committee to push the idea as fast as possible.[9] Jerry Sayre chaired the special committee and had the perspective of the grower since at the time he managed D&PL.

In 1958 the election of Harry Baker as the first president of the NCC from the West gave fresh momentum to the fight against synthetics. His recognition was unsurpassed. He had attended the founding meeting of the NCC in 1938 and had long been a delegate for the ginner segment. Baker served on the board of directors of the Bank of America and the California Manufacturers Association. He traveled in circles of influence, accompanying President Truman on his whistle-stop tour of 1948 through the San Joaquin Valley, and riding also with Republican nominee Thomas Dewey on his trip through the Golden State. Baker was well acquainted with senators James Eastland and John F. Kennedy. California growers remembered Baker for extending credit to them after World War II.[10]

In the meantime, Sayre's committee moved to the conclusion that a new system of fund-raising had to be employed beyond the regular mechanism used by the NCC; it needed "a fresh organizational initiative," according to one account.[11] The committee followed the suggestion made by Brooks that each grower would pay an assessment of one dollar per bale on each year's crop. Participation would be voluntary, but with the high percentage of producers already holding membership in the NCC, there was general assurance they would accept another assessment. Each participant would thus pay $1.20 per bale for belonging to the NCC and contributing to the campaign to fight synthetics. To administer the program, the recommendations included the creation of a separate organization for dispersing the funds raised by the special fee, but staffing should come from the employees of the NCC, who would be expected to perform double duty. The NCC's Field Service would handle the recruitment of participants, so by incorporating this last proviso, proponents claimed that little in overhead costs would be incurred. The plan placed the responsibility for fighting artificial fibers on the growers, with the supervision coming from the NCC.

Although the assessment hike made in 1957 brought revenue for the NCC to a new high, hitting $3,027,995 for 1959, it paled beside the monies spent on behalf of synthetic fibers. Besides the incentive payments for textile mills, synthetic manufacturers helped retailers with advertising costs if they carried polyester clothes. By 1959 cotton growers recognized that "they were trying to put out a bonfire with a water pistol."[12]

The NCC welcomed the proposal to create a new branch within its ranks. Indeed, it saw the move as a way to raise the funds that had never been available through regular membership dues. Behind the discussions among growers, the NCC provided consultation and arranged meetings. It identified growers who had influence and fund-raising experience and would likely participate in setting up a self-serving body for growers. Though westerners were the most adamant about such an undertaking, the NCC realized the importance of bringing in supporters from all regions of the Cotton Belt and did not overlook the power and influence of the Farm Bureau. For these reasons, the election of Boswell Stevens as president of the NCC in 1959 proved fortuitous.

Born and raised on a farm in Noxubee County, Mississippi, Stevens came from a family that migrated there in 1837. He had experienced the hazards of cotton farming, particularly boll weevils and low prices. He became active in the establishment of the Mississippi Farm Bureau and was elected its president in 1950; he headed the organization in his home state for twenty-two years. His association with the NCC began when Oscar Johnston requested his help in forming the organization in 1938. Stevens wore two hats, one as president of the Mississippi Farm Bureau and another for the NCC, and when in 1959 he called a meeting of growers from around the Cotton Belt to discuss the creation of a separate producer organization within the NCC, his experience and recognition in agriculture gave the concept much integrity.

The NCC created a committee to advance the idea and carefully recruited growers with wide acquaintances and recognition. Jimmy Hayes, president of the Alabama Farm Bureau, and Harold Ohlendorf, head of the Arkansas Farm Bureau, agreed to serve. D. W. Brooks joined along with E. L. Story of Missouri. The cooperatives were not overlooked: Jerry Sayre, now head of Staplcotn, Russell Kennedy of Calcot, and the highly influential Roy Davis from the Plains Cotton Growers Association in Lubbock. No one outranked Jack O'Neal, a producer in California, in name recognition. He had founded Producers Cotton Oil. Keith Welden and Russell Griffin, California producers, joined. O'Neal, Griffin, and Walden "were the three biggest in the United States" recalled Kirkpatrick. "If we can't get the folks with the most prestige, influence, money, and time, we are defeated to begin with."[13]

The momentum for a separate organization now moved faster. To the NCC board of directors, this committee with the "big three" recommended the creation of a separate organization to collect funds voluntarily from growers. Acting on this recommendation, the directors created the Cotton

Producers Institute (CPI) in 1960, which represented the most significant change in the structure of the NCC since its founding. It meant that growers now had the dominant voice in promoting research and advertising, considered to be the best weapons for fighting synthetic fibers. Proponents hoped to raise $10 million annually, and "the rallying cry was Dollar a Bale."[14]

Russell Griffin of California stood out as the logical person to lead the new CPI. Born near Fresno, he farmed one of the largest cotton operations in the United States, only a few miles from the Kettleman Hills and Black Mountain on the western edge of the San Joaquin Valley. In 1960 he had an allotment of seventeen thousand acres that produced over sixty thousand bales. He raised other crops, but "it was cotton that carried the load."[15] Griffin had done some fund-raising on behalf of the NCC but showed no further interest until 1960, when the NCC's Kirkpatrick, along with Jim Mayer from Producers Cotton Oil, John Benson, Fresno County cotton adviser, and Sherman Thomas, another large grower in the valley, visited Griffin in his home. After dinner, they explained how they were trying to organize the CPI to advertise cotton and conduct research to "reverse the trend of the market that we were losing to the man-made fibers."[16] They asked Griffin to chair the CPI and help raise $10 million for it. Undoubtedly his experience with fund-raising had drawn attention because large sums of money were the key to the plan. Griffin accepted the new responsibility; he quickly invited growers and related interests to a large meeting at the Tagas Ranch in Tulare County. About six hundred people came, and after explaining the proposed CPI to them, Griffin called for a hand vote. The hands in favor of the experiment were so overwhelming that he did not call for the nays. Griffin credited the NCC for the CPI. "It was their idea, not mine."[17]

The creation of the CPI was a bold step that went largely unnoticed. None of the other staple crop organizations had taken such steps to generate demand for their product, but wheat, corn, and rice had no alternative foodstuff threatening to grab their market and drive their farmers out of business. Cotton enthusiasts needed large sums of money for television advertising, and they resolved to rely on themselves through the "dollar a bale" program. However bold and innovative it might be, the CPI nonetheless had to raise $10 million.

Kirkpatrick and the Field Service would have to do the heavy lifting. They needed large-scale participation from NCC members, but since the Cotton Belt was so vast, the campaign had a three-year plan: it began in the West, moved to the mid-South for the second year, and ended the last year in the Southeast. Serving as chair of the CPI, Griffin worked in sync with the Field Service, holding meetings across the Cotton Belt, sometimes drawing a group of thirty to forty, or getting only a handful. Growers were assured their contributions would go solely for research and promotion and that all funds would be held in escrow and returned to them if the drive fell short. From California, Arizona, Texas, and New Mexico, the campaign got over $1 million. About the same amount came from the second-year drive in the mid-South, but the third year fell short in the Southeast. All cotton-growing states participated except for Georgia and Tennessee. To facilitate and administer the disbursal of funds for advertising, the CPI had a special governing body, the Cotton Board, whose membership consisted of growers from all regions of the Cotton Belt. The board awarded contracts for advertising that appeared on popular television shows across the United States. "The CPI television ads certainly were not bashful," reported one account.[18] They were well designed, classy, and effectively aimed at the consumer, touting 100 percent cotton for being comfortable, absorbent, and static free.

But the CPI had inherent flaws. To begin with, it advertised 100 percent cotton, and despite the ads' slick look, they could not overcome housewives' dislike for ironing. Homemakers wanted to reduce their household chores, not increase them by using all-cotton garments. When shirts and dresses could be more easily pressed with a blended fabric, housekeepers could not be expected to stand any longer than necessary at an ironing board for the sake of cotton farmers. And two advantages of cotton, its ability to breathe and its absorbency, meant less by the 1960s owing to the widespread use of air-conditioning in homes and offices. That new convenience had spread widely in the late 1950s, and the ability of clothes to absorb perspiration no longer meant as much to shoppers. Blended fabric put 100 percent cotton on the defensive.

Worst of all, the CPI never had enough money. Because of the expense of television advertising, the ads appeared only about three times per week, not enough to hold consumer attention. Repetition was a key tactic of television advertising; it was necessary to employ the "saturation principle." Consumer awareness of cotton improved, but not enough to overcome the preference for blends. Synthetic manufacturers were still winning the battle over the airwaves with more frequent advertisements "that best captured the convenience-fixated, push button spirit of that era."[19] Cotton would have to step up the fight or else succumb to defeat. America's cotton farmers were not willing to withdraw from the arena, but they knew money had to be raised on a much larger scale. "Cotton needed a lot more than advertising by a bunch of amateur cotton farmers," recalled Morgan Nelson.[20]

Griffin regarded voluntarism as the chief flaw of the CPI. Small-plot agriculture dominated the Southeast, so support there was minimal. In west Texas, the enthusiasm of the initial enrollment waned, and some ginners began to drop out of the program, which put ginners that continued to collect the assessment at a disadvantage. Some merchants and shippers used the refund as a competitive tool to buy directly from growers. Once that occurred, Rhea Blake explained, "The whole thing fell like a house of cards, and that's what happened."[21]

COTTON INCORPORATED

Both the CPI and the NCC realized they were not raising enough money. "We reached the point," Griffin recalled, "that we either had to go to the growers and say we can't do it for this amount of money and we'll just have to kill this thing and stop, or we are going to have to persuade the growers that there will have to be some more positive approach to collecting money."[22] In usual fashion, the NCC appointed a committee to examine the situation, while Griffin headed a similar study in the CPI. Together they saw volunteerism as the weakness and concluded that a "mandatory-volunteer" system of collection had to be adopted or else the fight against artificial fibers would be lost. In other words, all growers would have to pay the dollar-a-bale assessment at the gin, but any grower could apply for and receive a full refund. Few growers would be inclined to make a request, so went the reasoning, after they had paid the assessment. Congress would have to pass legislation authorizing any mandatory assessment.

The NCC tackled both tasks. In maneuvered the proposal through Congress, and President Lyndon Johnson signed the legislation in 1966. This new measure, the Cotton Research and Promotion Act, followed the CPI practice of requiring a Cotton Board to be appointed by the secretary of agriculture, which had the responsibility for collecting the assessment. Farmers would have to pay the fees at the gin but could apply for a full refund with no argument. The Cotton Board would rely on the grower organization, the CPI, to award contracts for research and promotion projects, but the law required a referendum of the country's half million cotton farmers, who had to approve with a two-thirds majority for the plan to go into action. Thus the USDA had oversight of the program but had no particular authority or discretion to carry out the measure. Provisions for the refund had to be included because of the strong dislike for government mandates among cotton farmers, an attitude particularly strong among growers with membership in the Farm Bureau. Many of them, including Boswell Stevens, split with the NCC.[23] To win approval, the NCC "carried the ball," citing the sixteen-million-bale surplus and the steady loss of market to synthetics. Mac Horne told the Memphis Rotarians that the Memphis economy was in jeopardy because of the looming collapse of cotton farming.[24] In December 1966 farmers approved the program by 68 percent. "The old CPI became the new CPI," wrote Albert Russell, "with its own charter and bylaws," but it remained connected with the NCC by relying on its personnel to carry out the program."[25]

How well did this revised program meet the competition of artificial fibers? By 1969 the CPI raised $10 million and devoted two-thirds of the money to advertising, but the total expenditures, both public and private, made on behalf of cotton were no measly figure. One estimate placed the annual figure at more than $30 million, but that number included USDA and land-grant school research on farming practices, insect control, seed research, irrigation, and the like. Across the Cotton Belt, Experiment Stations tackled the woes of farmers, gin companies invested in new technology, and implement manufacturers developed bigger and better machines. Federal subsidies remained in effect, with the Agriculture Act of 1965 setting price supports for 1966–68 at 9.42, 11.53, and 12.24 cents respectively per pound.[26] Mandatory-volunteer assessments did not undermine enthusiasm for the program. "We fully support the Research and Promotion Act of 1966," the Arizona Cotton Producers Association resolved, "and urge all cotton producers to support this program."[27]

But the demand for an organization totally separate from the NCC continued in 1968 with the creation of the Producer Steering Committee, made up of grower representatives from across the Cotton Belt. The new committee remained in the NCC but became the repository and advocate for proposals to obtain better representation for growers, and the next year committee chair Griffin assembled a small group of business-oriented growers to come up with ideas. They decided to hire the consulting firm Booz, Allen and Hamilton (BAH) to make a study and furnish recommendations. They expected BAH to reinforce their conviction that growers should strike out on their own and form an organization completely independent of the NCC. They got what they wanted.

For the most fundamental change, the consultants thought the CPI would have to leave the NCC because the NCC represented all segments of the industry. Aggressive marketing had to be undertaken, and only a separate organization with its own funding could give growers the muscle they required to fight effectively in the marketplace. It would be impossible for the NCC to provide the time and money needed to expand the cotton market. The consultants envisioned a promotional operation unlike anything seen in agriculture, something akin to an advertising agency. Such an undertaking would furthermore require a highly motivated and assertive leader.[28]

By this time, a malaise had spread among growers, who vented their frustration on the NCC. "You couldn't put your finger on it," Russell wrote, "but harsh and unfair criticism of the Council began to creep into conversations among cotton people in some sections of the Belt."[29] Likely this attitude thrived in the West, where interests thought the NCC remained too close to the mid-South, that the NCC made the CPI ineffective because the two bodies worked so closely together in Memphis, and that even the older organization had outlived its usefulness. But discontent originated from more than dissatisfaction with advertising. The price of cotton had steadily spiraled downward, dropping from 31.52 cents per pound in 1960 to 22 cents by 1969, while the cost of living, the level of wages, and the general price index moved upward. Planting restrictions via allotments led to a drop in the annual yield. In 1966 the United States produced almost fifteen million bales, but in 1967 and 1968, the yields dropped to 9.5 million and 7.5 million respectively. Yet the price of raw lint remained below the levels of the 1950s. Mac Horne and Dabney Wellford made a special study in which they explained the cotton puzzle as a three-legged stool with research and promotion as one leg and price and supply as the others. For cotton to remain competitive with synthetics, they pointed out that the price must not get too high, or else textile mills would further concentrate on artificial fibers. Growers needed higher prices, but Horne and his fellow economists never offered figures, only keeping price and supply in focus. They shied away from political issues but saw the CPI as a step forward.[30] Leaders in the producer segment acknowledged the point about prices, but overpriced cotton seemed more like a fantasy than an irritating reality.

Like the multiheaded Hydra, cotton's problems increased. Critics began to lambaste growers because some large landowners received hefty sums of taxpayer dollars, a consequence of the Agricultural Act of 1965, which provided direct payments to farmers. In June 1967, after the first year of the new program, Senator John Williams of Delaware pointed out that some growers in the San Joaquin Valley had received over $2 million in direct payments from USDA. His list included Griffin's operations at Huron and the J. G. Boswell Company at Corcoran. For Williams, this was inexcusable at a time when small-plot farmers were leaving the land. He proposed a $10,000 limit on subsidy payments. But the most adamant critic was Congressman Paul Findley of Illinois, who spoke frequently against subsidy payments. He identified some growers as wealthy beneficiaries but also pointed to the unintended effect of subsidies by citing the example of the Texas correctional system, which received $288,911 in 1966. Findley wanted to see subsidies phased out over several years for crops in surplus and had introduced a bill for that purpose. Secretary of Agriculture Orville Freeman defended the payments on the grounds of reducing surpluses: "Commodity programs are not welfare grants." *Time* magazine sarcastically replied that the recipients are not "exactly welfare cases."[31]

"Mr. Findley is a very worthy adversary," recalled Macon Edwards, NCC's Washington lobbyist. "He knows how to use the publicity of the press."[32] The arguments against subsidies presented a challenge to the NCC, so as the CPI struggled against the powerful synthetic manufacturers, the issue of subsidies came up, and growers felt threatened and worried that subsidies might be taken out of farming. Further pressure came with the election of President Richard Nixon, who saw the use of payment limitations as a means toward a balanced budget, so the NCC had to accept the inevitable when the Agriculture Act of 1970 established a maximum payment of $55,000 for each farmer. It was in this unsettling and shaky atmosphere that more and more growers came to believe they needed an organization separate from the NCC and free of oversight or influence from cotton's related segments—ginners, textiles, brokers, and shippers.

Growers felt they were putting up the money and not seeing results. Griffin wanted the CPI to get off "dead center" and have some "new eyes" look at it.[33] He met with Blake and his staff in Memphis and said it might be time "to change managers." Specifically Griffin meant the NCC's managers for research and promotion, who along with the NCC staff still thought in terms of 100 percent cotton. Griffin and Blake then attended a meeting in Lubbock, where sympathy for Griffin's view was strong. At the Lubbock meeting, the decision was made to put a new person in charge of research and promotion.[34]

DUKE WOOTERS

"I could see that Mr. Wooters was a little bit brash, and that he wasn't very impressed with some of these older southerners that were reluctant to make changes."[35] So Russell Griffin described the personality and management style of CI's first president. Growers, particularly westerners, wanted a self-confident and aggressive executive who would shake things up, and they found their man in Duke Wooters. In 1970 the CPI hired Wooters to lead the new venture, which proved to be, Kirkpatrick remembered, the beginning of "trying times" and "normal growing pains."[36] It would be an era in which CI began to change the public's association of cotton with old South farming to thoughts of stylish and comfortable clothing made from a product of the natural environment.

Wooters was a native of New England with the advantages of good upbringing and solid schools; he graduated from the Harvard Business School. He knew nothing about cotton and referred to it as "that fluffy white stuff."[37] He has been described as "bright and quick, imaginative, high-energy, hard-driving, and possessed of a deep and arresting voice."[38] Wooters's natural world was New York's Madison Avenue, the nation's advertising center, and he was working as the marketing vice president for *Reader's Digest* when the CPI search committee interviewed him. He traveled among a fast-paced business set who could be irritating to some southern growers, but Wooters understood marketing and had a reputation for being effective. If "cotton need shock treatment," as suggested in the BAH report, he was the man.[39]

Wooters took over the leadership of the CPI in 1970 and with it the responsibility to restore cotton as the fabric of choice. He disliked the name Cotton Producers Institute and in 1971 pushed through a replacement, Cotton Incorporated. The new organization still answered to the Cotton Board, made up of growers, which collected the assessments from farmers and released them to Wooters's organization. Wooters had broad leeway, partly because he seized it, but growers saw the need for a broad-minded approach to promotion, so they left him alone. Some grumbled about his management style, but others liked his fresh approach.

Federal appropriations began supplementing the assessments when in 1970 the Agricultural Act authorized $10 million to be added annually to the program. The act furnished another $10 million for the secretary of agriculture to use as a discretionary fund for support of research and promotion. These supplements were made available for two years and then were to be cut to $3 million annually. Known as 610 funds, these appropriations ended with passage of the Cotton Research and Promotion Act of 1977.

If anything, Wooters moved fast and treated the new organization as a profit-making company with a responsibility to the farmers whose livelihoods were at stake. Increasing the consumption of cotton was the order of the day, and he felt time should not be wasted. He acted like a missionary compelled to bring cotton back to its preeminence among fibers. He gave himself the title of president, which ruffled some feathers in the South because the NCC president had always served without compensation. His salary was high because the CPI had thought it necessary "to pay competitive Madison Avenue salaries," recalled a member of the CPI. "It was a wise decision, and it caused considerable trouble."[40] Wooters located the headquarters of the new operation in a tony high-rent building in New York City in the midst of the fashion center. The offices were plush. It "raised eyebrows" in Congress, the USDA, and the Cotton Board, wrote CI's historians, but Wooters wanted to welcome designers and buyers in flourishing and prosperous surroundings and present the operation as a class act. He established the organization's new research branch near the location of textile mills at Raleigh, North Carolina. For him, common sense meant researchers needed to be close to the mills for the "fluffy white stuff" to rekindle interest. He made these decisions without consulting the NCC or cotton leaders. To ignore the traditional guardians of the cotton kingdom in such heady matters engendered resentment, and when Wooters did consult with growers, he turned to his allies in the West.[41]

Wooters realized that CI needed a symbol that would instantly be recognized by the public, one that conveyed a positive and admiring image. Like Will Clayton, who complained about the South's "barefoot standards," Wooters wanted to "get rid of the image that cotton had of blacks in the field with a hoe."[42] In other words, cotton needed a logo. Through a friend at BAH, Wooters learned about Walter Landor, a designer in San Francisco. Landor showed little interest in creating the perfect logo for an agricultural commodity, but a year later he agreed to produce a design for the hefty sum of $50,000. A few months later, Wooters and two of his executives chose a logo designed by Landor's daughter, Susan. It was a boll based on top of the word "cotton." In 1973 the logo became the trademark of CI and became a hit as consumers associated it with the soothing comfort of a naturally grown fiber. This identifying mark alone did not create a new image of cotton, but it succeeded in becoming an instantly recognized symbol.

Cotton Incorporated quickly proved to be an effective combatant on behalf of the cotton farmer. It exploited the popularity of blue jeans, the most American article of clothing, by advertising the natural feel of jeans made of 100 percent cotton. It encouraged consumer loyalty by identifying jeans with youth. Sales of blue jeans rose steadily, reaching over five hundred million pairs in 1981 alone.[43] No item in retail clothing did more for the farmer than these 100 percent cotton indigo pants.

But the craze over blue jeans was greater than the management skills of CI or the foresight of cotton growers. All benefited as cash registers rang with the sale of jeans. To be sure, the slick ads coming from cotton's new promotional arm contributed, since they emphasized denim as a natural fiber, but the popularity of wearing jeans went beyond the efforts of trade organizations. Mothers liked blue jeans for the same reason they preferred blended shirts: ease of care. Blue jeans, especially for children and teenagers, needed no ironing and were unsurpassable for long-lasting wear.

Indeed, blue jeans became synonymous with youth; they served as raiments of identity. Cultural historians associate them with rogue independence, citing the example of actors like James Dean in *Rebel without a Cause* (1955), Marlon Brando in *The Wild One* (1953), and Marilyn Monroe in *Clash by Night* (1952).[44] The singer-songwriter Neil Diamond expressed this connection in the song "Forever in Bluejeans," and in Britain a rock group took the name "Bluejeans." The growth of a teenage culture after 1945, dependent on rising affluence, drove this market, and the sight of young screen actors or sitcom regulars dressed in jeans provided the best advertising.

Not even youth's rebelliousness, however, explains the overwhelming popularity of blue jeans. All age groups wore them, as seen in *The Misfits* (1961), starring Clark Gable and Marilyn Monroe. By the new century, jeans were ubiquitous in life and on-screen. The growth of leisure time in the last half century created a demand for all-purpose durable clothes, and for relaxation consumers wanted easy-to-wear garments. Jeans fit that purpose. With the rise of the relaxed look, including "casual-dress Fridays" in business offices, sales increased. The country-and-western craze also spurred sales, and blue jeans entered the world of haute couture when name designers began to produce their own brands. Expensive skintight jeans, some embroidered with rhinestones and other finery, became the rage. But the craving for denim included common wear as skirts, shorts, dresses, and jackets. Artificial fibers went through the polyester craze of the 1970s but had nothing to match the popularity of blue jeans. In 2005 the Internet sale of jeans worn by the Hollywood star Leonardo DiCaprio in the Celebrity Jeans Auction ("How much would you pay for a chance to get into a celebrity's pants?") raised money for the National Multiple Sclerosis Society. Referring to jeans as the "American uniform," the cultural historian James Sullivan wrote that "the classic pair of blue jeans might carry more implications about the American consumer than anything else we consume."[45]

Cotton Incorporated scored a hit with the "Fabric of Our Lives" campaign, which debuted Thanksgiving 1989. Researchers noticed that people rubbed their arms when they described the soft feel of cotton next to their skin. "Body and mind apparently worked together," they reported.[46] From these observations, CI hit upon the theme of "cradle to grave," meaning that cotton was commonly used throughout life from beginning to end. The campaign featured no brand names but promoted the use of cotton for various age groups in the events of life. It was subliminal advertising, a soft-sell generic pitch with pleasant scenes of weddings, anniversaries, and the like, focusing on women in the eighteen to thirty-four age group. Ogilvy and Mather, the advertising agency that CI employed, kept the ads updated and won a 1998 Silver Effie Award for television advertising. The campaign successfully established cotton as a desirable item in American consumer culture, and along with the logo and the Seal of Cotton, the "Fabric of Our Lives" was recognized "as a masterpiece of the sloganeering art."[47] In 2003, however, CI replaced Ogilvy and Mather with DDB New York and retired "The Fabric of Our Lives," but the new agency intended to keep the same target audience of young women.[48]

For all its success, CI went through some bumpy years. In 1976 the USDA Office of the Inspector General (OIG) began investigating whether Wooters had arranged a consulting contract for $60,000 per year with America's largest cotton producer, J. G. Boswell Company in the San Joaquin Valley. The press reported allegations that the "agreement amounted to a conspiracy to circumvent the intent of Congress by avoiding a law limiting Wooters's salary in 1975–76."[49] In 1975 Congress had ruled that no person employed by CI could have a salary above that of the secretary of agriculture, which at the time was $63,000. Wooters had agreed to a cut and went without pay for nine months, although his contract permitted him to have consulting assignments if they did not impair his work as president. The CI board had approved the consulting agreement, but only two years later. Suspicions were further aroused because the Boswell Company applied for and received a refund of $60,000 from the Cotton Board soon after making the agreement with Wooters. By law the California grower was entitled to the refund, but the timing of the agreement raised suspicion.

When the powerful voice of W. B. "Billy" Dunavant, a broker in Memphis, demanded the resignation of Wooters, the matter acquired deeper significance. Dunavant's anger began when he and members of the American Cotton Shippers Association held a closed-door meeting with officials of CI. After the meeting, Dunavant was quoted in the Memphis *Press Scimitar*: "I have felt threatened for our association because of this man [Wooters] and members of his staff."[50] The Shippers Association wanted to know if CI had been suggesting the names of suppliers to foreign mills, a charge that the Office of the Inspector General planned to investigate.

Old suspicions between southerners and westerners rose again. Wooters's lavish offices on Madison Avenue had raised eyebrows, but he wanted advertisers to see cotton as modern and sophisticated and thought the headquarters should not convey a shabby image. He understood marketing and knew the old image of cotton with "barefoot standards" had to be overcome. Only if the public associated cotton clothing with fashion could the losses to synthetics stop. His supporters embraced this reasoning and had no objection to his actions, and from the Cotton Board he received a formal statement of support commending his willingness to take a temporary cut in salary; it was backed up with a further resolution that his actions had been in the best interests of growers.[51]

Morgan Nelson exemplified the western perspective. He was a member of the Cotton Board, but his influence went further. Nelson was a prominent legislator in the New Mexico House of Representatives, where he served as chair of the Public and Military Affairs Committee and also the Board of Educational Finance. He had long tenure on the 1517 Cotton Association, the New Mexico State University Cotton Advisory Committee, and the state's Crop Improvement Association. He was well known in educational and agricultural circles as a heavyweight, and his voice did not go unheeded. "The inquiry into any compensation Wooters possibly could have received from private sources is meddling into his private civil rights and is beyond the legitimate functions of government."[52] Nelson saw Wooters taking cotton in the right direction.

Wooters had nonetheless miffed powerful southerners. They found themselves left on the sidelines when CI moved to New York with its enormous budget. Illinois congressman Paul Findley attacked Wooters, and there was speculation in western circles that disgruntled southerners had fed information to the congressman.[53]

Soon afterward, however, the Department of Justice dropped its probe without giving reasons. Wooters felt exonerated. Members of the Cotton Board were pleased that no criminal investigation would now develop, and Dunavant thought the decision would help CI survive the upheaval.[54]

When the Office of the Inspector General filed its report in August 1979, it cited CI for misusing funds in activities related to foreign sales and for using information about growers, furnished by the Cotton Board, to discourage requests for refunds of bale assessments. Such actions violated federal law, and the report recommended that the USDA would have to implement tighter budget controls and better oversight or else the federally mandated assessments would have to stop. A special committee of CI had been formed to look into personnel issues, but it stood behind Wooters. Paul Findley called for Wooters's resignation. The Office of the Inspector General thought the USDA and the Cotton Board had not provided effective oversight, which had enabled CI to develop "a degree of autonomy not envisioned under the act, the implementing order, or the contract with the Cotton Board."[55] Investigators reported that the Cotton Board had disclosed the names of growers who requested refunds to other producers, a claim that was denied.

Other issues included the use of funds raised by the old CPI to entertain textile executives at the 1976 Olympics. The Office of the Inspector General reported some cheating on expense accounts and recommended that all accounting systems within CI be brought up to standard and audited yearly by an independent certified public accountant. On the matter of foreign sales, the report revealed that staff at CI had recommended foreign mills to buy from specific suppliers and had notified those suppliers of the mills' interest, but there was no evidence the organization had taken orders to sell cotton on "behalf of any firm."[56] Investigators faulted the USDA for poor oversight and recommended that the USDA Foreign Agriculture Service (FAS) initiate procedures to ensure that CI's foreign activities remain in compliance for foreign sales. Likely it was this activity that disturbed the American Cotton Shippers Association. In 1981 CI and Wooters agreed to resolve the dispute over his consulting contract by paying the USDA $120,000, which would return to the Cotton Board. They denied any wrongdoing but chose to settle to avoid further litigation. Wooters did not like to compromise, a "characteristic that made him a lightning rod for Cotton Incorporated's critics," stated one account. In 1982 the Cotton Board did not renew his contract.[57]

Only the activities relating to foreign sales caused resentment among growers. "I personally am not in favor of their trying to go overseas," stated NCC president Chauncey Denton.[58] Through the Cotton Council International (CCI), an arm of the NCC, efforts to sell cotton overseas had been under way since 1956, and any intrusion on that program worsened the criticism. Much of the discontent stemmed, however, from personality clashes. In 1976 the USDA had arranged for an outside evaluation of CI by a New York consulting firm that reported: "In trying to divorce itself from traditional industry practices and relationships to concentrate on its promotional mission, Cotton Incorporated's management had alienated friend and foe alike."[59] Throughout the industry, the study discovered strong approval of the "objectives, organizational concept, and overall program of CI" but found the organization "insensitive to how its activities may be perceived." There was a consensus that the new organization ignored or overlooked opportunities for cooperation that might have "avoided the hard feelings which have developed." However, the evaluation pointed out that cotton's slide in market share had slowed and had "resisted competitive inroads better in the last three years than at any time in the past two decades."[60]

How should Wooters be judged? Nearly all agreed that he was brash, arrogant, and insensitive. He particularly offended southerners and felt they were mired in the old cotton culture, conservative to a fault, and locked in their own inertia. Wooters saw himself as saving cotton from its own mishaps and worried little about bruising feelings along the way. But he misjudged the southerners. To be sure, westerners had shown much leadership in establishing a separate organization for growers, but southerners like Jerry Sayre had made major contributions. Indeed, the Memphis-based NCC had made the original commitment a generation earlier to the same objectives Wooters now sought. In the CPI's early years, the NCC had drawn from its limited finances to keep the experiment going, and while the NCC welcomed a new partner, it had to tread softly among the unhappy southerners. The responsibility to protect the funding arrangements for CI from congressional interference continued to rest with the NCC.[61]

To what extent did Wooters see correctly? The CPI had foolishly continued to advertise 100 percent cotton in spite of the popularity of blends when he came aboard. He understood how a large part of the population still saw cotton farming as backward and immersed in poverty. He knew that to carry out his task it was necessary to overcome this perception and to give cotton a more respectable place in the mainstream of popular culture. He wooed and courted textiles and pulled them closer to cotton after they had drifted to synthetics, and though artificial fibers understandably remained an important part of textile production, the turnaround in the falling consumption of cotton spoke loudly. Much of the contribution achieved by CI came after Wooters left, but his vision remained.

The bumps in the early years did not impede CI. Growers saw the facts unveiled by the Office of the Inspector General as a worthy correctional measure, a spanking for misbehavior, but they did not demand an end to the organization. Dissatisfied growers could still obtain a refund of their assessments. Drought or excessive rains, which led to low yields, were often responsible for refunds. Bill Pearson of Mississippi remembered that the bale assessment was automatically deducted from the sale of his crop like a payroll income tax withholding. "We gave it little attention," he said.[62] Growers instead had to think about maintenance of machinery, insect control, diseases like wilt, and the threat of nematodes, the microscopic parasite that attacks plant roots. After the departure of Wooters, CI continued to pursue its original objective of raising consumption and public awareness of cotton. In 1988 it began a cooperative program with textile mills known as the Engineered Fiber System (EFS) that improved the ability of mills to obtain the grade of cotton they desired for a particular market. This program has significance because the quality of cotton varies within a bale as well as among bales. Since synthetic fibers consistently retained the same quality, expectations rose for cotton, a product of nature whose characteristics of strength and length are conditioned by soil and weather. Just as flowers in the same plant bed will vary in the size and beauty of blooms, so will cotton have some variations within a five-hundred-pound bale. EFS proved to be popular with mills, and CI expanded the program to Europe and started an EFS program in China in 2006. Through EFS, foreign mills tend to buy more American cotton because it has consistent grade and quality.[63]

In 1990 Congress revised the Cotton Research and Promotion Act to end the opportunity for growers to get a refund. For CI this would both increase and stabilize revenue, which it had long wanted. The measure also put a mandatory assessment on the importers of cotton textiles because imported clothing had become a large force in the retail market, and growers wanted importers to bear a share of the burden of promotion and advertising. The inclusion of importers softened the growers' objections to a nonrefundable assessment, and they ratified the new legislation in 1991. This action slightly lowered the supplemental assessment, first started in 1977 at .4 percent of each bale and raised to .6 percent in 1985. Now Congress set the supplemental figure at .5 percent. If he deemed it necessary, the secretary of agriculture could call a referendum each five years; the new measure also established a mechanism for growers to initiate a referendum should the secretary ignore their request for one. No referendum on the 1990 measure had been held by 2006. Much authority continued to rest with the Cotton Board, and the legislation specified that only "cotton producer organizations" could make nominations for the thirty directors of the board. Certified organizations include the Farm Bureau, cooperatives, the NCC, and various producer organizations. If a farmer has no membership in such an organization, he is voiceless, though he must pay the assessment. By 1998 CI had a budget of $61 million, with the importers' share reaching 25 percent.[64]

Table 2. Cotton and Synthetic Fibers: U.S. Mill Use, 480-Pound Bale Equivalents, 1965–2005

Year	Cotton	Synthetic	Cotton's Percentage of Total
1965	9,595,000	2,267,885	80.9
1975	7,249,667	3,762,567	65.8
1985	6,412,861	3,534,810	64.5
1995	10,647,329	2,981,412	78.1
2005	5,871,318	1,121,092	84.0

Source: Leslie Meyer, Stephen MacDonald, and James Kiawu, *Cotton and Wool Situation and Outlook Yearbook* (USDA Economics Research Service, November 2008), 47, http://www.ers.usda.gov.

Cotton's share of the world market was less than in the United States. In 1989 cotton held 48 percent of the world market, which amounted to 83 million bales. Synthetic fibers, and particularly polyester, amounted to the equivalent of 91 million bales. Over the next decade, however, cotton consumption rose by only 3 million bales, but the use of artificial fibers rose to the equivalent of 139 million bales. It left cotton with a world share of 39 percent.[65]

CONCLUSION

By 2000 Cotton Incorporated's research and promotion arm had begun to think globally. It had allowed its logo to be used with imported goods over the objections of domestic textile mills. Through the Engineered Fiber System, CI attempted to nudge foreign mills to buy more U.S. cotton, but serious concern arose whether the organization had peaked in the domestic and international markets. The new organization faced the same reality that had confronted the cotton kingdom for two hundred years: each triumph led to another barrier, as if the future amounted only to an endless path of resistance.

Cotton Incorporated vastly improved the image of cotton. By law it could not involve itself in political matters; it could not lobby Congress or otherwise seek to influence policy. These constraints allowed CI to escape nearly all the controversies over pesticides, fights over water rights, issues over trade policy, and the thorny question of farm subsidies. The organization had an enormous budget compared with the NCC, which enabled CI to entrench itself in the glamorous world of advertising and fashion. When NCC lobbyists delicately tiptoed through the cloakrooms of Congress, agents of CI waltzed through the salons of fashion designers. But glamour was a dimension that Wooters had understood, and he realized it would be necessary to undertake initiatives that might be disturbing in order for cotton to penetrate the world of chic designers and style makers. Therein lay CI's contribution. Safe from political involvement and richly funded, CI moved the fruit of common farming into an expanded market.

CHAPTER 12

THE TEXAS PLAINS
America's Cotton Patch

On the South Plains in Texas lies the most intensive cotton farming area of the world, the High Plains and Rolling Plains, which the inhabitants proudly call the "Cotton Patch." King Cotton still reigns on the three million acres, where the cultivation of the royal plant continues to affect social and cultural outlook just as it drives the economy. Here is the most developed cotton infrastructure in the United States, which processes and handles 25 percent of the country's crop each year. In the center sits Lubbock, the self-proclaimed "cottonest city in the world." On this windswept plain survives a culture that combines an admiration for farming with the resilience of living on a stark and harsh land. Indeed, the land dominates, forcing man and beast to accommodate and adjust to a prairie blown by a relentless wind. Residents in the state's eastern urban centers often ridicule and deride the area, but the recipients of such scorn have a strong attachment to the land and even the harshness that comes with it. The Cotton Patch could be compared with the Mississippi Delta in sense of place, where attitude and history along with the natural environment establish an identity for both areas. But the history of the Cotton Patch is much shorter, having begun in the twentieth century and coming to fruition only after World War II. Without question, the emergence of the Texas Plains stands as one of the important developments in contemporary agriculture.

The Cotton Patch forms a rectangle bordered on the north by Interstate Highway 40, which follows the old Route 66 across the Texas Panhandle, and bounded to the south by I-20. The New Mexico state line marks the western edge, and the 100th meridian draws the eastern edge. Within this general area lies the Llano Estacado, bordered on its eastern edge by the Caprock Escarpment. Around twelve million years ago, streams originating in the Rocky Mountains carried soil southward and deposited it on the plains. Over time the Canadian and Pecos Rivers eroded the area but left a high plain that Spanish explorers named the Llano Estacado, or the Staked Plain. The Caprock contrasts sharply with the lower plains, rising abruptly from two hundred feet to one thousand feet in some locales. On this alluvial plain of sandy loam, elevation reaches over three thousand feet, justifying one description of it as "an isolated plateau that sits fortress like amid the surrounding lowlands."[1] Town names like Levelland and Plainview hark to the area's geography, and towns like Cotton Center reflect the importance of cotton farming. Memphis, Texas, calls itself the "cotton capital of the Texas Panhandle." Caricatures of the region generally describe a flat and treeless landscape with dry or near-drought conditions, where severe storms carrying hailstones or causing flash floods occur not infrequently, and "blue northers" bring bitter cold. Large spans of open land and big sky are visible in any direction, where Comanches, Apaches, and Kiowas once rode. The region provided much of the setting for Larry McMurtry's *Lonesome Dove*. On the northern edge lies Palo Duro Canyon, where one of the last Indian battles took place. Nowhere in the United States is farming more risky because of recurring droughts, hailstorms, early and late frosts, and blowing sand and dust that cut tender plants. In May 1951, plains farmer William DeLoach poignantly noted a sandstorm in his diary: "A hard sand storm blew all day. Lots of land changed hand. No deeds issued. The wind does not require a deed."[2]

The harshness of the land with its remoteness and solitude shapes a culture known for its stoic doggedness, where farm families exhibit independence and self-reliance, and the determination to persevere resembles the spirit of nineteenth-century pioneers. Hard work and the ability to endure are taken for granted, and the willingness to live with hardship is expected. Crop failures have to be accepted, and "local wisdom has it that if a cotton farmer can make an exceptional crop every five years," reported a writer for *Texas Monthly*, "he can withstand three unimpressive years and one disastrous one."[3] This agrarian life under adverse conditions encourages the sense of community just as it fosters individualism, causing farm families to give priority to church membership and children's school activities like football and 4-H projects. Sons and daughters compete in livestock shows and home demonstration clubs that in reality are family affairs. More than anything the threat of a failed crop, where growers stake so much into a risky venture, bonds families and fellow crop men in a way not understood by wage earners in eastern cities.

Growing cotton on the Texas Plains offers advantages that help to combat the region's many drawbacks. The vast spaces of flat land are ideal for mechanization, which explains why growers used tractors and mechanical harvesters there sooner than in the South. Since large-scale farming lowers the cost of production on a per-acre basis, growing cotton becomes attractive with large acreage. After World War II, production on the High Plains rose in amazing numbers for over fifty years, and the sheer number of bales alone accounted for the region's importance. In 1945 production amounted to a measly 139,683 bales but hit 4,877,600 in 2004.[4]

But the achievement of America's Cotton Patch came only because of a dogged determination to overcome a slate of obstacles, some inherent in the natural environment and others imposed by historical tradition within the cotton industry. And it is necessary to recognize the sense of second-class citizenship heaped on plainsmen by the residents of the state's eastern and urban half, who see the remoteness and harshness associated with the Llano as undesirable and not suitable for their tastes. Indeed, in the nineteenth century, the area was thought to be uninhabitable and was described on maps as the "Great American Desert." Settlement picked up about 1900, and the economy rested on ranching and general farming that depended on the scarce rains. Crop production was low, and self-sufficient farming, or conditions near it, prevailed. The Depression worsened economic conditions, and the population declined in several counties. In 1940 less that 10 percent of the farms had electricity and running water. Towns were small, with Lubbock mustering a population of only 31,853. Texas Tech University, founded in 1925, had a reputation in eastern areas as a "cow college." Starting in 1945, the plains grew in population and developed economically, and the single most important reason was cotton farming.

STORM-PROOF COTTON

Growers on the plains faced a steep uphill battle. They relied on a variety known as Georgia Half and Half, meaning that it produced an equal amount of seed and lint, whereas most varieties had several hundred more pounds of seed than lint. Half and Half tolerated dry weather and produced an extremely white color, but the short staple, usually 13/16 inch, had coarse lint disliked by textile mills. West Texas cotton generally had some discoloration owing to sand, so these characteristics gave it a poor reputation, and growers received a discounted price. To worsen matters, the Half and Half had a large boll that made it vulnerable to high winds, and the infamous windstorm of Thanksgiving Day 1926 blew much of the area's cotton onto the ground. In Dawson County 75 percent of the crop blew out of the fields.[5] L. C. Unfred remembered moving as a child to the plains and driving up a steep road to the Caprock on that holiday. His family thought they had come upon a snowfall but then realized what they saw was cotton.[6] After the storm, H. A. Macha went through his fields in southern Lubbock County and found only a single stalk that held on to its lint. He gathered seeds and started growing them in his home garden in hopes of getting enough to plant. In 1927 the Experiment Station in Lubbock began trying to develop a storm-proof variety, but a fire destroyed all the records in 1934. When the Lubbock researchers advertised for seeds resistant to wind, Macha took his seeds to them. From that stock they developed a storm-proof variety known as Macha that replaced Half and Half. The new cotton produced fewer seeds per bale but had a small boll that did not generate high yields of lint.

To grow better cotton on the plains, seed varieties had to be further improved. One breeder thought the small boll on Macha resembled pecans. Derivatives of Macha were commonly planted along with strains from Anderson, Clayton and Company's Paymaster or Five A developed at the Lubbock Experiment Station. An improvement developed by Joe Lambright, a cross of Macha and Five A, resulted in a large boll with wind resistance that came to be known as Lambright 123. Other improved varieties included Lockett and Lankart, but nearly all cotton grown on the plains had some lineage to Macha. Rex Dunn, a seed breeder, estimated that growers lost fifty dollars per bale due to the poor reputation of West Texas cotton. Among crushers, seed grown on the plains carried the nickname "popcorn" because it was small, with little oil, and drew the "Dallas differential," a lower price than other Texas cottonseed. "Cotton from our area had a horrible reputation in the textile industry and, unfortunately, most of it was accurately perceived," recalled a plainsman. "We had absolutely horrible cotton."[7]

Dunn attributed much of the problem to farmers because he felt they failed to understand the importance of variety as a factor in production. "Is there really a difference in cotton?" he quoted one farmer who had raised crops for forty years.[8] As late as 1955, Half and Half still grew on the plains. To teach growers the importance of planting better cotton, seed breeders, private and public, along with the Extension Service, started an educational campaign in the 1950s that Dunn described as an "educational revolution." He found younger men more willing to experiment.

THE RISE OF IRRIGATION

Growers typically planted several varieties in the same fields, which increased yields but meant the cotton would grade differently. Textile mills wanted consistency in the lint, and with the rising market in artificial fibers, they could be persnickety. Volume of production continued to be the response to the discounted price, partly accounting for the area's dislike for acreage allotments. Seed breeders continued to search for storm-proof varieties with the desirable characteristics of large bolls and longer staple length. Early maturing also remained an objective because the northern limit for growing cotton rested only about seventy-five miles north of Lubbock. By the early 1980s, plains cotton had greatly improved, and Dunn gave credit to commercial breeders, university researchers, and independent operators. Seed improvement never reaches a final point and remains a constant objective in all areas of the Cotton Belt, but for West Texas, the pursuit of better cotton had special meaning, since the area had to overcome its reputation for growing inferior grades. But the low rainfall restricted yields, and the dry spells worsened matters.

The answer to scarce rain came in the Ogallala Aquifer, the largest underground source of water in the world. The aquifer begins in South Dakota, stretches southward through the Texas Plains, and ends near the Permian Basin oil fields. Farmers on the Caprock began drilling into the aquifer in the mid-1930s, but not until the end of World War II did irrigation grow significantly, owing much to improvements of the submersible pump for submarine warfare.[9] That advancement, along with the Ogallala's readily accessible water table, made pumping inexpensive once the infrastructure was installed. In the postwar era, initial expenditures per well ran about $4,000 to $6,000, and once installed, irrigation cost little, since the landowner had the water rights. Fuel costs for pumping were low, thanks to the abundant natural gas in the area.[10] With irrigation, per-acre yields generally doubled, rising 250 percent in some cases, so an irrigation boom got under way with a 1,000 percent increase in the number of wells for between 1945 and 1957.[11] As tractors replaced horses and mules, farmers converted feed acreage to cotton. Implement dealers often accepted teams of mules as a trade-in for mechanical equipment and shipped them to wholesale mule traders in the South. This disposal of animals partly accounted for the postwar surge of mule trading in Memphis, where mechanization trailed behind farming on the plains.

Irrigation proved advantageous once Congress imposed acreage restrictions after the Korean War, since watering enabled growers to increase yields and offset smaller acreages. During the next decade, however, as cotton prices remained low, plainsmen put a lot of their crop "in the loan" to finance the installation of irrigation systems. Dryland farming did not disappear, because some landowners continued to rely on rain while others practiced both types of farming, but generally cotton instead of grain received the favored treatment. Banks encouraged irrigation, since it was the key to further economic growth in the area. In 1948 the High Plains had 8,356 wells, but the number grew to 42,225 by 1957. Irrigation became an ancillary industry in the infrastructure of cotton farming, and to serve the wells there were drilling companies, equipment dealers, installers, and mechanics. Lubbock emerged as a supply center for pumps, equipment, and services. Farm size grew.[12]

With an adequate supply of water seemingly ensured for an indefinite time, landowners put more land under the plow, whether to raise cotton, sorghum grain, or other crops. During the era of the irrigation boom, from 1945 to 1970, some counties on the plains lost population, but those with irrigated farming grew as green fields stimulated the economy in cities and towns along with the rural environs. Irrigation raised farmers to a better level of prosperity and contributed, wrote an economic historian, "to the economic growth of the entire High Plains region."[13]

To overcome the obstacles imposed by the natural environment, more than irrigation was needed. Use of the mechanical cotton stripper took bolls, opened and unopened, plus any leaves left on the stalk, which made plains cotton trashy. This drawback was overcome when in 1946 E. L. Thaxton first observed in North Carolina that calcium cyanamide, the "black mammy" used for fertilizer, caused the leaves to drop off the stalk. Along with researchers at the Lubbock Experiment Station, Thaxton saw an advantage for mechanical stripping and began studying defoliants for use on the plains in 1948. They moved past black mammy to better chemicals such as sodium cyanide. Thaxton and others learned that defoliants needed moisture to react, so growers kept the strippers out of the fields until 10:00 a.m. to let the morning dew moisturize the chemical. Thaxton and other researchers developed desiccants that dried the leaves but left them on the stalk. Desiccants required no moisturizing, which made them attractive. Chemical companies saw the potential in this research and began commercially producing both defoliants and desiccants, but the desiccants proved to be the most popular. Growers in the cooler northern counties of the High Plains did not adapt the new chemicals, since they needed to maximize the growing season, so the acreage devoted to use of leaf removal was greater in the southern counties.[14]

Among the hardships of living and farming on the plains, none proved more irritating than blowing sand and dust. Townspeople complained that it penetrated under doors and around windows, and housewives loathed the fine coat that covered furniture and shelves. Sand and dust damaged young cotton plants and stripped away topsoil. A breakthrough came with chisel plowing, which involved the use of a special plow, narrow and with a chisel point, which ran ten to twelve inches below the surface. Chisel plowing required a powerful tractor, but it brought clay to the surface and held moisture better than sand. A clodded field also reduced the sweeping effects of wind. When combined with irrigation, this practice significantly reduced airborne sand and dust.[15]

Because of the capital expenditures involved with irrigation and the opportunity to further profits with greater production, growers tended to buy more land and engage in "factory farming," known as industrial agriculture. Profitability in cotton farming steadily required more land, and on the plains that meant the use of irrigation. Successful growers had to increase investment for acreage expansion and install wells, so risk and indebtedness climbed, too, and the culture of family farming succumbed to commercial operations based on volume production with heavy reliance on machinery and technology. Single proprietors continued to dominate the pattern of ownership, but the definition of a farm changed. Acreage became larger, machines replaced mules and workers, and owners carried much debt, a pattern that recurred throughout the Cotton Belt.

HOME MODERNIZATION

During the postwar growth on the plains, cotton generated enough new wealth that a boom in rural housing occurred. "Cotton builds fine farm homes," reported the *ACCO Press*. Across the Cotton Belt, construction of more-spacious dwellings with two or three bedrooms was under way, particularly on the Texas Plains.[16] With a sense of pride and accomplishment, farmwives displayed their new kitchens and children's rooms. With the massive construction program of the REA making electricity available in rural areas, wives no longer had to rely on kerosene lanterns and outdoor bathrooms and maintain homes with no refrigerators. "The mother of the family can vouch for automatic refrigeration, automatic washing machines that do just about everything but mind the baby," the magazine continued, "and home freezers full of fryers and choice cuts of beef, and a host of garden-fresh vegetables and home orchard fruits." One cotton wife stated: "Just look across the road to see that little place where we lived, and you'll understand why we appreciate the roominess of this new home."[17]

In 1949 the extension agent of Lubbock County arranged a farm home tour. The new houses were larger and had the latest features of design such as extra storage space, a sewing center, abundant kitchen cabinets, and a washroom for husbands and children. Many of the women had requested the installation of special features that had not been available in their smaller houses. Shelves for displaying favorite objects were popular, and the latest rage, the "family den," appeared in several homes. Housewives budded with self-satisfaction because in the past they had to concede fine living to their city cousins, but the new homes erased the differences. Lubbock County boasted an air of cordiality and a fresh enthusiasm for home life, and one reporter noted the booster spirit: "Cotton has built some fine homes on Texas' South Plains."[18]

THE RISE OF COOPERATIVES

Even as conditions improved with irrigation and seed breeding, growers nonetheless recognized their cotton had to be more marketable. For one thing, short-staple lint broke more frequently in the spinning process, and textile mills would generally blend it with a longer variety. Plains cotton did not compete with Acala in the Far West or the traditional varieties of the South, so much of the West Texas cotton went into CCC warehouses or sold at a discount. The CCC even refused to accept one grade common to the area, light-spotted (which referred to its coloration). Plains cotton would remain a minor feature in the market unless growers could overcome these drawbacks. With the renewal of planting restrictions in 1954 and the quarrels over allotments with the older growing areas, a feeling of crisis crept onto the Llano and drove the growers to act. Just as their grandfathers had banded together to fight prairie fires or Comanches, the instinct to be self-reliant led them to form cooperatives. In another respect, Earl Sears testified, the remoteness of the region left them no alternative.[19]

Table 3. High Plains Cotton Production	
1945	139,683
1950	858,480
1960	2,018,800
1979	2,750,800
1990	2,950,900
2004	4,877,600
Source: Plains Cotton Growers, http://www.plainscottonorg/esw/stats (accessed August 30, 2006).	

Cooperatives provide several services for members, with marketing nearly always being the most important. Before the rise of cooperatives, buyers typically traveled to gins and offered prices that growers generally accepted without knowing comparable prices a few miles down the road.[20] Through a cooperative, farmers could band together and raise capital to build storage facilities and empower the cooperative to sell their crop when prices were strong. A particular feature of this concept involved the employment of experts whose knowledge and information of commodity markets, one of the riskier and more complex in trading, would ensure that members had experienced traders working on their behalf. Co-ops also helped with the buying of fertilizer, fuel, and other production inputs. Cooperatives, in other words, offered the opportunity to gain advantages and make every dollar squeak.

From this perspective, a group of growers formed the Plains Cotton Cooperative Association (PCCA) in 1953. It began as a grassroots undertaking when the executives of the Plains Cooperative Oil Mill (PCOM) recognized the compelling need to overcome the disadvantages of cotton farming on the plains. Roy Davis, manager of the PCOM since its founding in 1937, led the move, but he had the assistance of the mill's directors. Lack of capital blocked any initiatives, but Davis offered $12,000 in uncashed checks that had been made to members of PCOM, only with the understanding that the directors had to honor the checks if the recipients demanded them. This offering became the venture capital of PCCA. "Roy Davis didn't create PCCA, but he was the midwife who delivered the organization into the world," one writer described his role.[21]

Probably no figure shone brighter on the plains than Davis. He was born in McLennan County in the heart of the Texas blackland strip, but his family moved to Lamesa in 1905. He was child number six in a family of ten children. He first picked cotton at the age of six and at one point lived in a plains dugout as a child. He went to Texas Agricultural and Mechanical College and graduated in 1927. Davis always worked in agricultural education or businesses, first as a county agent in Gaines County, then as manager of the Hale County Dairy Association, and finally for the Houston Bank for Cooperatives before accepting the position as head of PCOM.[22] In the course of his career, "Mr. Roy" received many accolades and served as president of the NCC in 1968. He embraced the concept of agricultural cooperatives, and along with his work as manager of the largest cottonseed oil mill in the United States, his contribution to PCCA made him renowned.

The new cooperative was a fledging operation, but it quickly improved sales for members. The Davis family became identified with the organization when Dan Davis, Mr. Roy's son, took over as manager in 1956. He concentrated on marketing in the early years, and the cooperative won recognition for improving the incomes of its members and generating more annual income for the Lubbock area.

In 1963 PCCA acquired two warehouses, one in Altus, Oklahoma, and another in Sweetwater, Texas, to hold cotton off the market until prices improved. In the beginning, the two facilities compressed bales as did privately owned compresses, but they dropped this feature when gins began installing universal density compressors in the 1970s.[23] The two warehouses offered storage and shipping and saved members enough money to issue a dividend of $10.13 per bale over a five-year period. PCCA cut costs by following the economies of scale.[24]

Plains cotton nonetheless generally brought a discounted price, and grower interests there believed they faced unfair treatment. They wanted to find a scientific and objective way to establish the grade and color of lint, which they felt would work to their advantage in combating the bias against their product. This objective led to the early work in developing high-volume instrument (HVI) testing, and when in 1960 PCCA manager Davis hired Emerson Tucker to be chief engineer, he gave him an assignment: find a way to test all bale samples with scientific accuracy instead of relying on the standard practice of manual grading. Tucker was a native of Connecticut who grew up in Brownfield, Texas, and graduated from Texas Technological College (renamed Texas Tech University in 1963). He combined the sharp intellect of an energetic Yankee with a naturally acquired familiarity of the Texas Plains. Before joining PCCA, he had worked in a textile mill and had learned the importance of a fiber's spinning qualities. He got a break in his assigned mission when he discovered that a Dallas cotton buyer used an instrument invented by Motion Control in Dallas to grade cotton, but the instrument had a slow rate of speed and had no system for testing staple length or the strength of fiber for spinning.[25]

In Dallas, Motion Control had developed the Fibronaire to measure fiber fineness, or micronaire, which provided an indication of how cotton would spin. Davis assigned Tucker to work with Glenn Witts, owner of the Dallas firm. Tucker lived in Witts's house for two years to perfect an instrument; he recalled how they sat around the kitchen table each morning tossing out ideas. At one point they experimented with an instrument at PCCA and graded over four hundred thousand bales. The USDA saw the potential to test micronaire for the entire U.S. crop, but it wanted a high-speed device that could also measure length and strength. By 1965 Motion Control had a faster length tester, and the Stanford Research Institute, with funding from the CPI, provided a speedy strength tester, but each instrument had to be used separately. By 1967 Motion Control had a single instrument that performed all three services, and PCCA put it into operation. In 1972 the USDA installed the instrument in the classing station at Lamesa, Texas.[26]

This new method of classing accounted for one of the most important developments in the cotton industry since 1945. It removed the subjectivity of human judgment and established a uniform standard. It gave textile mills a better grip on the spinning quality of each bale and let them avoid bales with varying characteristics and thereby spin fabric with more consistent quality and make fewer adjustments to machinery. HVI testing provided the answer to the advantage of fiber consistency enjoyed by synthetic manufacturers.

For growers on the plains, a major breakthrough had come. They could now assure buyers of the quality of their bales through instrument testing administered by the USDA, which enabled growers to identify their better cotton and sell it without a discount. Plainsmen now sold cotton at prices based on scientifically determined quality rather than human prejudice. In the words of PCCA's Tucker, "it modernized the cotton industry."[27] But it also ended the aura associated with cotton buying markets such as Cotton Row in Memphis because no longer did classers appear on streets and coffee shops with lint clinging to their trousers. The "snakes" of discarded cotton disappeared from sidewalks. Data obtained from HVI testing now appeared on computer screens, and one dimension of the romance of the cotton kingdom disappeared.

PLAINS COTTON GROWERS

To establish identity and authority in the Cotton Belt, plainsmen needed political muscle. In 1956 the Plains Cotton Growers Association (PCG) went into operation to fill the gap. For the mid-South there was the powerful Delta Council, and an assortment of other organizations like the Texas Cotton Association addressed the needs of local interests throughout the Cotton Belt, but growers on the plains had no organization to speak on issues requiring a political voice. Necessity pushed the establishment of the new cooperative, which began when the Texas Agricultural Stabilization Committee of the USDA intended to shift some acreage allotments from West Texas to East Texas, about two hundred thousand acres. And bills had been introduced in Congress to make one-inch staple the average length for cotton stored through the CCC program. The short staple grown in the Cotton Patch measured 15/16 inch, and since much of the West Texas cotton generally graded as light-spotted, which placed it in the lower CCC standards, an increase in the definition of average length would have worsened the penalties on it.[28]

To ward off these threats, growers in twenty-three counties established the PCG. They sought to protect their interests through political lobbying and to undertake programs for enhancing production. The new organization acknowledged the inferior qualities of plains cotton and soon began an educational campaign in cooperation with the Extension Service and Agricultural Experiment Station to make farmers aware of the importance of growing better-quality fiber. Some of the acreage lost to East Texas soon went back to the plains, and George Pfiffenberger, first executive director of PCG, lobbied the CCC and persuaded it to remove light-spotted from its list of inferior cottons. The new organization also helped finance the installation of humidifying equipment in the USDA classing stations at Lubbock, Brownfield, and Lamesa. This move caused the area's cotton to grade slightly better, at least in cases when the excessive dryness of the fiber had worked against it.

Bad luck loomed over the Caprock, however, when the boll weevil appeared on the lower Rolling Plains. Ironically, the absence of weevils and the expectation they would never reach the plains had been responsible for the growth of the area, but in 1959 a confirmed outbreak occurred in two counties east of Lubbock, Crosby and Dickens.[29] Further outbreaks came a few years later in Brisco and Floyd counties. News of scattered infestations continued until by 1963 the boll weevil reached the edge of Lubbock County. "These infestations were nearly beyond comprehension," reported one account.[30] Growers had believed the Caprock Escarpment would stop the weevil migration. Crosby County's Don Anderson recalled that the arrival of boll weevils caused a near panic on the plains.[31] Anderson, Pfiffenberger, and a handful of growers invited entomologists from Texas A&M University to the plains to offer advice. J. C. Gaines, head of the university's department of entomology, doubted if the migration could be stopped because no proven system of eradication or control had been developed, but entomologist Perry Adkisson suggested the new concept of diapause control. Adkisson made clear that the method was unproven, but he thought it might slow the further spread. Anderson and Pfiffenberger drummed up support for an experimental program among growers, and in 1964 they created the Texas High Plains Boll Weevil Suppression Program, with Adkisson as head of the technical group.

Plains Cotton Growers had organized the program and split the costs with Texas A&M, Texas Tech University, the USDA, and the Texas Department of Agriculture. Airplanes sprayed only targeted areas where boll weevils were found. It was a costly undertaking because an "army of flaggers" had to stand at the edge of fields to guide the pilots.[32] Anderson remembered that when crop dusters were in short supply, C-47s with U.S. Air Force pilots were used. This effort stopped the weevil migration and was described by a team of writers as "the first cooperative attempt in the U.S. to control the boll weevil on an area wide basis."[33]

Whether lobbying members of Congress or USDA bureaucrats or simply carrying its message to the general public, the main task for the PCG remained in the political realm. It resembled the NCC in this respect, though the Texas organization limited its activities to issues pertaining to the plains. It sent spokesmen to congressional hearings and assisted with special hearings conducted by congressional committees at the local level. When an issue arose among plains growers that required federal assistance, the PCG carried the ball. It maintained a close relationship with the congressional delegation for the area and often went to George Mahon of the Texas Nineteenth District for help. Mahon managed to get President Lyndon Johnson to make a special budget request to Congress to fund a special federal matching program for the PCG Boll Weevil Suppression Control Program.[34] When Mahon became chair of the House Committee on Appropriations, he gave the growers a distinct advantage; Larry Combest followed Mahon in 1984, and Randy Neugebauer replaced Combest when he retired in 2003. Charles Stenholm of the Seventeenth District often fought on behalf of cotton growers during his congressional terms from 1979 to 2004. In 1997 the PCG established an e-mail service that provided members with daily news about developments pertaining to their interests, ranging from political matters to announcements of meetings and similar items.

THE COTTON ECONOMICS RESEARCH INSTITUTE

Established with the Department of Agriculture and Applied Economics at Texas Tech University in 1997, the Cotton Economics Research Institute (CERI) became renowned for its studies of cotton markets and economic impacts of policy, both proposed and enacted. The institute's origin stretched back to 1975, when the USDA moved its Economic Research Services (ERS) to Lubbock, but the department closed the office in 1981. The university wished to keep the program and employed Don Ethridge to continue the studies, but the program needed funding, so representatives of the grower constituency managed to get line-item budget status for it in 1985 by the Texas state legislature. Leading in this effort were S. M. True of Plainview in the Texas Farm Bureau and Bob Poteet of the Texas Cotton Merchants Association. In 1997 the legislature consolidated many budget line items in a general appropriation that stabilized the funding for cotton economics research at the university. The annual appropriation, ranging near $130,000, funded projects undertaken by faculty in the Texas Tech Department of Applied Economics and graduate students' work. The impetus for creating a special institute within the department came a year earlier from Bob Albin, associate dean for research in the College of Agriculture. Interest in a separate institute also came from PCCA. In 1997 the university formalized CERI and appointed Ethridge as director.

Ethridge earned a doctorate in economics from North Carolina State University in 1970. He worked first at the University of Missouri with a quarter-time appointment in the school's Water Resource Center. The CIA recruited him for economic studies, where he stayed for over two years. Ethridge joined the USDA-ERS and moved to Lubbock in 1975. He took a position on the Texas Tech faculty in 1981 with a clear assignment: build a research program in cotton economics. From 1981 to 1997, he raised external funds for graduate students. When he took over CERI, he became department chair of applied economics. His mission remained the same, to build a research program in four areas pertaining to cotton: economics, water, risk management, and international trade.[35]

The institute conducts industry-related research but does not confine itself to the plains. For growers on the Llano, however, CERI has made economic impact studies of optimum application of water and fertilizer, the impact of modules on ginning economies, and the feasibility payoff of new harvest systems. It made studies of HVI classing. One accomplishment was the Alternative Price Analysis and Reporting System available for growers. This project, which took two years to develop, received funding from the USDA before Cotton Incorporated took over responsibility. The system, based on an econometric model, furnishes growers with data as they make decisions to sell their cotton.

A particularly valuable service provided by the CERI involves the international dynamics of cotton. Through analysis, it estimates the impact of government policy on the various segments of the cotton industry and consumers. In its Global Fibers Model, an econometric simulation model, the CERI conducted a study of the U.S. proposal for cotton at the 2005 Hong Kong meeting of the WTO Doha Development Round of negotiations. The institute examined the expected impact of the proposal on world production, consumption, textile production, and prices of synthetics. It made this study for each of the twenty-four countries that grow and import cotton. The CERI also assists the NCC when it needs economic analyses. The two organizations cooperate closely but maintain separate roles.

LATINO CULTURE

Development of the Cotton Patch involved more than improvements in farming and marketing, for Latinos made a contribution even though the Anglo-dominated ownership culture prevailed. Farmers on the Texas Plains used to rely on Mexican laborers who came from the San Antonio area and the Rio Grande Valley to pick cotton each fall. Crew chiefs, the jefes, carried out the orders of the growers. Migrants typically were at the mercy of the jefes, who owned the transportation and arranged the employment. Growers furnished small shacks in the fields that workers used as campsites. Migrants generally cooked outdoors on an open fire and slept outdoors unless weather drove them inside. Their hot meals were simple, consisting of the basic tortilla with peppers, chilies, onions, and vegetables such as potatoes and some meats such as chicken. Migrants would travel to the nearest town on Saturday afternoons to buy supplies and socialize with fellow workers. Neuman Smith, a rancher and grower, described migrants as trustworthy and hard workers and sold them milk and eggs.[36]

Latinos faced much hardship throughout Texas, whether on the Llano Estacado, the South Plains, or the Rio Grande Valley, but labor unrest rarely occurred among migrants in the Cotton Patch, because relatively few lived there permanently.[37] A few outbursts, however, took place. In October 1959, two hundred Latino pickers appeared at the gin of O. C. Heard located five miles west of Levelland, Texas. They demanded their pay and transportation to Mexico. They were peaceful but refused to disperse. Militancy was so uncommon among the laborers that this appearance provoked a reaction. Two Texas Rangers and five policemen went to the site, but no violence occurred, and no arrests were made. The workers were unhappy because they were kept out of the fields until 10:00 a.m., which shortened their day and the opportunity to pick a maximum poundage of cotton. Landowners wanted to let the morning dew evaporate, since dry cotton ginned better. No further action occurred, but the laborers chose to return to their homeland.[38]

Growers had more serious dissatisfaction with the U.S. Department of Labor. They had begun to assign braceros to drive tractors, though the practice violated the terms of the Bracero Program. In 1958 New Mexico and Texas Plains farmers appealed to the Dallas regional office of the Department of Labor for permission to let migrants operate tractors on the grounds that local labor was too scarce. The El Paso Valley Cotton Association made the formal request and pointed out that it had placed ads in newspapers for one hundred drivers and got only thirteen responses. The organization lost the appeal, and the growers had to cope without bracero drivers. Growers indicated a need for more braceros because local labor would no longer work on farms owing to the low wages or a new preference for nonagricultural jobs. Landowners in the El Paso Valley appealed to Senate majority leader Lyndon Johnson for help, but the Bracero Program had begun to be regarded as exploitative and unfair to workers.[39]

Migrant labor on the plains passed into history once the mechanical stripper and herbicides became common. In this respect, it resembled black labor displaced in the South by mechanization. Machinery changed the status of Latinos, however, for while it ended the large flows of migrants each year, growers employed a few as permanent employees to operate machines and serve as mechanics. This new permanency encouraged Latinos to become U.S. citizens and put their children in public schools. Latinos then began having an impact evident in food, music, and language. Growers, for example, nearly always acquired some speaking ability in Spanish as a matter of necessity. Though Anglo culture dominated, a powerful Latino subculture emerged on the plains, recognized through Lubbock's Viva Aztlan Festival and numerous Cinco de Mayo celebrations. Cotton farming alone did not account for this development, but it provided much of the original impetus.

MUSIC OF THE COTTON PATCH

Just as the blues grew out of the fields of the Mississippi Delta, the Texas Plains developed a version of a rockabilly sound. While the blues traced their origins to slavery and the life of black sharecroppers, West Texas rockabilly came out of the Anglo experience on the Llano Estacado after World War II. Music and place went together in both cases, and while the sounds were different, the messages were similar. However much the two forms of music parallel or diverge, they had a common root: the annual cycle of cotton farming.[40]

Rockabilly, described by a cultural geographer as "a uniquely Texas strain of music that falls somewhere between country, rock and folk," reflects the flat West Texas landscape with few trees, little rain, and blowing sand. Texas rockabilly originated in the 1950s among artists like Buddy Holly and Roy Orbison, but in the 1970s it enjoyed a revival on the plains and bespoke how cotton featured so prominently there. In "Because of the Wind," country- and-western musician Joe Ely remembers a former lover, the memory of whom flows through him like a West Texas wind. In his "Dry Land Farm," Butch Hancock sings about a thunderstorm on the horizon bringing wind and sand that "burns like a blazing blowtorch when you're living on a dryland farm."[41] Unlike the blues musicians, balladeers of the Caprock lamented the pain of place: sand, flatness, drought, and wind. The harshness of the physical environment was a common theme. West Texas rockabillies complained about the conservatism and religiosity of the people, so like blues, escape was a frequent topic. They expressed the oft-heard frustration over a conformity that squashes creativity and tolerates no rebellious behavior. They complained of a stifling environment imposed by nature and man alike, whereas blues of the South reflected social injustice. This made escape, nonetheless, common to both. Despite their protestations, the artists had an identity and love for the area. Lubbock native Mac Davis sang that "happiness is Lubbock, Texas, in my rear view mirror" but concluded that "happiness is Lubbock" and "you can bury me" there "in my jeans."[42]

The group that best expressed the connection of cotton farming with music, the Flatlanders, formed in 1970, had an "old timey acoustic sound" that did not follow the strings and electronics stylish in country music at the time.

They recorded an album, *More Legend than Band* (1972), that displayed the "band's geographic origin." This music, in the words of a scholar, was written by people "who knew what it was like to sink up to their ankles in mud and wash dust from their faces."[43]

Butch Hancock wrote some of the songs on the album, and his songwriting background came straight out of a cotton field. Born in 1945, he grew up in Lubbock and started driving a land leveler in 1968. He spent long hours on the lonely fields of the Caprock and later described how the sound of a tractor helped him write music. "Second gear at two-thirds throttle on a John Deer tractor—I found that that speed and gear was the key G and you could play any song you wanted to in it."[44] But the Flatlanders had a short life span, and their album was not released until 1990.

During the interim and through the 1990s, the members of the band went their own ways, and Joe Ely, Hancock, and Terry Allen achieved some recognition as individual artists. Hancock released two albums, *West Texas Waltzes* and *Dust Blown Tractor Tunes*, based on his experiences on the plains. In 1996 Allen released *Flatland Boogie*, which reminisces about cotton fields, coyotes, and other features of the Texas Plains. Ely lived in Lubbock as a child and remembered that Mexicans who went to his father's used-clothing store during cotton picking season liked accordions, so he incorporated them into his songs.[45]

Rockabilly and country-and-western music were not confined to the cotton-growing area of the Texas Plains. They belonged to a broad genre of music in the era after World War II. But the plains contributed some of the most popular and influential singers of their time who had roots in the rural environment of West Texas. The famed country-and-western singer Bob Wills came from Turkey, Texas, about one hundred miles northeast of Lubbock, near the Caprock Canyons. His music appealed to rural folk, and his *Cotton Patch Blues* resonated with them. Waylon Jennings came from Littlefield on the Llano. Their music went beyond the cotton culture, but it was rooted there just as blues traced its origin to the cotton South. Willie Nelson grew up on the Texas blacklands but stated that he was "raised and worked in the cotton fields . . . with a lot of African-Americans and a lot of Mexicans, and we listened to their music all the time. . . . There was a lot of singing that went on in the cotton fields."[46]

The sound of music on the Texas Plains had an appeal similar to the blues. Country-and-western and rockabilly musicians developed a message rooted in their own experiences, a childhood spend in a cotton culture that emerged after the emancipation of subsistence labor. Music was an escape. "No one had to tell the musicians," went one account, "that picking the guitar beat picking cotton."[47]

CONCLUSION

A strong sense of place continues to reside in the hearts of plainsmen for their land. They give it fierce loyalty and unashamedly boast of weathering the wind and sand, of enduring droughts, storms, and more wind. They show pride in overcoming the harsh and oppressive conditions that nature imposes on the area as they see towns and cities grow, schools and shopping malls appear, and patches of green interrupt the monotonous sandy brown landscape. An esprit de corps lingers that is visible among all, whether Anglo or Latino, and drives them onward season after season. Their common denominator is cotton, the plant that offers a livelihood, however harsh it may be, and gives meaning to the work ethic and the resilience of men and women. For the first generation after 1945, the cultivation of cotton on the vast open spaces meant a chance, perhaps even a second chance, to own land, make a home, and raise children. While some longed to escape, particularly the second generation, others stayed and expanded their operations, and the area became more developed as a consequence.

The Texas Plains took their place as the giant of the Cotton Belt, and more cotton grows on the plains each year than in any other place on earth. Here King Cotton can still be sensed, and the sight of raw cotton lying on the ground, or the appearance of cotton stalks growing along sidewalks in small towns, is taken for granted. Rows of cotton stretch to the horizon, and gins run all night during picking season. Module trucks speed along roads and through fields, and semitrailers loaded with cottonseed hurry to oil mills from whence the sweet smell of cottonseed oil permeates the air. To produce a crop on the plains involves more than cultivating the soil and tending the precious plants. It is a display of coping with the challenges of farming on the plains, and it is a show of independence from the old South. Here families thrive on the Caprock and prosper with their own work ethic and sense of purpose. Here the drive to succeed still flourishes. However they make their contributions, whether as educators and researchers, gin operators or cooperative managers, or the sunburned farmers, they are a cotton fraternity. Like westerners they admire the big sky and the vast expanses of space. "In Lubbock, there is nothing between you and the clouds or you and the earth," flatlander Jimmy Dale Gilmore stated. Easterners remain puzzled over this fondness for the unbroken horizons, but as expressed by one country-and-western musician, "You have to train your eye to see it."[48]

CHAPTER 13

THE QUESTION OF SUBSIDIES

Subsidies for cotton began in 1933 as temporary assistance during the Depression but continued through the farm bill of 2007. The purpose for federal assistance was simple: to better enable growers to survive the economic pressures of farming and remain on the land. Maintaining the beloved family farm served as the justification for the support programs, and crops such as wheat, corn, sugar, and rice were also subsidized. Since their inception in the early New Deal, assistance programs for cotton have varied widely, but the essential ingredient, a guaranteed floor price, remained in effect. Through the years the machinations and intricacies of the programs became complex, often befuddling growers themselves. Criticism of agricultural subsidies appeared as soon as they were introduced in 1933 and continued through the years, but the most serious attacks on cotton appeared after 2000.

Use of subsidies rested on the premise that farming reinforces values set in motion as far back as the founding of the republic. Known as agrarianism, this justification remained a part of U.S. culture after World War II. In 1947 a member of the House of Representatives from Minnesota stated, "I have always considered the farmers of this country as the backbone of American democracy."[1] A literary editor wrote a generation later that "the importance of agrarian thought in the development of American culture cannot be overestimated."[2] But this idealism dated to ancient Greece and Rome and continued through Europe into the Middle Ages. Aristotle wrote, "For the best material of democracy is an agricultural population; there is no difficulty in forming a democracy where the mass of the people live by agriculture or the tending of cattle," and the spirit of his statement thrived in North America, where the cultural extensions of Europe took hold. Here the ideals were expressed through the admiration for the yeoman farmer, the idealization of the rural classes as the nursery stock of America, and the popular association of innocence and simplicity with the pastoral life. In his *Notes on the State of Virginia*, Jefferson wrote: "Those who labour in the earth are the chosen people of God." Such ideals easily survived into the twentieth century, and southern writers like the Vanderbilt Agrarians boldly stated, "The theory of agrarianism is that the culture of the soil is the best and most sensitive of vocations."[3] William Faulkner liked to claim that he was a farmer and only wrote incidentally.

Don Paarlberg, noted agricultural economist and adviser to President Eisenhower, explained that agriculture had a uniqueness or a special place in U.S. culture. "Farmers were considered God-fearing citizens, stalwart defenders of the republic, and a stabilizing element in the society."[4] They were considered to have a special way of life and often enjoyed special status such as being exempt from the military draft or labor laws like collective bargaining rights and workmen's compensation. Indeed, Paarlberg continued, farming was a way of life, and agriculture had "unique worthiness."[5]

Subsidies were meant to guarantee growers a minimum price based on parity, a confusing term referring to a balance of agricultural and industrial prices. When parity payments began in 1934, policy makers identified 1910 to 1914 as the period when the relationship was equal. Since agricultural prices had not kept pace with nonfarm prices after World War I, support payments based on parity represented an attempt to strike a balance. Defining parity proved difficult, but Paarlberg used the following illustration: "If a bushel of wheat would buy a pair of overalls in 1910–14, then, to be parity, a bushel of wheat should be priced so as to buy a pair of overalls today."[6] Many drawbacks were associated with the concept, so that a true and accurate support price for cotton, or any crop, proved impossible; but the concept nonetheless continued to be incorporated into federal agricultural programs until it was removed in the farm bill of 1970.

Because farm prices on the open market often fell below parity, the subsidy made up the balance. Numerous schematics were developed since 1933 to carry out the assistance programs, incorporating concepts such as target prices, countercyclical payments, loan redemptions, and deficiency payments. To qualify for support, however, producers had to reduce their acreage planted in cotton, generally known as allotments. Government requirements for reduced planting thus accounted for much of the drop in the U.S. total acreage devoted to the fiber, but the restrictions had the most adverse effect on owners of small plots, even though the general public wanted to help them.

THE OPERATION OF SUBSIDIES

Complaints were common. In 1958 Milton Brown, a ginner from Arkansas, wrote Congressman Brook Hayes that allotments had brought suffering to the state, that he knew of owners allowed to plant only fifteen acres on a hundred-acre farm. Such small production did not warrant investments in machinery, and other crops did not yield enough revenue to offset the loss of cotton income.[7] An Arizona grower in the Sulphur Springs Valley complained that on his eighty-acre farm he had to accept a cotton cutback from seventy to twenty-one acres, while his costs of operation and living expenses did not go down.[8] His complaint brought up another effect of the subsidized cutbacks: in an adjacent county, a farmer who owned six thousand acres received an allotment of sixteen hundred acres. So although large-scale operators had to accept cutbacks, their greater size enabled them to continue. In the case of the smaller farmer, subsidies would not offset the loss of income from his normal production of forty-nine acres of premium Acala 1517.[9]

Subsidies were meant to supplement income when the price of cotton fell below parity, so growers did not collect them every year. Table 4 illustrates the variation in the market price compared with the parity support.

An example of the mixed results of cotton subsidies occurred with the Soil Bank program established through the Agricultural Act of 1956. It was popular with the general public because it was touted as an environmental measure. It took land out of production through two plans: acreage reserve and conservation reserve. For participating in the first, growers received payments for removing land from production in addition to the acres already diverted by their cotton allotments. The other reserve arranged long-term diversion of cropland to conservation uses. Setting land aside for protection appealed to the nonfarm public and gave cotton farmers a way to get payments for unproductive land not already set aside in the allotment program. Some reservations existed, however, over the effectiveness of the program. A. T. Jefferson, a grower near McDonough, Georgia, did not anticipate much change in reducing the surplus via the Soil Bank, since he expected the least-productive land to go into it. Only increased consumption and foreign sales would alleviate the woes of cotton farmers, in his opinion.[10] In 1957 and 1958, however, production fell noticeably, but the price of cotton fell for those years because the Commodity Credit Corporation unloaded stocks.[11]

Through the CCC, growers could place their cotton grown on the remaining acres into government storage and receive a "nonrecourse" loan. Each farmer could take his crop out of the CCC facilities and sell it if the open

Table 4. Average Price Support Levels and Average Prices Received by Farmers for 7/8 Middling, 1945–1960

Year	Percentage of Parity	Price Support Loan (Cents per Lb.)	Average Market Price (Cents per Lb.)
1945	100	20.90	20.72
1946	92.5	22.83	22.51
1947	92.5	26.49	22.51
1948	92.5	28.79	32.63
1949	90.0	27.23	31.92
1950	90.0	27.90	30.38
1951	90.0	30.46	28.57
1952	90.0	30.91	39.90
1953	90.0	30.80	37.69
1954	90.0	31.58	34.17
1955	82.5	31.70	32.10
1956	78.0	29.34	33.52
1957	81.0	28.81	32.27
1958	80.0	31.23	31.63
1959	75.0	30.40	29.46
1960	82.0	28.97	33.09

Source: National Economics Division, Economics, Statistics, and Cooperative Service, USDA Department of Agricultural Economics, Texas Tech University, *The Cotton Industry in the United States: Farm to Consumer*, Publication T-1-186 (April 1980), 12.

market price advanced beyond the guarantee, but if the price remained unchanged or went lower, he was allowed to leave his cotton with the CCC and keep the loan. Thus began the practice of "farming for the loan" rather than the market. The CCC generally held the forfeited cotton until the price went up and sold it at a profit.

To qualify for Soil Bank assistance, growers had to reduce their acreage planted in cotton, but the supply continued to exceed demand because per-acre yields climbed. Herein lay an important dimension: production of cotton in the United States did not fall with the reduction in acreage. From 1954 to 1963, and from 1964 to 1970, the yield of lint per acre averaged 431 and 486 pounds respectively, a real jump compared with the average from 1941 to 1953, which was 271 pounds.[12] So although federal programs were intended to remove the excess supply and provide subsidies as compensation for loss of acreage, advances in technology, the use of irrigation, and the efficiency of large-scale mechanized cultivation drove up the per-acre yields.

OPPOSITION TO SUBSIDIES

In 1956 President Eisenhower questioned why hefty payments went to large landowners while small family farms fared poorly. In 1954 the USDA had reported only seventy landowners, regardless of crop, receiving over $100,000; some cotton payments went as high as $1,250,000 with wheat coming in second at $354,000.[13]

Cotton subsidies caught public attention when in 1957 *Life* magazine featured a full-page editorial, "King Cotton: The Royal Nonsense." Renowned for its photographs, *Life* was the most popular magazine in the United States and was read by millions each week. The editorial lambasted farm subsidies, but particularly cotton subsidies: "The most absurd of these absurdities is reflected in cotton." The editorial claimed that growers no longer needed assistance because cotton farming had become "mechanized, automated, and incorporated. The original purpose of cotton subsidies—to help the small farmer—no longer makes sense." *Life*'s writers chose Mississippi's D&PL as an example of a large commercial operation drawing taxpayers' monies. The piece hit hard and provoked sharp responses.[14]

B. F. Smith, executive vice president of the Delta Council, fired back. "The editorial on cotton . . . is a masterpiece of inaccuracies, twisted facts, half truths, and outright misstatements." Hodding Carter II of the *Delta Democrat-Times* responded with an editorial, pointing out that the subsidy to D&PL, $1.4 million, which *Life* claimed was a giveaway, had been recovered by the government with a profit. Carter brought up subsidies that went to shipping, the airlines, and the second-class mail rates that benefited *Life*. If cotton supports ended as *Life* proposed, Carter warned, it would upset the southern economy and "cause even the author of that editorial to sweat through his cotton shirt." Jerry Sayre, head of the Mississippi Agricultural Experiment Station, congratulated Carter for his response and attributed *Life*'s attack to merchandisers in the futures exchanges campaigning to get the cotton loan level lowered. Gerald Dearing of the Memphis *Commercial Appeal* stated that the editorial's claim that subsidies injured the foreign markets for growers was true, but it was not true that cotton growers exploited the government. Dearing best reflected the attitude of growers, who felt that farm policies of the past had caused the surplus of cotton, and they would prefer to operate independently of government assistance, but until a thorough and fair revamping of farm policy occurred, farmers needed assistance.[15]

Congress imposed no limitations, but in the late 1960s the issue reappeared when cotton's principal opponent, Representative Paul Findley of Illinois, fought for a $20,000 maximum payment. His proposal passed the

House in 1969, but the Senate shot it down. Within the House the Cotton Bloc was strong with Bob Poage as chair of the Agriculture Committee, George Mahon as head of Appropriations, Olin Teague of Texas as chair of Veterans Affairs, Mendel Rivers of South Carolina as head of Armed Services, and Mississippi's William Colmer as chair of the Rules Committee. But Findley had been persuasive. The *Nation* called for a limit on subsidies, referring to recipients of large payments as "fat cat farmers" and "subsidy men" who raised subsidies rather than crops. "The whole $5 billion setup of farm regulation and subsidization needs going over."[16]

In a special article in *Farm Journal*, Poage and Findley debated subsidies. As explained by Findley, the general public was losing patience because large landowners received extravagant payments. Such payments, he continued, accelerated the trend toward bigger farms; they also encouraged an addictive dependence on government assistance. Findley claimed that cotton growers obtained about half their income from supports compared with wheat farmers, who received one-third. He wanted a five-year transition period to end payments, but recommended that special arrangements be made to help marginal farmers earn supplemental income in nonfarm employment, or provide assistance for them to withdraw completely from farming.[17]

Poage saw payments as fair compensation to landowners because the government had taken portions of their acreage out of production. Payment must be made, he replied, in proportion to the amount of land taken, or else the agricultural program would amount to confiscation. If large landowners received a payment less than the productive value of their land, they would not participate in the program, and great surpluses would reappear and cause a Depression-era fall in farm income. He pointed out that mule farming and walking plows would not provide the United States with enough food and fiber and that large-scale operators had evolved naturally through the open market. As long as Americans wanted abundant supplies of food at reasonable prices, Poage concluded, large-scale operations would be imperative.[18]

Farm organizations fought limitations. For the NCC, the subsidy program kept a great industry alive. Dislike for limitations was strongest in the western states because a ceiling on subsidies would disproportionately affect large-scale growers. In Arizona the average allotment in 1969 stood at 187 acres, compared with the national average of 33 acres, which the Arizona Agricultural Extension Service estimated would severely impact 31 percent of the state's cotton growers, but only 8 percent across the Cotton Belt. Subsidy checks for Arizona growers averaged $21,980 in 1969, though the Extension Service reported that 50 percent of the checks fell below $10,000. To protect their interests, growers in Arizona vehemently opposed payment limitations.[19]

Farm organizations not related to cotton opposed limitations. For the National Association of Wheat Growers, restrictions "would be unfair and discriminatory and would encourage widespread noncompliance and disrupt effective supply control." The National Wool Growers Association thought limitations "would create economic hardships," and the National Sugarbeet Growers Federation opposed them, too. If large-scale operators did not participate in the crop reduction program, declared the National Grange, surpluses would reappear. For the Farm Bureau, a five-year phaseout should be implemented, during which large landowners would continue to receive payments as before. Only the National Farmers Union offered a specific schedule of limitations: full payment to the first $25,000, 75 percent of payment for the next $10,000, and 50 percent to the second $10,000. Hence the maximum would be $37,500 for farmers qualifying for $45,000 or more.[20]

MARKET-ORIENTED ASSISTANCE PROGRAMS

Despite resistance, in 1970 Congress limited payments and set a maximum of $55,000 on all crops covered in the federal program. How much of the payment each farmer received depended on his historical planting acreage. Large-scale operators typically overcame the limit by dividing their land among family members or carving several small farms out of the original operation.[21]

In 1973 Congress passed the Agriculture and Consumer Act. Legislation for agriculture, generally known as the farm bill, had to be renewed every few years because of changes in the world market that could be induced by weather, insect damage, the accumulation of surpluses, or the rise and fall of foreign competition. The new measure introduced the use of "target prices" and "deficiency payments" to replace price support payments. A target price incorporated costs of production and would enable a grower to earn a reasonable return on his investment. In one sense, it replaced the old parity price used in former years. Farmers received deficiency payments whenever the market price of cotton did not reach the target price; the payments covered the difference. Deficiency payments would make it possible to keep the loan rate for CCC below the world price, a desirable objective, and still supplement a grower's income. If the open market exceeded the target price, growers were free to take advantage and expand production, but when the market price fell below the target, the production controls with payments went into effect. This geared cotton toward a market-oriented policy. But deficiency payments proved unnecessary in 1974 and 1975 because the market price surpassed the target. In 1976, however, the target had to be adjusted to 43 cents, and to 47.8 cents in 1977. The measure further stipulated that Congress would henceforth reexamine the program each four years, but not during the year of a presidential election. Again in 1973 Congress reduced the payment to $20,000. With the new farm bill of 1977, however, Congress raised the payment limitation to $50,000.[22]

In the early 1970s, developments in the world economy affected cotton prices. In 1971 President Richard Nixon floated the dollar through his New Economic Policy, which came to be known as the "Nixon Shock." It ended the 1944 Bretton Woods system of monetary control because the United States had begun running a sizable deficit with its balance of payments. In other words, foreigners and other governments held dollars at such a level that U.S. gold reserves were insufficient to permit the convertibility of dollars held overseas. The U.S. role as global hegemon had stretched the balance of payments to uncomfortable levels, and though the United States did not face a crisis, it needed to adjust its position through international monetary exchange. The cheaper dollar instigated by the shocks stimulated exports, and cotton benefited. To further encourage exports, Congress enacted the Agriculture Export and Trade Expansion Act of 1978, which established export credit programs, opened trade offices around the world, raised agricultural attachés in U.S. embassies to the rank of counselor, and extended eligibility for export credit to China.

Through the Rural Development Act of 1972, Congress had authorized the Farmers Home Administration (FmHA) to make loans to rural residents for small businesses, to guarantee loans made by commercial lenders to rural residents for small businesses, to increase authorization for water and waste disposal, to raise to ten thousand the population in towns included in the FmHA system, and to raise the lending limit for loans made for farming operations. Obviously such assistance did not bear directly on commodity prices, but it helped enhance rural life and had some effect in encouraging opportunities for nonfarm employment.[23]

But the switch to direct payments had made cotton growers more visible and vulnerable. For one thing, the supports for cotton proved to be more costly than for any other crop, and opponents of farm subsidies liked to cite examples of large-scale growers receiving huge payments. Urban congressman Ray J. Madden from Gary, Indiana, wanted his colleagues in the House to acknowledge formally that taxpayers demand "that this unfortunate raid on the public treasury be terminated."[24] Other opponents cited cases of banks, insurance companies, and even state prisons receiving cotton payments. In the House discussion of the 1973 measure, cotton became such a point of disagreement that it had to be treated separately from other crops. At one point when a narrow vote preserved cotton subsidies, Silvio Conte of

Massachusetts stated, "It was a great victory for the boys in the Cotton Belt, but... a defeat for the taxpayers and consumers."[25]

More threatening to the welfare of cotton, however, were the rising costs of farming. Fuel went up with the oil embargo of 1973, machinery became more expensive, fertilizer prices climbed, and the expense of fighting insects and weeds rose. Feeling they were caught in the middle between the attacks from urban interests and spiraling costs, cotton growers wanted relief. Inflation slowed in the 1980s, but large crops overseas, along with the improved value of the dollar, forced exports down, and surpluses began to reappear. Spending for farm programs went up although President Ronald Reagan wanted constraints in the federal budget. This set of conditions was the backdrop for the 1985 farm bill, the Food Security Act.

It remained necessary to keep U.S. cotton competitive with foreign producers and hold down the buildup of surpluses in CCC warehouses, so the 1985 measure introduced marketing loans. A grower could still put his crop in CCC storage for the nonrecourse loan, but now he could repay that loan at less than its original value, thereby getting a marketing loan. The value of the loan would be calculated, however, on the world price of cotton instead of the domestic price. Devised by NCC economists, a system known as Step 1-2-3 that incorporated cotton prices in northern European markets determined the price used for making marketing loans.[26] Exports grew, hitting 44 percent of the U.S. crop in 1987 and generally staying above 40 percent through 2000; exports accounted for 64 percent of the 2006 crop.[27] Use of marketing loans managed to keep down stocks in the CCC and cost to the U.S. Treasury. In 2001 the NCC described the 1985 nonrecourse marketing loan as the cornerstone of the program because it made cotton more competitive on the world market. The 1985 measure set a payment limitation of $50,000 per person.

Subsequent farm bills dropped and added features such as planting flexibility and changed the limitations, but the farm bill of 1996, known as the Federal Agriculture Improvement and Reform Act (FAIR) introduced a new set of provisions. Instead of customary income supports that were made via target prices and deficiency payments, seven-year production flexibility contracts were introduced. Growers could plant crops other than cotton or place their land into conservation without losing their subsidies. In the past, if a farmer planted a crop other than cotton, he endangered the right to future subsidies, a feature that kept more acres in cotton than the market warranted. Nonrecourse marketing loans continued to be available for farmers who used production flexibility contracts, but for cotton the total spending made through the Step 2 payments could not exceed $701 million over the next seven years. And there was a total payment limit set at $230,000 per person. The outstanding feature of the measure lay in the plan to end supports by

2002 through the gradual reduction of payments. For this reason the 1996 legislation got attention as the measure that would end subsidies; it was touted as the end of the safety net and the path toward farming as a self-reliant undertaking. Agricultural interests throughout the United States, however, were skeptical. Phil Burnett, executive director of the NCC, predicted that Congress would continue to assess and consider a new program for 2002 and that it was too early to conclude that support for commodity programs would be lost forever.[28]

Subsidies did not end, however, because low cotton prices and unfavorable weather drove many growers to the edge of bankruptcy. The collapse of the Southeast Asia "Tiger economies" in 1997, which came to be known as the "Asian flu," cut exports for nearly all crops. Congress responded with emergency payments that reached $22 billion in 2000 for all farmers. The *New York Times* concluded that the hope for ending support payments had vanished and that the next farm bill would safely embed subsidies in the agricultural program.[29]

Subsidies continued with the 2002 farm bill, so the expectations that assistance might end came to naught. No longer was there anticipation that subsidies might stop at some point, because the new measure brought back target prices and introduced countercyclical payments (CCP) that gave the grower the difference between target and world prices. Congress meant for the CCPs to remove the need for the emergency payments that had become necessary from 1998 to 2001. Payment limitations were redefined with the maximum set at $360,000. To help make subsidy funds go only to bona fide growers, any landowner with income over $2.5 million would not be eligible unless 75 percent of the income came from agricultural production. Press reports of banks, insurance companies, and nonfarming landowners receiving large payments were behind this stipulation.[30]

Have subsidies been a friend or foe to cotton growers? Will Clayton had criticized price supports when they began in 1933 because they handicapped exports. Fleming, who had been tutored by Clayton and took over leadership of ACCO in 1940, took the same position. In 1957 Fleming published a pamphlet under the auspices of the company in which he set forth the rationale against subsidies between 1945 and 1970. Since small-plot agriculture remained widespread in the South, Fleming repeated the oft-heard point that government assistance kept inefficient farmers in business while they plowed exhausted soil. "Old cotton plots, handicapped by Nature for adaptation to mechanization or in other respects, have been kept in cotton," retarding the natural shift to western lands better suited for modern farming.[31] The activities of the CCC in selling cotton not redeemed by growers interfered with the business of private merchants and brokers. Fleming added that government supports put a floor under cotton prices and discouraged buying or selling futures. Perhaps the New York or New Orleans Exchanges would be forced out of business, he predicted, and when in 1964 the New Orleans Exchange closed its doors, his words proved prophetic. He pointed out that the export subsidies inaugurated in 1939 and used in various fashion in the 1950s violated the U.S. trade objective of preventing the use of subsidies or other devices to obtain an advantage in world markets. Other nations "are deeply concerned by the contrast between the U.S. profession of principle and the U.S. actions in drastic subsidization of our exports of cotton."[32]

Other nations see our actions, Fleming warned, as indifference to their well-being. His words were again prophetic in 2003 when the WTO discussions at Cancun broke down over this question. He also felt subsidies kept cotton prices at artificial highs and encouraged the manufacture of synthetic fibers. And though subsidies and cotton allotments were meant to protect small farmers, particularly southerners, "the brutal fact is that the revolution of mechanization has made many traditional cotton lands obsolete for cotton production."[33] Fleming believed society should help the struggling marginal farmers, but the help should come separately and distinctly apart from the general cotton problem.

Fleming had succinctly stated the objections to cotton subsidies for his era, but the question over the shift to western lands and the perpetuation of small, inefficient farms in the South lost importance with the passage of time. By 1970 small family farms were rare; cotton acreage had fallen drastically in the Southeast, and the West had fully established itself as a cotton growing area.

NEW ATTACKS ON SUBSIDIES

A new line of questioning appeared with the rise of environmentalism: the idea that subsidized farming led to wasteful use of natural resources for a crop in surplus. Water became one focal point for this objection, and the crescendo of unhappy voices rose with urban growth in the arid West. Critics often stated that growers grew their crop with subsidized water and sold it at subsidized prices, while urban interests paid much higher prices for water. Environmentalists anticipated that unless water could be distributed more evenly, urban growth might slow, and the agricultural use of water would come under even more pressure.[34]

That subsidized farming led to unnecessary use of insecticides and herbicides became another criticism. In the name of producing a crop dependent on subsidies and in abundant supply, went this charge, growers polluted soil, water, and air. A bloated program of supports for production, assistance with exports, help in fighting insects, and federally funded research projects represented a misuse of resources that should be devoted to protection of the environment. These charges effectively gained notice and forced congressional defenders of cotton to compromise with sponsors of environmental protection when designing various bills. In this respect, environmental concerns have accounted for the inclusion of the conservation reserve program and provisions for protecting wetlands since the mid-1970s.

A popular and accurate charge in the era before the 1970s accused growers of producing cotton to put "in the loan" rather than sell on the market. This referred to the practice of storing a year's crop with the CCC for a loan and later forfeiting the crop and keeping the money. When cotton prices were low, farmers followed this practice, but they were also responding to the USDA requirement that to remain eligible for assistance and not lose their allotments, growers must plant cotton. Congress eliminated this drawback with flexible planting contracts that did not jeopardize direct payments to landowners if they planted another crop. Lawmakers intended to encourage growers to respond to market opportunities rather than produce cotton for the CCC safety net. Nonrecourse loans remained available, but the introduction of planting flexibility contracts alleviated some of the problem.

Critics charged that subsidies inflated farmland prices. Community banks were more likely to loan to growers participating in the federal program and receiving assistance, so the value of land tended to remain strong if the owner retained the base allotment. Defenders of subsidies acknowledged this effect. Critics added that high land prices favored the commercial operators who could afford to expand, while smaller farmers were blocked out, resulting in more large-scale operations that qualified for bigger subsidies. From this perspective, the purpose of taxpayer support became distorted, helping the rich get richer.

Indeed, the sharpest attacks targeted the fact that larger farms received the larger subsidies, particularly in regard to nonfarming corporations. As reported by the *Atlanta Journal-Constitution*, the USDA program "props up big farmers at the expense of small growers both here and abroad."[35] In 2005 the newspaper reported that $23 billion went to agriculture and that cotton farmers accounted for $3.4 billion. "The government pays if farmers grow too much, or nothing at all." Allegedly, the program turned farmers, "once symbols of self-reliance—into government dependents." The newspaper's special study revealed that approximately half of the total subsidies for agriculture, $10.5 billion, went to 5 percent of the eligible landowners. Four percent of the eligible farmers in Georgia received half of the state's subsidies, averaging $200,900 per farmer, while the other 96 percent received $8,300 on average. The *Atlanta Journal-Constitution* cited one grower who marketed seed plantings but grew cotton to supplement his business with subsidies. "If you unplugged my USDA column," he told the newspaper, "I'd be way south of break-even." One cotton farmer stated: "Shoot, we're on welfare. We're no different that people driving up to the welfare center and getting food stamps."[36]

Interest zoomed in the subsidy program when U.S. cotton became the focal point of the dispute with Brazil and African countries. Suddenly growers found themselves having to justify the assistance. Press accounts and television documentaries attacked them as a wasteful and greedy self-interest. "No other crop is subsidized to such an outrageous degree," stated an editorial in the *New York Times*, "enriching so few at a cost so high to millions elsewhere."[37] The newspaper invoked the old association of cotton with the South, saying, "King Cotton, the evocation shorthand for the supremacy of cotton in southern culture, still ranks high among the hierarchy of Washington's power lobbies." Cotton subsidies, it continued, are "a potent source of anti-Americanism."[38]

THE DEFENSE OF SUBSIDIES

Most of the responsibility to defend subsidies falls on grower organizations, and cotton interests stand together on the subject. Joining them are the farm press, agricultural economists, members of Congress from farming areas, and assorted agricultural interests. After 2000 they faced a tough challenge, the most threatening since the inauguration of federal assistance in 1933. Interests such as pharmaceuticals want to foster trade, so they see agricultural subsidies as outdated obstacles that endanger negotiations. But cotton still has assets that keep it formidable.

Since the Cotton Belt spreads from the Atlantic to the Pacific, it has a greater range of constituents than in the older days, when cotton was associated with southerners walking behind mules and chopping weeds with a hoe. Now growers are businessmen, banded together in area, regional, and national organizations that know how to speak politically and make alliances. An overlooked feature accounting for the persuasiveness of cotton lobbyists is their knowledge and use of data. Armed with information supplied by the Cotton Economics Research Institute (CERI) at Texas Tech University or the staff economists with the NCC, cotton spokesmen challenge the accusations made against subsidies, and even if they do not completely block the efforts to remove supports, they manage to soften the attack and retain some benefits. The close contact between organizations and their members, evidenced by the NCC Field Service and the grower associations on the Texas Plains, means that cotton lobbyists come armed with firsthand information and a sense of authority based on facts.

In April 2007 the CERI published a study requested by the Southwest Council of Agribusiness on the use of agricultural subsidies in other countries. The study showed that input subsidies were commonly used by China, meaning that farmers obtained seeds, fertilizer, and electricity at subsidized rates amounting to $5.17 billion. In the United States, growers purchased these items at market prices, though they received assistance in marketing. In 2006 China's supports for general agriculture amounted to $43 billion, while Brazil and Pakistan had projected assistance of $26 billion and $2.7 billion respectively, and foreign nations made greater use of agricultural tariffs. When compared with other agricultural exporting countries, the United States, according to the CERI study, fell into the middle of the pack with price supports; but when input and indirect subsidies were taken into account, the United States fell further down.[39]

In the 1990s, almost two-thirds of a century after farm subsidies started, the justification for government assistance continued to rest on the fundamental premises that agriculture must be kept viable for the sake of the general welfare and that cotton is essential to well-being just like food. Since products using cottonseed oil were found in the family refrigerator, spokesmen liked to say that cotton was both food and fiber, though clothing accounted for its principal value. Other considerations include the importance of cotton for manufacturing smokeless gunpowder and other explosives. The United States must be self-sufficient in this respect and not have to import an ingredient so vital to national security. From the perspective of all segments in the industry, the justification was self-evident and rested on the common assumption that a viable agriculture, not exclusively cotton, is vital to the general welfare.

In a more direct sense, cotton impacted a significant portion of the U.S. economy. For 2003 the Congressional Research Service (CRS) reported cash receipts for lint and seed at $5.5 billion. Corn had $18.7 billion in receipts, soybeans had $15.7 billion, and wheat was $6.8 billion.[40] The United States continued to be the top exporter of cotton, accounting for about 42 percent of world exports of the fiber in 2003. Herein lay a principal justification for subsidies: cotton exports helped to counter our deficit trade balance. "A sharp reduction in commercial farm exports as a result of restricting government payments," wrote the executive vice president of the Arizona Cotton Growers Association, "could result in a disequilibrium with international consequences."[41] As reported by CRS, direct payments and countercyclical payments did not raise the price for cotton but provided income assistance to growers when market prices were low. A severe swing in price had occurred between 1995 and 2001, when cotton dropped from 75 cents per pound to 30 cents. Other crops had also suffered, but such a drop in cotton had not been seen since the early 1970s. With the help of the subsidy provisions put into effect with the farm bill of 1996, cotton growers weathered the storm.[42]

Without supports, cotton and other crops would undoubtedly have price swings. Political unrest around the world also affects market prices. Subsidies thus prevented price volatility and kept smaller or medium-sized growers in business and helped make the United States a "consistent supplier to international markets."[43] During the Cold War, the argument of self-sufficiency in agriculture had extraordinary meaning.

THE COMMISSION STUDY

The Farm Security and Rural Investment Act of 2002 mandated a study of agricultural subsidies and the impact of payment limitations. Conducted by the Commission on the Application of Payment Limitations for Agriculture, the study examined the long and short-term impact of subsidized agriculture and focused on all subsidized crops. It solicited public comments and held a public hearing in Washington in June 2003 just before filing its report in August. The commission summarized the common points made on behalf of subsidies.

It stated the oft-heard defense that government assistance maintained the "family farm and vitality of rural communities." By helping keep farms financially stable, government assistance lessens rural outmigration and reduces "pressure on social services in urban areas." With the requirement that landowners who enroll in the government program must not cultivate eroding land or endanger wetlands, the farm policy promoted protection of the environment. Since other countries subsidized agriculture through a variety of tactics, price supports in our economy "merely put U.S. farmers on a level playing field." Indeed, the countries of Europe furnished assistance to their farmers on a larger scale. The commission included a summary statement defining agriculture in broad terms, stating that it encourages tourism and recreation, promotes conservation, and fosters "economic activity in rural areas."[44]

The commission also saw drawbacks. Subsidies purportedly enriched large-scale operators and enabled them to further expand their holdings by buying smaller farms. In some cases, nonfarmers who nonetheless own land received payments, which had never been the intent of assistance programs. Subsidized agriculture further led to surplus production and lower prices, renewing the demand for supports and driving up government spending. The commission included the loudest complaint, that a larger proportion of payments went to the wealthiest growers.[45]

Cotton interests acknowledged these points but replied that growers in other countries similarly received assistance. China initiated price incentives from 15 to 50 percent and allowed above-quota premiums for cotton procurement. In another instance, China's communist government stimulated production with preferential loan rates and advanced payments to farmers before planting that made China's yield jump by 16 percent in 1993 and 1994.[46] Delta Council director Chip Morgan emphasized that U.S. growers were expected to operate in an open market whereas some of the world's largest producers of cotton, like China, operated in a planned economy.[47] In the 1990s, he added, the role of proprietary interests like Monsanto has been responsible for advancements for which American growers must compensate, but foreign interests have acquired the advances without paying for them. In other words, there was a loss of intellectual property, which the Mississippi grower Bruce Brumfield described as a parallel to drug companies that made U.S. consumers pay a higher price for drugs than in other countries. On the merit of paying growers to raise cotton, Californian Jack Stone believed that implement dealers and seed and fertilizer companies would be too severely impacted by the loss of cotton farming for Congress to let the industry become insignificant.[48] Earl Sears liked to recall the tax breaks given to the synthetic industry to build plants during World War II.[49]

The cotton community nonetheless acknowledges its dependence on government supports and would like to be able to operate without them, but grower interests expect payment limitations to continue and requirements for land conservation to remain. They anticipate the U.S. cotton policy will be geared to compete in the world.

THE TEXAS BLACKLAND STRIP

If the intent had been to preserve small farms, subsidies failed. The severe drop of 1.5 million cotton farms in 1945 to twenty-five thousand by 2003 is overpowering evidence that federal assistance did not accomplish that goal. If the purpose, however, was to keep cotton farming a viable part of agriculture and supply the United States with a commodity essential to the general welfare, the assistance programs proved successful. Where did the hundreds of thousands of small farmers and their families go, and why did they leave the land? The twenty-six-million-acre blackland strip in Texas, where fertile soil had attracted thousands of settlers after the Civil War, demonstrated why cotton farming became unattractive for the bulk of small farmers after World War II.

In 1948 Anderson, Clayton and Company's corporate magazine published an article titled "What Ails the Blacklands," which recalled the earlier years when the waxy black soil, called gumbo, produced cotton up to 1.5 inches in staple length. "Old timers said there was no limit to the richness of this soil."[50] Bountiful farms flourished with cotton as the main crop, while grain and feed crops supplemented. Small towns proliferated, and cities were established on the strip. On the northern end, Dallas became a major cotton trading center, and Houston the port from whence Texas cotton went to Europe and Asia. The abundant agriculture sustained towns and cities, but starting in the late 1940s and throughout the 1950s, cotton farming began to fizzle. Although farmers received subsidies, they could not generate enough income to stay on the land. The article explained that intensive farming had exhausted the soil until it produced yields equal to only half that of its glory days. Worn-out soil had long been known to drive men off the land, but other factors accounted for cotton's demise on the blacklands.

The pattern of landholding there evolved with the migration into Texas from 1880 to 1900, when the state experienced a large influx of new residents. Rich new land at reasonable prices attracted southerners who established small self-sufficient operations and raised a few bales of cotton for cash. Few commercial operations developed, so that once mechanization came after World War II, the numerous small acreages could not justify the investment in machinery. Farm workers had begun to disappear, too, which meant that capital investment would be necessary far beyond the capabilities of self-sufficient farmers. The federal allotment program reduced cotton acreages, which left many landowners with no opportunity for meaningful production. And once the declining soil fertility affected yields, farming on the blackland would require expansion with capital-intensive practices. Not to be overlooked was the impact of root rot, a plant disease that forced cotton off "several hundred thousand hectares."[51] Farmers on the blackland strip were not competitive with other cotton-growing regions, and with the devastating drought of the 1950s, cotton farmers sustained losses beyond the ability of subsidies to overcome. Texas had also begun to industrialize heavily, so small landowners and tenants alike moved into the mushrooming urban centers, where they could earn better livelihoods without the risk of losing a crop to storms or insects. "Blackland's farm people wiped the black, waxy clay from their feet," wrote a rural historian, "and headed for Waco, Dallas, Austin, and points beyond."[52]

CONCLUSION

Subsidies did not rescue farms with small acreages. For the general public, this proved disappointing and seemed incongruous with the purpose of assistance, but the economics of producing commodity crops required enough acreage to warrant investments in costly machinery and the new technologies associated with insect and weed control. Each advance increased costs, which meant that capital outlays far exceeded the investments necessary to grow cotton in former years. Opportunities to earn a livelihood in urban centers appealed to younger men and women, which had not been available to their fathers. They also enjoyed the social and cultural attractions in the city. In the swirling mass of human goals and economic realities, cotton farming logically declined in the central zone of the Lone Star State; rural sociologists had long insisted that a surplus population lived in the older cotton-growing areas. From this perspective, the subsidy program since 1945 should not be criticized for failing to preserve an outmoded way of life.

How should the United States view subsidies if they failed to keep the mass of cotton farmers in operation? During lean years, subsidies provided the margin for success, but in fat years growers fared better through the open market. The supplemental income from the U.S. government ensured a steady supply of a commodity vital to the United States, though a large portion of the annual crop increasingly went overseas. Generally the Cotton Belt produced about fifteen to eighteen million bales annually, and the United States consumed the same amount, but imported clothing accounted for approximately 60 to 75 percent of the consumption, owing partly to the decline of textile manufacturing in the United States. Because more U.S. cotton went into international markets, growers were pitted against developing countries, which produced charges that Americans engaged in unfair practices. Joining this global battle were artificial fibers, a threat that cotton alone must bear. Subsidies kept cotton viable under these pressures, and with viability came jobs in rural towns and taxes for schools. Ideally subsidies would not be necessary, but they were so widespread in cotton and other segments of agriculture that even a gradual and equitable reduction would likely bring adjustment and hardship to a large number of people.

The question of agricultural subsidies had a new dimension when in 2008 the U.S. Treasury and Federal Reserve System extended loans and buyouts to some of America's largest financial institutions, even agreeing to absorb some of their losses. A sharp ideological issue surged to the forefront: should the U.S. government attempt to rescue private banks and investment firms with public funds? Congress provided $700 billion in the first rescue plan, or bailout, to world-renowned Wall Street firms and central banks. Justification for this action, which prompted emotional outbursts of criticism and protests, was based on the grounds that by the sheer size of their operations ("too big to fail"), recipients of the aid should not collapse because severe repercussions might occur throughout the economy. For the sake of employment, retirement incomes—indeed, the solvency of the country's financial system—government supports were deemed necessary. Estimates of the expected total cost to taxpayers for rescuing all the troubled private firms ran over $1 trillion. In this context, federal assistance to cotton growers seemed more justifiable and explained the comment by a Mississippi grower, "There's no perfect solution."[53]

CHAPTER 14

CROP LIEN TO FUTURES
Financing Cotton

The impressive body of Depression literature on southern agriculture focuses heavily on social injustice. Writers attributed the crushing poverty in the cotton-growing areas to a culture embedded with racism, narrow social attitudes and outrageous behavior, and outdated farming practices. A topic that frequently appeared in the literature was the system for financing farm operations known as the crop lien. This referred to the practice of creditors, sometimes known as "factors," making loans to landowners who then lent money to their sharecroppers and tenants. Interest rates were reputedly unfair, but the most damaging feature was the creditors' insistence that the loans be used to raise a cash crop, cotton. Such practice left no opportunity to escape one-crop farming and the numerous ills associated with it. In his classic *Human Factors in Cotton Culture*, Rupert Vance likened the crop lien to "legalized robbery" and quoted an author who felt that borrowers "have been but the prey of sharks and harpies, bent upon keeping them in a state scarcely better than that of slavery."[1] Delta native Frank Smith recognized that cotton farming did not generate enough income for small landowners and that the gap between the poor and prosperous "became acute when a major portion of the wealth was drained off through the credit system."[2] Over and over, Depression analysts blamed credit practices for the social and economic woes of the cotton kingdom.

The lack of capital had plagued the South as far back as the colonial era, making the region dependent on bankers of the North before and after the Civil War. Numerous factors accounted for the South's slower rate of economic development when the northern states became the industrial heartland of the United States, but the shortage of capital in Dixie stood out. Money flowed into the North, giving rise to the southerners' claim that they had a colonial economy. In 1938 the National Emergency Council devoted one section of its famous report, *Economic Conditions of the South*, to the shortage of capital and the nefarious crop lien.[3] By World War II, progress in industrial development had been made, but cotton farming, with its reliance on mules and hand labor, had changed little from its nineteenth-century methods and customs. The first steps toward improving the sources of credit came when the Roosevelt administration of the 1930s recognized the need to overhaul the system of financing agriculture and initiated several steps to attack the problem. At the end of World War II, when Oscar Johnston and his colleagues set out on their crusade to save the cotton industry, the practices of furnishing and acquiring credit were in a state of flux that incorporated both the old and new.

FEDERAL CREDIT

In 1916 Congress passed the Federal Farm Loan Act, which established twelve regional banks operated as loan cooperatives under the direction of the Federal Farm Loan Board. These banks provided only mortgage loans at a relatively low rate of interest, intended to encourage farm ownership. Mortgage loans did not cover production expenditures or provide assistance for emergencies such as drought, floods, and the like. Though commercial banks played a major role in cotton financing, they shied away from short-term loans, particularly for nonlandowners. Further legislation came in the 1920s such as the establishment of intermediate credit banks, but federal credit for emergency purposes amounted to little. When the Depression hit, circumstances on the farms worsened, and the need for federal assistance grew.[4]

In the cotton culture, the creditor, or factor, stood out. Factors might be equipment dealers, storekeepers, or private investors who provided the smaller loans for year-to-year expenses and put a lien on the crop. In the case of planters, the money trickled down to sharecroppers through the old fifty-fifty share system. For the thousands of sharecroppers, characteristically living in strenuous circumstances, credit meant an account at the plantation commissary or a general store backed by the landlord. In some instances, factors made mortgage loans to landowners, but the infamous crop lien nearly always dealt with loans for planting, harvesting, fertilizing, equipment, and other production activities. After World War I, when farming became increasingly unable to keep pace with the rising consumer economy, particularly for the small family farm, lenders became more insistent on protecting themselves with a lien on a crop that could readily be sold for cash, and that meant cotton.

During the 1930s, important changes came in federal credit assistance that impacted the era after 1945. For one thing, the thirteen Banks for

Cooperatives were established with the Farm Credit Act of 1933, which provided funds to be made specifically to agricultural cooperatives. Borrowers, or cooperatives owned by farmers, had to subscribe to the stock of the bank where they placed their loan, at a minimum of $100 for each $2,000 requested for operating expenses and $100 for each $10,000 in commodity loans.[5]

A dimension of public lending to farmers ended just when the cotton industry began to make itself viable. It was the well-known effort by the New Deal to extend assistance to the tenant class through the FSA that stopped in 1946, a move that terminated credit via federal channels for those at the bottom of the cotton ladder. But the tenant class had begun its migration off cotton farms and into cities, and "America's peasants," sharecroppers white and black, wanted to escape their environment rather than attempt to improve their operations with assistance. Thus the experiment at supervising and funding cotton's downtrodden ended in 1946, which meant that future debt relief and credit assistance would be aimed at the more stable landowners.[6]

With the end of the FSA, tenants and operators of small family plots had little resource for credit. Most of them fell into the category of high-risk borrowers for lack of collateral or their lack of volume to warrant loans. Both public and private lenders knew, too, that many tenants and owners of small plots were locked into outdated farming practices, had too little equipment, and generally lacked the resources to be more than a risky client. Certainly commercial bankers avoided such borrowers, and even the public lending agencies saw them as a gamble. The Cotton Belt had the greatest proportion of such cases at the end of the war owing to the region's higher rate of tenancy and the patches of cotton too small to remain viable. Like all farmers, the poorest faced the squeeze for more capital as mechanization became the answer for the loss of labor. Their mules and plows, though low cost, could not keep pace with machines, so that as production costs rose and the profit margin narrowed, volume production was not available to the small farmer. Creditors clearly understood the dilemma and agreed with the assessment made at the Cotton Conference of 1944 that small-scale operators would find it difficult to survive. Requirements for capital only heightened as technology advanced, and the natural vagaries of weather further reduced the chances for survival. "The earning prospects of these farmers even in good years," wrote one analyst, "are not such as to make them good credit risks for ordinary lending agencies."[7] This underling class disappeared between 1945 and 1970, and credit became a resource for the more established and solvent growers.

In 1948 Chester Davis, formerly head of the Agricultural Adjustment Administration in the 1930s, spoke to the Cotton Mechanization Conference held in Lubbock when he was president of the Federal Reserve Bank in St.

Louis. He saw mechanization as a fast-approaching reality for cotton farming and predicted that small-plot growers would continue farming by mule and hand because of the cost of machinery but would not likely survive because of their inability to compete with the exploding number of larger operators in the Cotton Belt. Davis also saw larger operators incorporating machinery to offset the shortage of labor. For him, the question was how to furnish capital for the mechanization of the Cotton Belt.

Davis reported no shortage of capital in the South as had been the case in the past. He saw a five-year transition from mules to full mechanization as an average for landowners, so they were expected to mechanize their operations on a piecemeal basis rather than in one quick transition, which would ease their requests for capital and the ability of the nation's credit system to furnish it. Much of the process "will be accomplished along conventional farm machinery credit procedures without encountering any serious credit problems as such."[8] Terms for repayment, he continued, would likely fall "between the conventional intermediate type of loan and the long-term mortgage loan." For machinery chattel loans, three to seven years should be expected, but credit for installing irrigation equipment and field leveling might extend to ten years or more and be serviced as mortgage loans. The practice of implement manufacturers furnishing credit to promote sales had already begun to end. "The manufacturers have virtually gone out of the finance business, and I believe that commercial banks, production credit associations and others have sufficient experience and willingness to carry the load from here on."[9] Davis encouraged mechanization. "This trend is going to continue; it is inevitable. It means better homes and a better life for those who remain on farms." He had no reservations about the supply of capital.

ANDERSON, CLAYTON AND COMPANY

An unusual feature of capital expansion that spurred the movement of cotton into the West came through ACCO that Under Secretary of State Will Clayton had founded with his brother-in-law Frank E. Anderson and Frank's brother, Monroe, in 1904. They started in Oklahoma City before Oklahoma became a state. At that point the brokerage firm dealt only in merchandising, mostly in Oklahoma and part of Texas. It sold cotton to mills in Europe, and within one year the firm leased several gins and built an oil mill in Elk City, Oklahoma. The company supplied the British and European markets in Liverpool, Bremen, Le Havre, and Milan, though it conducted sales with east European countries on a lesser scale. But the British, German, and French buyers dominated the global cotton trade with their established textile plants, large warehouses, and great resources of capital and lines of credit. World War I, however, ended Europe's dominance of cotton marketing.

Clayton and his partners saw new opportunities and moved the offices to the port city of Houston in 1916. They proceeded to expand their operations by building storage warehouses and compresses there, but more importantly for growers, the company began purchasing and building gins to service them. Further expansion came in the next twenty years with oil mills and more gins, and a seed-breeding operation in Abilene, Texas. During the interwar years the company expanded into Mexico, Brazil, Argentina, Paraguay, Peru, and Egypt and opened trading offices in Japan and Europe. Clayton detested the New Deal price support program because it kept the domestic price of cotton above the world price, which had prompted the company to open operations overseas. Before World War II, however, it continued to build gins, oil mills, compresses, and warehouses in Texas, New Mexico, Arizona, and California. By 1945 ACCO had become the largest private cotton organization in the world and was poised to further the expansion of the cotton kingdom into the West.[10]

Because Clayton and company president Lamar Fleming pushed for the westward expansion of cotton, they invested resources there after 1945. For young farmers in west Texas and the far western states, ACCO furnished crop financing but left mortgaging to the Federal Land Banks. During the era of the greatest westward expansion, 1945 to 1960, ACCO furnished credit at various times to approximately one-third of the growers, particularly new landowners who had installed irrigation. In the late 1950s, the company reported that it had about $75 million in crop production loans during an average year.[11] Farmers used the funds to buy seed, equipment, fertilizer, and to pay labor and cover energy cost for irrigation pumping. Growers' contact with ACCO was the gin manager who arranged the contracts and often furnished management advice. ACCO did not try to profit from the loans but did require the grower to use the company's gins. Loans to farmers were extended through stages from planting to picking rather than a lump sum at the beginning of the season. Crop financing by ACCO was responsible for the transition of much raw desert land into cotton fields, followed by more farms and population growth, particularly in portions of Arizona, California, and the Texas Plains. Financing by ACCO was also responsible for transforming the west coast of Mexico into a cotton growing area.

An intriguing example of ACCO's capitalization of the westward expansion of cotton farming occurred in Arizona's southeastern corner, the Sulphur Springs Valley. It was here that the famed Apache leader Cochise kept his headquarters in the Dragoon Mountains on the western side of the valley. Only a short distance away lies the historic town of Tombstone. It was an area where blood ran in the old West. Because the region received only about eleven inches of rain per year, the earliest settlers had dug shallow wells and irrigated small patches of alfalfa and feed grains for livestock. Once the technology of drilling arrived, well depths went to six hundred feet to find plentiful water, and newcomers from New Mexico, Oklahoma, Texas, and even a group from Kansas went there to plant cotton. In 1947 the first cotton gin was erected in the valley, which "gave new life to cotton hopes," and deeper wells over one thousand feet provided large amounts of water. Cotton acreage expanded rapidly from three thousand acres in 1948 to eighteen thousand by 1952, and the acreage was visible from the Cochise stronghold. But the southern end of the valley offered the chance for further clearing of land for fields. "New land," reported the *ACCO Press* that year, "is being cleared so fast that many persons predict it will be the garden spot of the valley in future years." Forecasters predicted up to seventy-five thousand acres of cotton would be planted.[12] Most of the cotton grown was Acala 1517, and with little Mexican labor available, approximately 80 percent of the crop was gathered with mechanical pickers. Aphids, thrips, lygus bugs, fleahoppers, and bollworms were worrisome, but not overpowering.

Anderson, Clayton and Company financed the growers in the valley. In San Simon the company built a new gin, and another at the Kansas Settlement near the Cochise stronghold. Wells, irrigation piping, land-clearing equipment, and land-leveling machinery received financing from the company, and growers obtained funding for cultivation expenses, seed, and other items as well. The growth of this locale occurred in the midst of the westward expansion of cotton, and with new landowners putting virgin land under the plow and receiving strong prices, a sense of the El Dorado spirit was visible. So attractive did cotton appear that a group of Kansas farmers moved there and planted cotton even though they knew nothing about cotton farming. It was this group that established the Kansas Settlement underneath the Apache headquarters.[13]

For ACCO the expansion of cotton farming meant good business. Profits came from merchandising the cotton and also from ginning, crushing, warehousing, and even shipping. At the end of World War II, the company was flush with cash because it went public in 1945 when the estate of Monroe Anderson sold his shares to build the M. D. Anderson Cancer Hospital and Research Center in Houston, a philanthropic move that produced much goodwill for the company. Monroe had died in 1939, and the proceeds from his stock were used to buy land for present site of the M. D. Anderson Center. With the readily available capital, ACCO built thirty-six more gins in Arizona, fifty in west Texas, thirty-five in California, and four in New Mexico. It also had fifteen oil mills in the same states. It operated seed farms in Aiken,

Texas, and Iguala, Mexico, and its Paymaster Feed mills in Abilene and Dallas. In 1949 it opened a fiber-testing laboratory in Houston and started a small textile mill, mostly for testing purposes, that made baby blankets. In 1951 ACCO built an oil-seed-crushing mill in Lubbock, the world's largest. To accommodate the cotton it bought for shipping overseas, ACCO built a massive thirty-two-acre warehousing area in the Houston shipping yard and started its own shipping line, States Marine.[14]

This impressive infrastructure, along with its merchandising operations, made ACCO the giant of the industry. Thanks to its size and diversity around the globe, the company was a creditor whose solvency was ensured, and its enthusiasm, unquestionably self-serving, for the expansion of cotton contributed to the westward movement of the Cotton Belt. The food products division became more lucrative than cotton by the late 1950s, and the profits from foodstuffs backed the gins and mills that were starting to become only marginally profitable. In its 1959 annual report, the company reported that the traditional cotton-merchandising operations had losses that were sustained by its foreign operations and food business. It further reported that ACCO had nonetheless "kept its oil milling and ginning business in the United States in step with the Western migration of cotton production."[15]

By the early 1960s, signs of trouble began to appear. Simply put, ACCO's merchandising business, the backbone of the company, generated relatively little profit. As early as 1958, the executive leadership claimed that the U.S. farm program of price supports "had rendered wholesale raw cotton merchandising on a large-scale far too risky for the low profitable obtainable," so wrote S. M. McAshan, Clayton's son-in-law and head of the firm starting in 1959. The two-price system for cotton wreaked havoc on U.S. textile mills, since it put American cotton at a disadvantage in the world market. For large-scale merchandisers, this left no adequate hedging medium in marketing cotton. "Cotton futures . . . no longer followed normal supply and demand," McAshan wrote. Orthodox hedging, vital to cotton brokers, had been rendered useless and had become "as risky as not hedging at all." The company felt it could no longer "perform its function of marketing huge volumes of raw cotton from the farmer to the mills of the world." In other words, the old business of buying and selling cotton was no longer viable for operators as large as ACCO.[16]

McAshan did not take ACCO in a new direction as much as he concentrated resources into the more-profitable food business. By 1961 ACCO's bankers complained that its large investments in cotton yielded relatively small returns. That year the company arranged for the consulting firm Booz, Allen and Hamilton (BAH) to study the company's operations from the perspective that ACCO's cotton business was not sacrosanct.

The consultants saw the company's cotton business as a poor investment because the profit margins were too low. Several factors accounted for this condition, with the emergence of farmer-owned marketing cooperatives at the top. Cooperatives such as PCOM in Lubbock or CALCOT in California had seriously eroded ACCO's profitability in merchandising, ginning, oil milling, and warehousing. World cotton consumption was growing at a snail's pace, too, as synthetics accounted for much of the increase in fiber use. About half of ACCO's gins had shown no profit for the previous year, and the company's vast warehouses would be more profitable, the report continued, if used for something other than cotton. The report recommended the elimination of ACCO's baby blanket operation in Houston. Seed breeding, however, looked promising because of the small amount of capital needed and the expected growth in the commercial production of seeds. For the decline of profitable merchandising, the consultants blamed the federal cotton program. To remain viable, ACCO should eliminate any gin, oil mill, or cotton buying office not returning a profit. Even the ACCO commercial farm, Vista del Llano near Fresno, demonstrated the lack of profit opportunity in the cotton business. Competitive pressures demanded the company to get more efficiency from its cotton operations.[17]

These observations and recommendations were reaffirmed by another consulting firm contracted in 1961, though it put stronger emphasis on cooperatives as a new force taking over the functions of the company. Marplan, a division of Communications Affiliates Inc., conducted a survey and concluded that ACCO's declining share of the cotton market had been due to the cooperative movement. To an extent, the consultants added, crop and capital financing by banks had been a factor, but more and more farmers were taking their business to cooperative gins and mills, and therein lay the serious challenge.

The Marplan survey provided a view of attitudes among western growers on crop loans that generally came from the following: ACCO gins, independent gins, community banks, and the Production Credit Association. To begin with, Marplan noted that co-op gins were becoming more common and replacing the role of ACCO gins in financing. Farmers stated that "ACCO put us in business" and "they have always given us anything we wanted," or "ACCO pulled us through a difficult crop period." Growers liked banking with an organization that understood cotton farming, but the consultants warned that some old-line ACCO borrowers had moved to community banks for several reasons. For one thing, financing through a bank was viewed as a status symbol. It demonstrated a good credit rating, since banks were more conservative and less likely to loan to a marginal grower. A bank loan also freed the farmer from having to use an ACCO gin. The banks further released the loan in full amount at once rather than the ACCO practice of forwarding loans in stages during the growing season. But ACCO gin loans nonetheless had appeal because some farmers did not feel comfortable in banks and preferred the familiarity of gin surroundings. Some growers feared that banks would foreclose during a bad year, and thought ACCO better understood the risky nature of farming. "It gives a fellow a pretty good feeling to walk in a place and come out with $30,000 and not even have to sign a paper," reported a longtime ACCO client. "Some of the other outfits want you to bring in the old lady, the sewing machine and even a setting hen to cinch the deal."[18]

The consultants noted that ACCO's overall record of financing growers was admirable, and despite the evidence that some borrowers had moved to community banks, there was no rush away from ACCO. "They'll give you loans on other crops and livestock," was a common reply. "They've even loaned me money to improve and build buildings on my farm," said another.[19]

Despite ACCO's record in extending credit to growers, however, it continued to encounter a new and stiff competitor that was encroaching into its markets: cooperatives. The Marplan study considered them "the major factor contributing to Anderson-Clayton's declining share." Too many farmers were taking their business to cooperative gins, oil mills, and warehouses, so ACCO's return on investment fell. This practice occurred across the West, with the Bakersfield area being the most prominent. Cooperatives were newer in Arizona and had less impact, but they were expected to take more business from the company. This condition, recognized in the late 1950s, put financial pressure on ACCO, and when coupled with the loss of hedging, the Texas giant started to feel pain. When the New Orleans Cotton Exchange permanently closed its doors in 1964, it came as a shock to the cotton community, and Clayton blamed the government price support program. "High price supports decreed by government have placed a floor under cotton's price and the surplus induced by the high price has put a ceiling over the price so close to the floor that there is no room for trading."[20] The company's 1962 annual report identified cooperatives and price supports as threats to its operations. For these reasons, the annual report of 1964 raised the specter "to shift the traditional but no longer adequately profitable basis of the business to other activities."[21] For growers, this development meant the credit made available by ACCO might soon disappear.

Pressure continued to mount on ACCO's revenue stream. It was during this period that the company fought with Lubbock's PCOM for the oil-crushing business on the Texas Plains, a significant development, since the huge ACCO oil mill there generated dollars for making crop production loans. In 1964 longtime president Fleming, who had been so influential in the

NCC, died unexpectedly. In 1965 the company sold warehouses and its part in transoceanic cargo ships, and some vice presidents in the cotton departments voluntarily took early retirement. In 1966 Clayton died; he had seen ACCO change from a raw commodity merchandiser to a food-processing and marketing operation. Clayton had fought cooperatives, trying to convince friends in Congress that they had an unfair advantage. To Speaker of the House Rayburn, he urged that cooperatives pay taxes like any other business.[22]

Company president McAshan continued to dispose of unprofitable or marginally profitable cotton holdings, and the organization increasingly depended on foods. In 1971 it sold its facilities in Peru and Columbia, and in 1972 the nonagricultural sector accounted for 75 percent of the earnings. In 1973 ACCO formally withdrew from cotton merchandising and liquidated its remaining gins, mills, and other cotton services, and with the influx of cash, ACCO acquired small cheese companies. In 1976 McAshan retired as chairman of the board. The last gasp came in 1984, when Quaker Oats acquired the foods division, and America's best-known old-line cotton company ceased to exist.[23] No lack of credit for growers occurred with ACCO's demise, since by the 1970s most of them had established enough stability to find other sources.

PRODUCERS OIL COMPANY

Producers Oil Company in California also made loans to growers in California, Arizona, and New Mexico but did not operate on the Texas Plains. The company resembled ACCO in that it owned and operated gins, oil mills, warehouses, and compresses. It also operated as merchandiser. The company started in 1930, and Harry Baker joined it as a gin manager. He became head of the organization and after World War II took the same position as ACCO toward helping new cotton producers. "When the growers are prosperous and successful, so are the other segments," Baker stated.[24] He had worked for ACCO for two years before joining Producers Oil Company and partly attributed his view toward growers from that experience. So after World War II, Producers Oil became a major supplier of funds to the new cotton growers in California and Arizona. In one sense, it served as an intermediary for the Bank of America, which furnished much of the capital for the expansion into the West. The branch of the Bank of America in Fresno was known as the "cotton bank." Jack Stone recalled that as a young man he borrowed from Producers Oil along with other upstart growers.[25] From 1945 to 1960, these two companies closely resembled each other in their multifaceted operations and furnished capital for much of the westward expansion, operating over half the gins in postwar America. "It was cotton financing that developed the San Joaquin Valley," one grower remembered. "When the banks wouldn't loan a red cent, the gins let us have the money we needed at 5 or 6 percent."[26]

The exit of the Texas giant along with the deaths of Clayton and Fleming marked the end of an era. After World War II, a new generation of growers felt more kindly toward cooperatives, which cut into the profits of the old-line merchandisers.

PRODUCTION CREDIT ASSOCIATIONS

After World War II, the attitude of farmers toward credit changed. Previously debt had carried a stigma and a forewarning of disaster, but the adoption of business management practices now made credit and debt acceptable. "The rate of progress in the revolution taking place in agricultural production and the accompanying increased use of agricultural credit," reported the Farm Credit District of Wichita, Kansas, "has been particularly rapid."[27] While funds came from private organizations or cooperatives, federal funding served as the foundation of farm credit, and the Production Credit Association (PCA) ranked high in this respect. Each PCA was charted under the Farm Credit Act of 1933. These organizations amounted to small credit institutions, since they were organized locally under the rules of the Farm Credit Administration. Local farmers act as the PCA board of directors, which usually employed a secretary-treasurer to handle business. Each PCA served several counties and tried not to overlap the jurisdiction of another. The capital was furnished by the federal government and also came from the farmers participating or borrowing from the PCA. To obtain loans, borrowers had to purchase a minimum amount of stock in the PCA, so that hopefully it would become a member-owned operation. By 1960, 456 of the total 494 PCAs in the United States were owned by members. This system of credit extended throughout the country and applied equally to producers of all crops. Cotton growers used PCA credit extensively.

Loans from a PCA could be used for any agricultural purpose: expenses for cultivation, livestock, and even living expenses. Most of the notes were secured by a first mortgage on chattel property, but borrowers had to show management skills and demonstrate the ability to repay the loan. Most PCA loans were for one year, but they could be renewed for a maximum of five years. Once accepted, the PCA forwarded the loan to the Federal Intermediate Credit Bank, and if accepted there, the money was transferred to the PCA. Interest rates followed market trends, and service fees were attached to the loan to cover operating expenses, but PCAs strove to charge only minimum fees. More than anything, the PCA short-term loans ended the dependence of growers on "furnishes" and the crop-lien system in the southern cotton-growing areas. "Most of the Production Credit Associations in the Yazoo basin," wrote a Delta native as early as 1954, "have paid back the original capital advanced by government and are today entirely farmer-owned and operated, all within a period of less than twenty years."[28]

COOPERATIVES

Cooperatives gained strength as they took a role in funding operations. They first tended to concentrate on building new gins rather than making loans, though before World War II, the California Cotton Cooperative Association, commonly known as CALCOT, had provided its members with production loans with capital that came from the federal Bank for Cooperatives, most of them for less than $3,000.[29] With the postwar boom in cotton, CALCOT aggressively began loaning funds for the construction of cooperative gins, getting twenty-five such gins into operation during the 1950s. This venture proved to be a sound decision, according to CALCOT's president G. L. Seitz, who remembered that all gin loans were repaid with interest. The organization built warehouses to store members' cotton until prices rose.[30]

In 1921 the Staple Cotton Cooperative Association, or Staplcotn, was established by Mississippi growers who wanted to market their own cotton. Two years later, the directors of the cooperative set up the Staple Cotton Discount Corporation because they saw a compelling need for agricultural credit within the state. It was the federal Rural Credits Act of 1923 that opened the door for co-ops to finances growers, and the directors began making loans for crop production expenses, farm equipment, gins and warehouses, and other agricultural undertakings. Staplcotn reported that its rate of interest averaged below the New York prime rate and that all borrowers received the same rate. Like any cooperative, the Discount Corporation returned dividends to its members from revenue earned during the year.[31]

COMMERCIAL BANKS

Despite the credit made available through semigovernment entities such as the PCAs, cotton farmers depended heavily on commercial banks. This meant that credit came simultaneously from several sources: merchants such as ACCO, PCAs, and banks. Banks had a highly important role in extending agricultural credit for several reasons: they were more numerous and convenient, they provided quick service with a minimum of red tape, they provided checking and savings accounts, and they provided counseling on financial affairs and farm management. Private banking further offered the advantage of being readily available.

Though the systems by which PCAs obtained their funds might mean little to their borrowers, commercial and investment banks along with private investors were the ultimate source. PCAs obtained money from the Federal Intermediary Credit Banks, which sold securities or debentures to large commercial banks. When an investment bank's reserves were high, it tended to buy federal securities, thus funding the federal agricultural credit system. Since the assets of commercial banks grow through the routine custom of deposits, bank loans, and further deposits, they increase the money supply, which meant they handled much of the financing of cotton's expansion into the West. Like all farmers, growers relied on a variety of sources for funding, perhaps switching from season to season among them for the sake of convenience or to obtain lower interest rates.[32]

The federal credit system fared well between 1945 and 1970, so well that by 1968 farm loans had been repaid to the PCAs and Intermediate Credit Banks. This development came partly because of the decline of small-plot farmers, many of whom left for urban employment, leaving farming in the hands of the commercial or sizable family operations. Being more solvent and a better credit risk, these larger growers met their obligations and accounted for the strong position of federal lending agencies. Because of the healthy nature of farm credit, and to anticipate future needs, the USDA established the National Service Commission on Agricultural Credit in 1969, and its recommendations led to the Farm Credit Act of 1971.

This measure amounted to administrative streamlining with a minimum effect on borrowers. It consolidated the PCAs from approximately one thousand to about two hundred across the United States and combined the Federal Land Banks and Intermediate Credit Banks. Known thereafter as the Farm Credit System (FCS) with the new Farm Credit Administration (FCA), it received no federal funding but obtained capital through sales of securities.

THE DEBT CRISIS

The 1980s brought a debt crisis to American agriculture, the worst since the Depression. Many farmers bought more land in the 1970s when commodity prices improved; farm debt tripled. Cotton prices jumped from an average of

28.6 cents per pound in the 1960s to 45.3 cents in the 1970s. Inflation soared, however, reaching over 6 percent per year from 1970 to 1978 and hitting double digits from 1979 to 1981. Economic growth slowed, and "stagflation" was the popular term used to describe the situation. The oil embargo of 1973 to 1974 impacted the economy, but rising competition from the Japanese and European economies also took its toll. A new term, the "Rust Belt," referred to the decline of the U.S. industrial heartland, and the government bailout of Chrysler Corporation symbolized the malaise of the decade. Economists differed on their favorite culprit: the oil embargo, monetary policy, or federal spending. The stage was set for the agricultural crisis of the 1980s.[33]

Cotton prices rose to an average of 60.2 cents for the 1980s, but many growers of cotton, as was the case with wheat and other crops, had borrowed in the previous decade for more land and equipment to take advantage of the strong prices. In the 1980s, the ratio of agricultural debt to assets was about 28 percent, and farmers throughout the country could not service their debt. Weather worsened the agricultural plight with a severe drought in the Southeast during 1986 that hurt cotton, corn, and other commodities. Farm land fell in value. Poultry growers lost thousands of chickens to the summer heat. "This last drought," stated a Georgia farmer, "it's a catastrophe."[34]

Unrest had been growing in the farm belt. In 1977 the American Agricultural Movement (AAM) began in Springfield, Colorado, and spread rapidly across the Great Plains and the South. The new organization felt that the new farm bill would not provide enough price guarantees. The AAM called for a farmers' strike, and in December held "tractorcades" by driving their tractors to state capitals as a form of protest. The next month, January 1978, approximately fifty thousand farmers from around the country converged on Washington and marched down Pennsylvania Avenue. Some had taken their tractors to the capital city. Congress passed some legislation on farmers' behalf, but in February 1979 another tractorcade converged on Washington. In this instance, the farmers angered Washingtonians, and the police arrested a handful of protesters and impounded some tractors. But a heavy snowfall hit the city, paralyzing traffic, and suddenly the unpopular farmers with their tractors were pulling cars and buses out of ditches, carrying medical personnel to hospitals, assisting the fire department and police. According to one report, AAM wives cooked in hospital kitchens because the regular personnel could not get to work. After a few weeks, the AAM farmers went home. They established an office in the city and continued to present their case to Congress over agricultural matters.[35]

The tractorcade drew attention to the debt crisis, of which cotton growers were a part. In Texas most of the dissidents came from the Plains, but AAM enthusiasts were found also in the Southeast. Cotton growers in the western states had the least enthusiasm, likely because of the greater proportion of large-scale owners there. Reserve funds for the Farm Credit System dwindled.

By the mid-1980s, the FCS faced insolvency unless Congress pumped money into it, and through the Farm Credit Act of 1987, the FCS was reorganized with a $4 billion backing. Improvements in commodity prices helped, though the average price of cotton for the 1990s fared only 2 cents per pound better than the 1980s. Still, the debt load fell in agriculture, and the FCS did not have to tap into the government guarantee. For the Farmers Home Administration (FmHA), which loaned to beginners or less-commercialized farmers, there was a $14 billion write-off from 1987 to 1992 that came at the expense of taxpayers. By contrast the loss for loans made through the Intermediate Banks and Banks for Cooperatives amounted to approximately $1 billion.[36]

Some conflict and disagreement broke out between the NCC and the AAM. Their styles of operations were vastly different because the NCC took a deliberate and methodical approach to advance cotton's interests. It relied on gathering data and information on each issue and presenting its case as informed lobbyists, purposely avoiding confrontations. It sought to gain favors through cooperation and by winning friends in political circles. The militant, confrontational style of the AAM was unproductive from NCC's perspective. For the members of AAM, however, the loss of homes and farms, which had indeed occurred, could not wait on prolonged deliberations and the promise of relief at a future point. For them the desperate situation required desperate actions, so the tractorcades seemed the only means to goad Congress into action.

AAM members attended the NCC's 1978 annual meeting, held in Houston. Jack Stone, president of the NCC, ordered that they be treated courteously, but he would not allow any disruption of business. No disruptions erupted, but AAM members presented their case during floor discussions. More serious action came, however, when some cotton gins on the Texas Plains defected from the NCC, and the staff of the Field Service faced heckling at meetings in the west Texas area. The NCC's historian Albert Russell reported that the AAM targeted Cotton Incorporated and caused a reduction in its funding for research and promotion. By the early 1980s, the AAM movement lost much energy, however, stopped its quarrels with the NCC, and adopted the tactics of lobbying Congress and the USDA. This new organization nonetheless remained independent as it pursued farm legislation that would have government purchases of crops and foodstuffs made at 100 percent parity, and it wished to prohibit the importation of any farm product selling below 100 percent parity in the U.S. domestic market.

NEW LINES OF CREDIT

Another new development for cotton growers since World War II involved the marketing of their crop, which did much to help them exert better control over their sales. The days are gone when a farmer sold his crop at the gin for the going price and hoped to have enough profit to make ends meet for another year. Several new ways to sell emerged, ranging from the popular forward contracting to playing the futures market. In 1976 Mississippi growers sold 67 percent of the state's crop through forward contracting. The frequency of using this method of selling varies, however, depending on market conditions, because the next year the percentage sold in Mississippi through forward contracting fell to 27 percent. That year there was a declining market with a fall over 20 cents per pound from March to December. Farmers avoid selling in a bearish market.

Forward contracting rose in popularity in the 1960s and 1970s, and one of the most active merchandisers in this respect was Dunavant Enterprises in Memphis. Contracts between grower and buyer varied, but the common "hog round" agreement specified that the buyer would take all cotton from a grower despite grade and staple. The contract would allow discounts for late delivery and excessive trash in cotton. Another variation was the "modified hog round," which contracted discounts for cotton with micronaire (fiber fineness and maturity) at 3.2 or lower. And some buyers like Dunavant offered the "escalator clause" to growers known for producing premium grade: the buyer offers a minimum price with an agreement to pay half the rise in price on the delivery date. Another method used by members of cooperatives such as CALCOT or Staplcotn are pool-sales contracts, which means the co-op provides a cash advance upon delivery, and later during the year when the cooperative sells the cotton it makes final payment. A large percentage of growers follow these patterns because of the advantages: lenders are more likely to extend credit for a contracted crop, and growers can reduce uncertainty and make better management decisions.

Still another alternative involved futures trading, which required knowledge and experience with a complex system of commodity trading. Most growers shied away from this practice because of their lack of knowledge and understanding. A farmer can get hurt if he fails to understand the futures market because he must cast his bread into shark-infested waters, the ruthless practice of commodity trading. He must compete with merchandising houses staffed with traders that keep constant watch on trading, and growers do not have the time for such scrutiny. Only the financially secure commercial growers successfully engaged in futures trading. Professor Carl

Anderson, Extension Service economist at Texas A&M University, recommended that growers learn the practice of hedging in futures or contracting options. He felt that cotton markets had become too volatile and uncertain not to practice some futures trading. "Failing to plan may be the same thing as planning to fail."[37]

To encourage growers to venture into futures trading, Professor Anderson hosted seminars in Texas. Cotton Incorporated also regularly conducted workshops for growers across the Cotton Belt. Often a trader from a well-known merchandising house such as Hohenberg Brothers in Memphis participated in the workshops. But growers did not rush to engage in futures trading because, as stated by the noted Extension Marketing specialist O. A. Cleveland of Mississippi State University, "Of all the marketing alternatives available to cotton growers, the futures market is the most sophisticated."[38] In 2001 the Cotton Growers Marketing Association in Georgia began offering a software and video program to teach growers how to hedge their crop in the market, a practice that is not speculation and reduces risk. But the skill of hedging is not easily acquired, so the program, developed by Glen A. Arnold, a commodity trade adviser, attempts to remove the technicalities and simplify the procedure. It includes an electronic newsletter and a ninety-day subscription to the Data Transmission Network (DTN) for rural users of the Internet. But growers nonetheless shy away from trading, with most of them placing their crop into a cooperative marketing pool like PCCA or CALCOT.[39]

The disappearance of yesteryear's crop lien with its accompanying high interest rates and its effect in prolonging one-crop farming should not be attributed solely to new marketing strategies or the presence of federal lenders. A variety of factors consigned the mules and gangs of cotton choppers to the history books: the migration of southern black and white croppers into northern and western cities; mechanization and the advances in insecticides and herbicides; and new technologies in cultivation. The greater dependence on exports in the face of greater global competition forced all interests to pay closer attention to marketing strategies. Gone now are the sharecroppers and small-plot farmers who grew a few bales a year on a "furnish" from a factor like a small back-office buyer or hardware store owner. Only the growers able to adapt to a ruthless world economy survived, and the business-minded commercial growers adapted best. The 1.5 million cotton farmers of 1945 shrank to approximately thirty thousand by 2000.[40] This smaller number had more in common with businessmen than any idolized class of yeomen, and the changes in the credit structure of farming were both a cause and effect.

CREDIT DISCRIMINATION

Farm credit did not go without controversy. In August 1997, Timothy Pigford, a black farmer from Virginia, filed a class-action suit against the USDA, alleging discrimination by the department against African American farmers when they applied for benefit programs. Specifically, the suit contended that blacks were denied loans or given smaller loans than white farmers who had similar credit histories. For these black farmers, the USDA served as the lender of last resort, since they had been shut out by private banks. Part of the suit centered on the absence of any recourse within the USDA for handling civil rights complaints since the department had abolished its Office of Civil Rights in 1983. The two parties quickly began negotiations for a settlement under the direction of a mediator. The USDA did not deny the allegations but sought to erect a structure of settlement satisfactory to its own interests. Two advocacy organizations represented the interests of the black farmers, the Environmental Working Group (EWG) and the National Black Farmers Association (NBFA).

In March 1999 they agreed to a settlement. Eligible farmers received a payment of $50,000, and the USDA agreed to forgive debts of farmers, which ranged from around $75,000 to $150,000.[41] But the plaintiffs remained unhappy because the settlement went only to thirteen thousand farmers and gave nothing to eighty-one thousand. The large number were left empty-handed because they failed to file a claim within a specified deadline, so the settlement did not quell the dissatisfaction, and the advocacy groups continued to lobby Congress to intervene and provide compensation. In July 2002 a group of black farmers staged a sit-in at a farm agency office in Brownsville, Tennessee, over allegations that they did not receive fair treatment. The incident caused the congressional black caucus to demand that Secretary of Agriculture Ann M. Veneman put forth more effort on behalf of civil rights for black farmers. She agreed to do so only if the protesters ended the sit-in. The black caucus further wanted Congress to provide relief for the disaffected farmers. "For generations these practices have continuously eliminated the hopes, dreams, and opportunities of black farmers," stated John Conyers Jr. to the press.[42] In the next month, a small group of black farmers staged a protest outside the USDA building in Washington to vent their anger over the slow process of extending the settlement payments. One farmer brought his tractor on a trailer rig, and another brought a mule. Another dissident brought two goats and three chickens. The NBFA criticized the large subsidy payments received by large commercial farms.

Most of the farmers involved in the dispute came from the South. Of the national total of 22,181 eligible recipients, 20,838 came from areas producing cotton and peanuts. Only 1,343, as reported by EWG, lived in noncotton states. The available data did not identify cotton growers among the eligible recipients, but a significant portion almost certainly produced cotton.[43]

The charges of discrimination against black farmers were originally made in December 1996 and led to the creation of the USDA's National Commission on Small Farms. The department's Civil Rights Action Team believed that besides racial discrimination, other government practices had discriminated against small farmers, regardless of race. In 1979 Secretary of Agriculture Bob Bergland had ordered a general study of agriculture, but no specific examination of small farms had since been undertaken. The new commission began in July 1998 with a public hearing in Memphis, but it conducted an all-inclusive study with no emphasis on cotton or any aspect of agriculture other than small farms.

On the subject of credit, the commission reported that lending practices favored large operators. Commercial bankers regarded small landowners too risky because their production was too low to generate sufficient income to service debts. Congressional action in the past, so the study found, had shifted farm credit from direct lending to guaranteed lending, which benefited bankers rather than farmers. And steps to make the USDA the lender of last resort for small operators, as set forth in the 1992 Agriculture Credit Act, had not been implemented. The commission thought credit should be aimed more equally at minority and beginning farmers, and loans should be granted based on a borrower's skills as a farmer rather than credit histories or pending loan applications. To further assist landowners, the commission recommended "the promulgation of regulations" to encourage loans under $50,000 as stipulated in the 1992 Agriculture Credit Act amendments.

Other recommendations included a moratorium on foreclosures while a farmer's case was under review, and a delay in debt collection until a farmer had exhausted all options for servicing his debts. The commission wanted the USDA to push for legislation that would remove the prohibition of further USDA loans to recipients who had been granted debt forgiveness. And procedures for applying and granting loans should be hastened so that a decision could be made within thirty days.[44]

CONCLUSION

Since 1945 cotton farming underwent dramatic changes, but few outranked the developments in crop financing. The crop lien that for decades had beset small southern growers with indebtedness and perpetuated one-crop farming disappeared. Credit agencies originating in the New Deal did much to alleviate harsh credit practices, but the migration of small-plot owners into towns and cities removed the most risky borrowers and left the business-minded large-scale growers. As farming became more like a commercial enterprise rather than a home-style operation based on loose and informal arrangements, in which a few bales meant a little cash, the banking and financing of growing cotton operated like general business, in which formal structure had to be followed. Since growers had to operate on economies of scale, they needed larger reserves of credit, which had to come from commercial lenders. The new culture of cotton left no place for the old-fashioned lien that required a crop easily converted into cash.

A variety of sources of capital were available, but the most unusual came from ACCO. Its role in financing young growers in the West during the post-1945 era hastened the expansion of cotton farming and contributed to the general development of some areas. In this sense, cotton drove the expansion into the West in the manner of nineteenth-century migrants seeking new land. But federal lending agencies and commercial banks were just as important, so that capital came from both private and public sources and demonstrated that expansion into the West and modernization in the South were a joint effort.

From the long list of conditions responsible for cotton being the "nation's number one problem" in the late 1930s, the exploitative credit structure that stretched back to the end of the Civil War was gone, in great part because of the new practices based on commercial credit operations. This aspect of the world of cotton has been overlooked, but it did much to push forward the new culture based on growing the fabric of life.

CHAPTER 15

THE ROLE OF TEXTILES

Cotton farmers and textiles mills have a symbiotic relationship, each depending on the other for survival. Raw lint must be spun and woven into cloth, which makes the mill a critical destination in the chain of growing and processing cotton until it goes to the apparel industry. Breakthroughs in the technology of spinning and weaving in Britain had accounted for the growth of the cotton kingdom in the antebellum era, but after the Civil War, mills in the United States became the principal market. Farmers became highly suspicious of the mills, however, believing that along with merchants and shippers, textiles were part of a market system built on exploitation of the dirt farmer, that textiles were the final beneficiary of a system that kept individual growers under the heel of poverty through manipulation of prices, rigged classing of cotton, and usurious interest rates. Rhea Blake recalled that on the eve of the creation of the NCC, the mills were seen unkindly. "We can't be anything but adversaries," so he described the farmers' view, "because we're on one price side and they're on another price side."[1] It was this lack of unity that drove Johnston and his cohorts to establish the NCC.

From the perspective of the NCC, this distrust had to be overcome, and when textiles joined the organization in 1941, both parties had the clear purpose of fostering cooperation between growers and mills. At the time, two trade organizations represented the textile industry, the Cotton Textile Institute, consisting of northern companies, and the Cotton Manufactures Association, composed of southern mills. With the membership of textiles as a segment, the NCC would become antiunion and would generally oppose increases in minimum wages. More than anything, the NCC regarded a healthy and vibrant textile industry as essential to the prosperity of cotton farmers, and believed the welfare of one complemented that of the other. Dr. Claudius T. Murchison, a former assistant secretary of commerce and well-known textile economist who headed the Cotton Textile Institute, retired in 1949. Dr. William T. Jacobs, who led the southern Cotton Manufacturers Association, died the same year. Executives within the textile industry decided to merge the two organizations into the American Cotton Manufacturers Institute (ACMI), since cotton still dominated the fiber market. Mill owners felt the need to establish a better rapport with growers, who generally remained suspicious. It was this consideration that led to the appointment of the NCC's Robert C. Jackson as the executive vice president of the ACMI in 1949. Jackson's experience with congressional lobbying and his recognized status within cotton circles appealed to mill executives. His fact-finding tour of Europe in 1945 and subsequent role in establishing the special cotton export programs had brought him respect and prestige. His appointment related unmistakably to his connection with the cotton-producing element.[2]

ROBERT JACKSON: COTTON'S AMBASSADOR

Jackson soon had an opportunity to serve two masters: growers and textiles. In 1949 British textile interests worried that Japan would soon threaten their dominance in the world market, a threat that had become apparent in the late 1930s but had stopped with the war. Now, with the likelihood of a renewed Japanese industry, British interests wanted their counterparts in the United States to join in establishing a cartel of world textile trade. Jackson and mill owners quickly explained that U.S. antitrust laws prohibited their participation in such a plan, but indicated they would be happy to send a delegation on a fact-finding tour of Japan's textile mills. This latter point appealed to both textile and grower interests that wanted to ascertain Japan's capacity for manufacturing fabric after the devastation of the war. Jackson thus became part of an entourage that visited Japan in 1950 before the outbreak of the Korean War. He went as head of ACMI, but his observations had value for the producers of raw lint, who wished to restore the market there.[3]

General Douglas MacArthur, head of the American reconstruction program in Japan, invited Jackson for a personal conference. Jackson recalled that MacArthur knew the subject well and could recite statistics and other pertinent information about the condition of textile manufacturing in Japan. MacArthur wished to encourage shipments of cotton into the country to get the restoration program under way.

The delegation learned that Japan's capacity for manufacturing textiles had indeed been set back, but owing more to the conversion of machinery into munitions than to Allied bombing. The potential to become a major player in world trade appeared clearly, which brought promise for the export of U.S.

cotton. The group established business contacts with Japanese mill executives that "turned out to be quite valuable," Jackson recalled. Later he recognized an important effect of the trip: "It was a new concept of industry to industry negotiations." It was the beginning of direct contact of mill executives from country to country, the exchange of information and the use of contacts and conferences instead of relying solely on formal negotiations conducted by trade ministers. Through the years, Jackson maintained his friendship with the Japanese executives, seeing them at international conferences.[4] During the initial trip in 1950, Jackson met Carl Campbell, who had been on General MacArthur's staff, though by that point Campbell was no longer in uniform. He would later work for the NCC and succeed Read Dunn as head of its international division.

Upon the return of the delegation, Jackson saw that the suspicion in agricultural circles toward textiles persisted. Mississippi senator James Eastland, champion of the dirt farmer and member of the Judiciary Committee, made an investigation of the trip, suspecting that some effort was under way "to divide up markets or something."[5] Nothing came out of the effort, and Jackson saw it as a warning not to endanger the market in Japan.

Cotton's international character and the importance of textiles in the well-being of growers appeared in the Rusk–Van Fleet mission to Korea in 1953. It began when General James Van Fleet and Dr. Howard Rusk, a medical adviser and philanthropist for the U.S. Army, created the American-Korean Foundation to raise funds in the United States for providing assistance to South Korea at the end of the conflict. General Van Fleet recruited representatives of industries and organizations for a trip to Korea to initiate reconstruction activities through private auspices. He asked Charles A. Cannon, chief executive of Cannon Mills and active member of the ACMI, to go on the mission because of the need to restore the textile industry there. Jackson went to Korea for Cannon. The delegation included the head of Monsanto Chemical Company, New York real estate broker Bill Zeckendorf, an editor of the *New York Times,* and Dr. William Carr, president of the National Education Association. Leonard Mayo, a member of the family connected with the Mayo Clinic, went along, as well as the president of Sharpe Dohme Pharmaceuticals. "It was quite an interesting and impressive group," Jackson remembered, and he felt "flattered to go along."[6] Soon after the armistice in Korea, the delegation went there and witnessed an exchange of prisoners of war at Panmunjon. To see soldiers physically and emotionally in bad shape, he said, "coming to the warm arms of their American counterparts . . . was one of the most touching experiences I have ever had."[7]

Jackson's purpose involved textiles, however, and he inspected several sites of destroyed mills. He quickly realized the Koreans wanted U.S. assistance in building plants to manufacture synthetic fiber. For him the proposition was unrealistic because of the lack of infrastructure for such operations; Korea had no chemical industry, energy supplies were unavailable or inadequate, and the transportation system lay in shambles. "It was just far-fetched to even think of building anything as sophisticated as a man-made fiber manufacturing facility."[8] He knew Korea grew limited amounts of cotton, but it was poor quality and "hardly suitable for consumption." Upon his return, Jackson arranged for D&PL, Coker's Pedigree Seed, and Aubrey Lockett's company in Texas to ship cottonseed to South Korea. Through ACMI offices there were arrangements to supervise the planting and cultivation of the cotton, but the effort amounted to little because the Koreans wanted to develop synthetic fiber. Korea became a player in the global trade of synthetics only years later when American investors and importers furnished much of the capital for the construction of synthetic mills. American importers wanted to buy large quantities of clothing made with artificial fibers at prices below the American prices.[9] This practice became common as U.S. importers would move from country to country, setting up mills to manufacture items to their specifications.

COTTON, TEXTILES, AND THE COLD WAR

Cotton farming, the textile industry, and U.S. Cold War policy began to mesh in the 1950s. To begin with, revitalization of the export market remained a critical objective of U.S. trade policy makers. With the communist seizure of China in 1949 and the Korean War, the United States extended its policy of containment to Asia, and the Eisenhower administration saw trade as essential to preventing the further spread of communism. Like the free-trade ideologists of the cotton South, Eisenhower believed the flow of goods across borders was vital to national security. In this respect, the objective of the cotton interests and the White House went hand in hand. During the 1950s, exports of U.S. cotton to Japan grew spectacularly, overshadowing the sales to other countries. For 1955 to 1956, the figure was an astounding 8,388,051 bales but leveled off to 1,530,840 bales the next year and 1,754,963 bales for 1959 to 1960.[10] While the raw-cotton industry welcomed such growth, it put domestic textile manufacturers into a global competitive squeeze because the Japanese sent much of the cotton back to the United States as clothing. Cold War strategists, however, saw Japan as a buffer to Asian communism and wanted to encourage the nation's industrial growth.

The ACMI, now headed by Jackson, had the responsibility to stop the flood of Japanese imports. He knew he would be violating the free-trade principles of the South, but he needed the support of raw-cotton interests. Jackson's past service with the NCC gave him an advantage, but he later recalled that he had to campaign diligently until the organization endorsed the implementation of restrictions on Japanese goods coming into the United States. He took the position that cotton's domestic market, the mills, would be endangered and would furthermore be encouraged to concentrate on synthetics if Japan grabbed the apparel market in cotton. After three years of Jackson's lobbying, the NCC joined ACMI in the fight and used its resources to persuade Congress. The White House did not want restrictions but accepted a compromise in the Agricultural Act of 1956.

Eisenhower directed Sherman Adams to negotiate a voluntary restraint system with the Japanese. Jackson's ACMI had prevailed over the opposition of the State Department, the press, and the free traders by urging a quota system broken down into categories of goods. The quota system kept the Japanese from inundating one segment of the U.S. apparel market, capturing control, and then moving to another. Cotton and textiles, via their national trade organizations, worked together and justified their stance by pointing out that the cotton-textile-apparel complex stood at risk.[11]

RETAILING FROM OVERSEAS

A new twist appeared, however, that challenged the traditional pattern of textile imports. Retail chains in the United States such as J. C. Penney arranged for Asian apparel makers to produce items such as shirts and dresses according to its specifications. The retailer imported as many as four million items of clothing and sold them under its name. This practice meant that overseas manufacturers had instant access to the U.S. market without having to develop a reputable brand name. A new influx of goods hit the market as chain stores moved rapidly among Asian apparel manufacturers, bouncing from Hong Kong to Korea or Taiwan. Again the U.S. textile and cotton interests, along with apparel makers, wanted protection. Toward the end of the 1950s, six million yards of material were moving into the United States, "equivalent to about thirty yards for every individual in the United States," Jackson recalled. "Its economic impact was just too large for it simply to be ignored."[12]

For help the textile interests turned to presidential candidate John F. Kennedy after he obtained the Democratic nomination. They enlisted the support of South Carolina governor Fritz Hollings, a friend of Kennedy. Jackson and Hollings visited Kennedy in his Senate office and explained the predicament. The future president offered his help and upon taking office assigned Mike Feldman, his White House counsel, to initiate a program. This resulted in the Kennedy Seven Point Textile Program, which led to an international textile agreement. It removed cotton textiles from the negotiations undertaken through GATT and erected a separate structure to handle trade in cotton textiles. A country could now put quotas on imports of cotton apparels without being subject to penalties. The agreement allowed the negotiations of bilateral agreements without the participating countries violating their other commitments made through GATT. "It provided... an umbrella under which bilateral trade negotiations," Jackson recalled, "could go on between the United States and anybody it wanted."[13] Agreements over the imports of cotton goods would last about five years until renegotiations had to be implemented. Textile interests liked the system and were particularly pleased that Kennedy agreed for industry advisers to accompany trade negotiators, which at the time was still a novel practice.

FURTHER IMPORT THREATS

Despite the imposition of quotas, the cotton-textile complex faced a new challenge, the influx of clothing made of synthetics. To avoid the restrictions on finished cotton goods, overseas textile manufacturers produced and sent apparels of synthetic fibers to the United States until by 1967 they surpassed cotton imports. To better combat this new threat, the ACMI changed its name to the American Textile Manufacturers Institute (ATMI) in 1962. Again the textile organization took the lead in getting relief, and Jackson joined a delegation of industry executives, Roger Milliken, Charles, Myers, and Fred Dent, which visited presidential nominee Richard Nixon at his headquarters in San Diego in 1968. Dent later became secretary of commerce. Nixon was sympathetic and promised to help if elected. After Nixon's victory and three years of globe-trotting, Jackson, textile executives, and U.S. trade representatives made the first multifiber agreement with Japan, Hong Kong, South Korea, and Taiwan in 1971. It included cotton, wool, and synthetic fibers and was negotiated through GATT.[14]

The origins of the cooperation between raw cotton and textiles went back to the origin of the NCC. Oscar Johnston had become convinced that the two segments had a common interest and could be beneficial to each other, so in 1941 he sent Jackson, working for the NCC at the time, into textile company offices to lobby and convince the executives of the advantages of joining the organization. C. A. Cannon became an enthusiastic participant in the NCC, and other executives took their firms into the organization. Despite their history of mutual suspicion, the two interests worked well together. "We were constantly in touch regarding our respective positions on various issues," Jackson recalled. "Almost always we were together on our positions."[15] When he left the NCC to head the ACMI in 1949, Jackson discussed the move with Oscar Johnston and Rhea Blake, who saw the advantage of the move to both industries. Jackson's relationship with cotton interests and textile executives like Cannon, who served on NCC's Foreign Trade Committee headed by Read Dunn and Fred Simms, contributed much to the amicable relationship.

Within the textile interests, however, there were cases of dissatisfaction with the NCC. By the 1960s some mills disliked Washington's policies on cotton support prices, acreage allotments, and the CCC loan arrangements. For some mill owners there were too many differences with the NCC over policy, so some left the organization. Jackson persuaded a group of disinterested executives to visit NCC headquarters in Memphis to observe its operations, and some of the mills renewed their membership. A few years later a similar development occurred, but again most of the defectors rejoined. Jackson always remained a strong proponent of membership in the NCC, and his loyalty to, and representation of, textiles never wavered. For him, understanding each other's operations and objectives was the key to teamwork because the textile industry gained the lobbying power of cotton.

THE TWO-PRICE SYSTEM

One of the most outstanding examples of the subtleties within the cotton-textile complex occurred over the two-price system. "It was probably the most dramatic situation that has come before the Council," Jackson remembered.[16] Conditions that led to the system began after the Korean War when cotton prices weakened and Congress had established a subsidy to assist growers. They tended to "farm for the loan" by putting their crop into CCC warehouses. American cotton suddenly cost more on the world market, so U.S. producers began losing sales overseas. To make U.S. cotton competitive in the international markets, Congress, at the urging of the industry, passed the Agricultural Act of 1956, which established a two-price system that provided a subsidy at six cents per pound for cotton sold to foreign mills. Nothing changed for growers who continued to receive domestic subsidies, while the federal government bore the cost of the export subsidy. In this arrangement, spinners originally agreed to support the export subsidy, but every effort would be made, explained Lewis Barringer, to either "lift the American export price, or equalize the American domestic price with the export price."[17] An unanticipated effect occurred: foreign spinners could buy American cotton more cheaply than could the U.S. textile mills. For eight years American mills endured this hardship and lobbied against it with the support of the NCC, but Congress refused to change the arrangement for fear of harming growers. When the Kennedy administration raised the domestic support price, it worsened the plight of the mills.[18]

The export subsidy went back to 1939 when it was set at 1.5 cents per pound, but it varied with world cotton prices, often being discontinued and then restarted. It ran continuously from November 1944 through 1950, when the subsidy was dropped because of the Korean War. With the reapplication of export subsidy in 1956, U.S. textiles found themselves in an untenable situation: they had to pay more for American cotton than did their foreign competitors. With the rise of textile production in Japan and other countries, imports of fabric began flooding retail outlets, and the two-price system, one for domestic and another for export, handicapped textile manufacturers. Growers saw the flood of apparel imports and wanted to protect their domestic market, which caused the NCC to oppose the two-price structure. Repeal of the system became a major objective.

At the annual meeting of the NCC in 1964, the subtle dynamics of farm interests were apparent. Delegates of the organization were prepared to endorse a repeal of the two-price system, but the delegates affiliated with the Farm Bureau opposed such a move. Walter Randolf, president of the Alabama Farm Bureau, managed to block a repeal measure by invoking the two-thirds rule among the producer segment. "Twenty-one people," reported Albert Russell, "killed the resolution," and kept the NCC from lobbying for the repeal. Textile interests nonetheless took over the responsibility to end the system. Albert Russell, head of the NCC, recalled that Lewis Barringer "practically moved to Washington to head up a lobbying of mill interests." President Lyndon Johnson supported repeal, and Congress ended the two-price system in 1964. "He [Johnston] had worked with us on this legislation very hard," Barringer later stated.[19] At the signing ceremony in the White House, President Johnson affixed his signature to the bill, but Barringer, the ghost, had slipped away.

The new law only changed the direction of the federal handout; it went to growers as a direct price support. In other words, the legislation transferred the subsidy to the growers because no other segment wanted to be identified with it. The consensus held that public acceptance would be more likely to support farmers than shippers or a single manufacturing industry. Leadership in the Farm Bureau warned, however, that direct payments to growers might lead to demands for limitations, a warning that proved prophetic.

Over the years, textiles and cotton generally stayed in tandem. Mills supported the calls for greater acreage allotments because a larger crop would ensure a supply of different varieties of lint. Textiles also wanted price stability in cotton, so they stood behind price supports. And they joined farm groups in opposing imports of raw cotton to encourage prosperity among the growers.

THE MULTI FIBRE ARRANGEMENT

Textile companies found themselves increasingly having to fight imports. The first Multi Fibre Arrangement (MFA) that the United States made with other countries operated from 1974 to 1977 but had to be superseded by MFA II from 1977 to 1981 owing to pressure from countries of the European Union (EU). But MFA II did not slow imports. Worsening the predicament was the growing demand by retail interests for fewer restrictions on imports. MFA III (1981–86) stiffened the criteria for developing countries to qualify for entry into the U.S. and European markets, but imports nonetheless mushroomed, almost doubling in a two-year period, so another extension was negotiated, MFA IV (1986–91). As this last agreement approached expiration, the GATT Uruguay Round of trade talks had been progressing for a few years. Additional extensions of the MFA were made through 1993, and once the Uruguay negotiations led to the WTO, effective January 1995, MFA agreements ceased. For cotton the MFA regulations had provided some relief from imported apparels, but not enough, as developing countries nonetheless sent cotton goods into the United States and further eroded the market of the domestic mills.

Grower interests got caught in the larger struggle to protect the textile-apparel complex. As it became clear that some countries shipping goods to the United States refused to open their markets to U.S. shippers, textile and apparel constituencies demanded stiffer regulations than those provided by the MFA. This development led to the fight over the Jenkins bill in 1985. That year Georgia congressman Edward Jenkins introduced a bill known as the Textile and Apparel Trade Enforcement Act. It would negate about thirty unilateral agreements made through the MFAs and reverse imports from Asian suppliers up to 30 percent. To lobby for the measure, a coalition of textile and apparel interests created the Fiber, Fabric, and Apparel Coalition for Trade (FFACT). Jackson's ATMI had an important part in the group, and the NCC formally joined the coalition and lobbied Congress on its behalf. Through its Communications Division, the NCC urged its members to contact their congressmen, and even the 1985 Maid of Cotton, Michelle Pitcher, described the situation behind the bill at her tour appearances. NCC's Gaylon Booker and Dean Ethridge made special appearances to explain the technical features of the issues.[20] Though Congress passed the measure, President Reagan vetoed it, and an effort to override his veto failed by only eight votes. This close vote demonstrated the power of the cotton-textile-apparel complex at the time, but it would soon face new and tougher challenges. To replace the MFA and make the transition to WTO, the United States made the Agreement on Textile and Clothing in 1995, which provided for a phaseout of quotas over ten years. Kitty Dickerson, a highly reputable textile analyst, described the end of MFA in 1995 as the beginning of "difficult times ahead."[21]

During the 1990s, the coalition of FFACT began to unravel. Retailers and importers saw quotas as age-old hindrances in the newly globalized economy and wanted to end them. A new organization appeared, the Retail Industry Trade Action Coalition (RITAC), to stop quotas. In 1990 a delegation from the organization had gone to the Uruguay Round discussions in Geneva to voice its dislike for quotas, and though they failed to achieve their objective, it was clear that the old unity had begun to erode. When developing countries that wanted open access to the U.S. market formed the International Textile and Clothing Bureau (ITCB), the traditional protection of quotas came under further attack. Most damaging, however, was the loss of support from the powerful retail sector.[22]

Free trade had been the clarion call for cotton interests for more than one hundred years, but the industry sought to protect its traditional adversaries, textiles, from foreign competition and joined in the general effort to obtain restrictions on imports of cotton goods. They acted because textiles and apparels were the final processors of the natural fiber. Wise action that emphasized cooperation had enabled cotton and textiles to tap into each other's resources for their own benefit, and the personal acquaintances and camaraderie among cotton activists and textile executives fostered their effectiveness. Cooperation and unity were a breakthrough, a reversal of the older era when mistrust and suspicion had characterized their relationship. As the twentieth century drew to a close, however, the relentless force of globalization overwhelmed textiles and forced many manufacturers to shut down, until the closing of another spinning or weaving mill became mundane. For 1997 through 2006, the National Council of Trade Organizations recorded 379 mills that permanently ceased operations in the United States. Jackson commented in 2005 that the weakened industry could not handle a challenge equal to the demand it faced in 1941.[23]

The contact between the grower element and mills became even greater, at least in research, when researchers at Cotton Incorporated's Raleigh center developed a chemical treatment to reduce wrinkling in cotton fabric. They also reduced fading by subjecting cotton to "dye-batch dying" and "beam washing of reactive dye stuffs." To avert shrinkage, the researchers engineered a system to measure shrinkage in cotton fabric, which helped the mills and apparel manufacturers, since they could predict the shrinkage values of cloth. Through a cooperative undertaking with Procter and Gamble, CI learned that cotton enzymes added to laundry detergent helped cotton fabric retain its colorfastness.[24]

ROTOR SPINNING

A major advancement in making yarn came with the development of rotor or open-end spinning. Invented by engineers in Czechoslovakia in 1967, this technique proved particularly advantageous for the producers of short staple. Ring spinning had been employed since the mid-nineteenth century, when it replaced the old spinning mule. Ring spinning worked well with cottons measuring over one inch, but shorter lint broke more frequently, so mills preferred the longer Upland varieties and paid a better price for them. Cotton of shorter length grown on the Texas Plains did not perform as well, which partly accounted for its disconnected price and explained why the Textile Research Center at Texas Tech University conducted research for improving the spinning properties of plains cotton. It was this comparably inferior spinning quality that had made growers on the plains rely on the Asian market.

Rotor spinning is faster, puts less strain on the fiber, and causes less breakage. Rotors turn at high speed, often more than 100,000 RPM. This method also employs the use of a pinned opening roller that further removes bits of trash and dust, an important consideration in preventing brown lung disease. Rotor spinning also eliminates the step of roving. Ring spinning, however, produces a finer yarn suitable for higher-quality fabric.[25]

Ring spinning put High Plains cotton at a disadvantage, but an effort to correct the situation began in 1973 when NCC vice president William Reid asked Don Anderson to observe rotor spinning in Japan. Anderson had just finished a tour of Uganda, Turkey, Pakistan, and Indonesia to examine cotton production. In Japan he visited a small mill in Osaka using rotor spinners to spin cotton baled in Tahoka, Texas, located in the midst of the cotton patch. It was short staple, but the small mill nonetheless spun the fiber approximately four times faster than with traditional ring spinning. Upon his return, Anderson explained his observations to Roy Davis at PCOM, and they agreed that the Textile Research Center at Texas Tech should undertake a research program to develop rotor spinning of plains cotton. Anderson led the fight to get an extra $1 million appropriation through the university trustees for the program, and the laboratory began the research in 1973.[26]

The Textile Research Center was well suited because textile research had been incorporated into the school's general mission in 1925. In 1945 the laboratory operated in a small department with only a handful of faculty and limited facilities. But as cotton farming expanded on the plains, growers looked to the university for help with improving the spinning characteristics of their fiber. Because of the lobbying efforts of organizations such as the Plains Cotton Growers and influential individuals among growers, the state legislature increased support for textile research, and the university stepped up the commitment by redesignating the department as the Textile Research Center in 1969. Raw cotton interests drove this undertaking; a portion of the bale assessment paid by members of the PCG went directly to the center. With the infusion of cash, the center better coordinated its research among the engineers, who had previously carried out projects on an individual basis.[27]

Roy Davis recruited Jim Parker to become head of the center in 1973. Davis saw promise in his recruit, and Parker recalled how the two worked together upon Davis's retirement from PCOM. Davis regarded textile research as another stage toward improving the fortunes of cotton farming and saw a need for a textile mill on the plains for the benefit of growers and job employment. Only by putting money into a mill, he would say, could cotton interests learn and understand the textile end of cotton.[28]

In 1993 the center was renamed the International Textile Center (ITC) and moved back into the university as part of the College of Agricultural Sciences and Natural Resources. It moved into a new building and was well furnished with equipment to spin and weave cotton into a finished product, which enabled the center to provide testing services for privately owned mills and seed companies like Monsanto and D&PL and to assist textile laboratories at other universities. The ITC conducted studies of "sticky cotton," raw cotton on which aphids and white flies leave a sugary deposit that gums up textile machinery. It developed a method by which a mill can measure the sugar on cotton and determine the likelihood that it will cause disruption of the milling process. In 1989 the center started its Texas International Cotton School, a two-week professional school for management trainees in various cotton businesses, mostly merchandisers. The school is similar to the nine-week courses taught by the University of Memphis, but the Lubbock school is shorter because it does not teach hand classing. Among its research projects, the ITC wants to develop an Advanced Fiber Information System that will be superior to high-volume instrument testing. The center sees itself as a liaison between growers and mills and follows a mission of making plains cotton as attractive and marketable as possible.[29]

Other universities have contributed to textile research through the National Textile Center, a consortium of eight schools in the Southeast and New England states. The ITC in Texas is not a member. Clemson University holds much prestige for its textile research that began in 1927 and strengthened its program in 1958 by establishing a Department of Research within its own Textile Center. The ATMI and the USDA Fiber Testing Laboratory conducted research in conjunction with the Clemson Center. Because of its location in the heart of the textile industry and its performance in assisting the mills, the program at Clemson has received substantial funding from private sources. Clemson University has a special collection of manuscripts and various materials from mills along with a large collection of books and journals pertaining to textiles. In 2000 the Eastman Chemical Company awarded Clemson $38 million for fiber technology research, the largest gift in the school's history. Among the textile programs in higher education, however, the ITC in Lubbock holds the distinction of being identified with growers, a development that evolved with the rise of cotton farming on the plains.

A TEXTILE MILL ON THE HIGH PLAINS

One of the least-expected ventures undertaken by the raw-cotton industry came with the establishment of a cooperative textile mill located in Littlefield, Texas, in the heart of the High Plains. Activists, led again by Roy Davis, believed that mills in the Southeast still did not show enough interest in plains cotton. The establishment of a mill wholly owned and supervised by plains farmers was their ultimate answer to a marketing problem.

For growers to construct and operate a mill seemed absurd. Among the numerous agricultural cooperatives in the United States, none processed and manufactured the crop in a stage as final as cloth ready for scissors and needle. Even citrus cooperatives did not make the concentrated juices that went into grocers' frozen-food section. Producers of cotton had no precedent or experience in textile manufacturing, and though oil mills crushed oil from cottonseed, it was a far cry from spinning, weaving, and dyeing fabric to high standards. Operating and managing a textile mill required a mentality different from that of farming; one was industrial and the other agricultural. Growers knew little about cotton after it left the gin; few could grade and class their own cotton. The notion of a cooperative establishing and supervising a textile plant defied logic and history, but the same skepticism had applied to the use of rural electric cooperatives in the mid-1930s. Interests on the Texas Plains were well aware that conventional thinking went against such a venture, but they remained defiant and determined to put their cotton on an equal footing. If this objective required a textile mill, so be it.

The date when plains interests got the idea to have their own mill fell somewhere in the late 1960s and early 1970s. Roy Davis was as usual among the first, and others included his son Dan and PCCA pioneers like Emerson Tucker and Joel Hembree, as well as growers like T. W. Stockton, Compton Cornelius, and Clyde Crausby. Jim Parker at the Textile Research Center belonged to the early group.[30] Doubting Thomases were common, and Davis, along with PCCA organizers like Jim Martin, prevailed on occasion by force of personality. The PCCA had tried to get into the textile business in the mid-1960s through a joint venture in Thailand known as Thaiamerican, but a switch in federal support for the project killed it, a development that Dan Davis blamed on the election of Richard Nixon because of his ties to southeastern mills. "The textile interests were against it even though we were not going to produce textiles for importation into the United States."[31]

When in 1971 BAH, in conjunction with Lockwood Greene Engineering, conducted a feasibility study of the textile industry in Texas, the general wish for a mill on the plains was in embryo. Through arrangements made by the Texas Industrial Commission, the study recommended three sites for the construction of a mill: Rio Grande Valley, the San Antonio–Austin corridor, and the High Plains. There was some speculation that Roy Davis was behind the study. The next step was the push by plains cotton interests to realize their dream. A mechanism to organize a textile plant already existed in the American Cotton Growers (ACG) cooperative that the PCCA had established in 1973 among ginners. ACG had built a "super gin" at Crosbyton that incorporated the new universal-density press and took into account the new clear-air standards imposed by OSHA. Only after strenuous campaigning and with good luck did PCCA organizers persuade the ACG directors to commit to the construction of a textile mill. The formal move occurred in 1973, but to obtain funding, as well as add legitimacy to the undertaking, there had to be a customer for the proposed mill.

Levi Strauss and Company in San Francisco, originator and maker of the famous Levi's blue jeans, stepped into the picture. These pants, prominently identified with American culture, were becoming so popular that Levi Strauss could not meet the demand. It wanted more denim cloth, but no textile manufacturers were willing to expand operations solely for denim. Levi Strauss contacted Roy Davis in 1973 after learning about the plans in Texas. Here the connection between the cotton grower and consumer is shown clearly, with a major apparel manufacturer going directly to farm interests. The Levi Strauss representative, John Hunt, recalled his first meeting, which included an enthusiastic Roy Davis, who stated that he wanted "to take cotton grown in Texas and make it into Levi jeans in Texas."[32] PCCA representatives stated the denim would have to be rotor spun, but ring spinning had always been used with denim, and Levi Strauss personnel questioned whether the plainsmen, despite their enthusiasm, could succeed in making cloth with rotor spinners.

But experiments at the Textile Research Center continued, and ACG moved forward with the mill. Funding came from the Houston Bank for Cooperatives, an arm of the Federal Farm Credit Board, and a commitment from ACG's cooperative gins to furnish backing. A condition of the federal funding required a five-year contract with Levi Strauss that it would buy the denim. When Levi Strauss executives saw denim samples supplied by the Textile Research Center, they felt assured of the quality of the material but reserved the right to reject unsuitable cloth.

After reviewing several locations, the ACG selected Littlefield for the site. The financing came from stock put up by ACG farmer members who had subscribed to the large gin at Crosbyton. These original stockholders were thus members of the PCCA, but not all PCCA members belonged to ACG. The Houston Bank for Cooperatives made a loan and construction began in 1975. After making adjustments to satisfy Levi Strauss and Company, the plant generated about twenty thousand yards of denim per year. On average the growers in ACG that supplied the mill made thirty dollars more per bale than selling their cotton or putting it into CCC loan. In 1987 the ACG members sold their interests to the larger membership of the PCCA, and the original members who had subscribed to the initial stock offering made a handsome profit. Ironically, many of the sellers used the profit to avoid bankruptcy in their cotton farming operations or to pay off loans. The practical effect of the sale was negligible because the mill continued to sell to Levi Strauss and hardly any change occurred with the employees.[33]

From the perspective of the growers on the plains, the denim mill provided another way to market their cotton and showed that a cooperative could indeed manufacture fabric to industry standards. That this success occurred with West Texas cotton made it more remarkable. A key to the success lay in the new HVI testing that the PCCA had pioneered, because instrument testing gave the cooperative the ability to identify which bales met Levi Strauss's standards and thus not endanger the denim fabric with low-quality raw lint. By the same token, the PCCA could sell bales exceeding or falling below the requirements for Levi Strauss in other markets. In other words, precise grading enabled the PCCA to identify bales that met the specifications for Levi's jeans and to market any superior bales at a premium to other buyers. This advantage made the team of PCCA and ACG highly effective.[34] Other cotton organizations were not involved in this bold experiment to establish a textile mill, which reinforced the conviction among plainsmen to be self-reliant.

THE CHINA FACTOR

Cotton farming became highly enmeshed in world trade with the rise of China in the 1990s as a trading partner of the United States. China had long been regarded as a rich market for the U.S. economy, but when communists seized control of the country in 1949 and intervened in the Korean War in 1950, Americans saw the country as a threat. When in 1972 President Nixon visited China, it was the beginning of an improvement in relations that led to U.S. formal recognition of the communist regime in 1978. In 1980 the United States restored the status of Most Favored Nation (MFN) to China, a move that would usually have increased trade, but protective American tariffs held down Chinese imports. Trade nonetheless increased, and China had a trade surplus with the United States that began attracting attention in the 1990s. The "China Factor" entered into the strategy of cotton traders. "Long gone is the time when the dynamics of supply and demand of cotton were national, hemispheric, or regional," reported *Cotton Farming* in 1996.[35] China became the world's largest producer of cotton in the 1980s, growing an enormous 28.7 million bales in 1984. Production fluctuated year to year, but China surpassed the United States as the world's top producer of cotton, and the growing trade power of China "was beginning to alarm members of Congress."[36]

China had set a course for modernization and intensified its production of textile goods, so that it became a net importer of cotton. American growers welcomed the expanded export market that accounted for the high price of cotton, reaching one dollar per pound in some spot markets in 1996. Such prices brought a jump in the world production of raw lint, so that a price decline began that bottomed in 2001 below forty cents. China continued, however, to import cotton to manufacture goods, so while growers saw an improvement in export sales and prices, the power of the communist country in the apparel market grew steadily. Quotas protected American textiles, but they were scheduled to expire in 2005, and textile manufacturers expected to be overwhelmed. "According to many," wrote one observer, "it will be the last nail for the U.S. industry in the sad story of plant closings and job losses that has lasted for nearly 50 years."[37] To combat the rising tide of imports, most of which came from China, the textile industry fought back with the backing of the NCC.

In 2001 China gained entry into the WTO, but the NCC feared the new member would not comply with terms of the world organization. Cotton and textiles allied with one another over the China trade, since the country would not fully open its markets to raw cotton just as it flooded America's retail shelves with clothing. In response to demands by developing countries, the WTO proposed the creation of a special subcommittee to focus solely on cotton in the name of removing trade distortions. From the perspective of the NCC, however, cotton issues should be handled as part of larger agricultural trade reform. American cotton, so it feared, would be singled out and treated unfairly. As seen by U.S. producers, the difficulty lay with China's refusal to reevaluate its currency, and when the U.S. import quotas expired in 2005, imports of textiles from China soared. Once the United States invoked some safeguards for textiles, China did adjust its currency by 2 percent, too little to have an impact. American trade negotiators offered concessions on cotton, which caused the NCC to warn that it might withdraw support for Trade Promotion Authority for the White House.[38]

The NCC had allied with the American Manufacturing Trade Action Committee (AMTAC) and other textile organizations in getting the White House to push China to limit its exports of knit fabrics, brassieres, and dressing gowns to the United States. In 2003 President George W. Bush negotiated such an agreement, known as a safeguard petition. It was "a first step in developing a much more comprehensive and rational trading environment with China," in the opinion of Roger Milliken of Milliken and Company. Former NCC executive director Gaylon Booker felt the "bipartisan coalition really helped push the safeguard petition over the goal line."[39]

ROGER MILLIKEN

Despite the general malaise in textiles, there remained successful and prosperous companies, and Milliken and Company stood at the top. In 2007 it had over sixteen thousand employees in South Carolina and several other southeastern states. The company had further operations in twelve countries, and the domestic and international plants together made Milliken and Company the largest private textile firm in the world. It manufactures a variety of products: finished fabric for apparel, carpet and rugs, and automotive fabrics. The company operates chemical plants for its own dyeing of fabrics and also markets dyes to other users. Within textile circles the company is recognized as a giant that wields influence on U.S. trade policy.

Behind the gargantuan organization is Roger Milliken, who inherited the operation from his father and took control in 1947. Regarded as an innovator and trade protectionist, Milliken's influence extends through South Carolina and Washington, D.C., and with fierce loyalty from his employees and the admiration of business leaders, he has a commanding presence that bespeaks firm convictions while maintaining a friendly manner. The recipient of numerous awards and honors, Milliken has been an avid philanthropist and environmentalist and helped organize the AMTAC and its fight to hold down imports. One author stated that "when Roger Milliken goes, things will be different."[40]

Several factors account for Milliken's success in a troubled industry. For one thing, he kept the company in private hands and refused to incur heavy debt for expansion, pumping profits into research on the grounds that innovation and new product development were necessary to remain viable. At the six-hundred-acre Milliken University in Spartanburg, South Carolina, headquarters of the organization, the research center dominates. Milliken destroys old machinery to keep it away from foreign competitors. Unashamedly a protectionist, Milliken wields influence in political circles. In 1956 he closed a plant in Darlington County, South Carolina, to avoid unionization, and the case was not settled until twenty-five years later when the Supreme Court ruled the workers or their ancestors were entitled to a $5 million settlement. But with the rise of textile imports and the loss of mills and jobs, textile unions have now allied with Milliken to protect the industry. Together they believe that in the new era of globalization, the United States must protect industries vital to self-reliance and economic security. Essentially this has been the stance of cotton interests since World War II toward textiles, though they have urged free trade. This discrepancy cannot be overlooked and fits the general pattern of U.S. trade policy to protect textiles while trying to break down trade barriers. Since 1933 the United States maintained restrictions on importing raw cotton, which made growers and textiles cooperative partners on the matter of imports.

TRADE POLICY

No area of U.S. trade policy is more baffling than textiles, and cotton gets caught in the swirling mass of contradictions, free trade versus protectionism, self-interest, bureaucracy, and politics, indeed, the whole array of factors that make up the murky economics of international textile trade. Since the welfare of cotton and textiles is often lumped together, they befriend one another and try to stand together on issues that affect them. Both have been in the forefront in coping with the new world economy that favors no country or interest group, and though individual countries may erect barriers for protection, they cannot withstand the onslaught of global forces, except temporarily. Producer organizations in the cotton community accept the reality of globalization and seek to establish and participate in trade initiatives rather than pull back and insulate themselves from international developments. Until the mid-1980s, the cotton-textile-apparel complex stood united, but in the following decade and into the new century, the coalition lost power owing to the break away by retail importers and the declining strength of textiles.

In Pietra Rivoli's analysis of global textile trade, *Travels of a T-Shirt*, she demonstrates the complexity, and futility, of trying to protect the U.S. textile industry from the avalanche of foreign imports. Her analysis is both critical and sympathetic as she follows the example of a T-shirt from its origin in a cotton field on the Texas Plains until it ends up in the hands of a small used-clothing retailer in Africa. Trade policies designed to protect U.S. textiles often fail, though Rivoli attributes the decline of jobs in spinning and weaving plants more to "mechanization and technological progress than by foreign competition" and predicts that "textile and apparel jobs . . . will continue to vanish, with or without protection from imports."[41] Rivoli combines textiles and related interests, which include cotton, as the "regime." Because the regime has not blocked the intrusion of foreign competitors, she suggests that cotton interests stand in danger, particularly in view of the rising attack on the use of subsidies provided to growers. For her the cotton organizations fighting to protect support programs resemble the futile effort by textiles to repeal foreign imports. Rivoli argues effectively and persuasively, pointing out that if African farmers suffer at the hands of U.S. cotton subsidies, they are also the victims of corrupt governments, political and economic instability, and lack strong property rights, what she summarizes as "bad governance." In the general effort to overcome the plight of "African farms and Asian sweatshops workers," she asserts that "cutting agricultural subsidies" would be among the "steps in the right direction."[42] Rivoli is sympathetic to American cotton interests, however, referring to their creativity and adaptability as they seek to improve their fortunes. But she warns that cotton might share a similar fate to that of textiles if growers seek safety through protectionism or rely too heavily on subsidies rather than competition.[43]

From the perspective of U.S. textiles, foreign countries engage in unfair trade practices that enable them to undersell American-made cotton goods by as much as 40 percent. Currency manipulation is the principal tactic that allows underselling, and U.S. textile interests consider China, the world's behemoth textile-apparel producer, to be the chief culprit. Other allegations against China include illegal export tax rebates and direct government subsidies of mills. One textile organization calls these practices "China's marriage of government power with manufacturing output."[44] For those product lines in which the United States dropped quotas in 2005, China took about 74 percent of the market. It was this avalanche of Chinese imports that prompted textile power brokers to persuade the White House in 2005 to impose quotas on the most sensitive textile and apparel categories. The United States and China made a bilateral agreement that reinstated a quota control system for certain items through 2008, which bought time and made textiles hopeful that the Doha Round of negotiations would lead to a more permanent program of quotas and the creation of a special group within WTO for negotiating imports.

Textile manufacturers in Europe also complained about imports from China. Europe was the second-largest market for China, and the flood of imports there set off demands for restrictions from France, Portugal, and Italy. But retailers in Europe opposed any action because they realized better profits with imports. Agreements between the United States and Europe over trade have been difficult to achieve, as they get caught in issues of anti-Americanism, foreign policies, and the popular distrust of globalization in Europe.[45] In the United States antiglobalization feelings run high, as seen in December 1999, when rioting protesters stopped the WTO talks in Seattle.

The threat of terrorism worsened the plight of textiles. Developing countries, of which many are Islamic, rely on textile manufacturing to improve their economies and want access to the U.S. market. Since American antiterror policy recognizes the importance of fostering prosperity in poor countries, trade ministers operate under pressure to make concessions. In 2001 after the 9/11 attacks when the United States needed the cooperation of Pakistan, and Pervez Musharraf, the country's president, asked the United States for debt relief and greater access to its textile-apparel market, there was a natural inclination to concede. But U.S. textile interests opposed concessions and managed to persuade the White House to stand firm. President George W. Bush agreed, however, to assist with debt relief and support a pending IMF $135 million loan to the country. Textile spokesmen complained that Pakistan would not agree to lower its tariff wall blocking access to its markets. Spokesmen also complained that textiles should not be used as a bargaining chip in trade negotiations or the antiterror fight.[46] "The textile and apparel sector has been on the front lines of the trade debate between poor and rich countries for decades," wrote an analyst for the Brookings Institution.[47]

CONCLUSION

For the denim mill in Littlefield, Texas, the tidal wave of Chinese imports had ramifications. For the mill to compete after the end of quotas on January 1, 2005, its six hundred employees took a pay cut. "We have to get very lean and mean," stated Wallace Darneille, head of the PCCA in Lubbock, which owns the mill.[48] In 2001 the Textile Division of the PCCA had already eliminated 650 jobs in its New Braunfels plant owing to foreign imports. From 2002 to 2004, the Textile Division lost $18 million, but it expected to survive the end of quotas in 2005 because it had little long-term debt to service and was renegotiating its electricity rates, cutting warehousing costs in the city of Shallowater, and seeking tax abatements. The PCCA reported that the closings of textile mills in the United States from 1997 to 2003 were higher than during the Depression years of 1929 to 1932.[49]

There had been speculation in the press that consumers would see a reduction in the price of clothes with the low-cost Chinese imports. Little savings occurred. The most practical result was the closing of U.S. mills or their transfer to other countries, but the denim mill in the Texas Cotton Patch kept its doors open and continued to turn High Plains cotton into clothing after the end of quotas. The mill lost much of the guaranteed sales to Levi Strauss and Company in 2005, when the popular jeans maker began placing orders with mills overseas to take advantage of lower prices. The cooperative mill diversified its markets and replaced 300 rapier looms with 160 more efficient and less costly air jet looms.[50]

Few achievements compared with the unity and cooperation of raw-cotton interests and textiles. In the past they had mistrusted each other, with growers accusing the mills of manipulating to cheat them out of a fair price. On the other hand, textiles saw farmers acting as a special interest trying to achieve an artificial price level through political lobbying rather than relying on the market. This mutual suspicion had been sharply embedded as a dynamic of the broader cotton culture and stood among the reasons why it had come to be seen as disorganized and hapless.

In 1941, however, they joined forces and soon began to combat imports; but despite their efforts, textile mills steadily fell to only a portion of their former number. For growers this development meant a reduction in the traditional domestic market and greater reliance on exports. These trade shifts also brought losses of textile jobs. To protect themselves in a global market, growers and textiles believe they must maintain their unity. "Unity has become," Jackson observed, "a condition of survival."[51]

Unloading bales in Memphis. Courtesy NCC.

"Snakes" of leftover cotton on Front Street, Memphis. Courtesy Memphis University.

The royal barge arrives with the Cotton Carnival's king and queen. Courtesy Memphis Public Library.

Beauty and glamour salute King Cotton at the Memphis Cotton Carnival. Courtesy Memphis Public Library.

Popular television star Ed Sullivan visits the Cotton Carnival king and queen. Courtesy Memphis Public Library.

Customers waiting at Schwab's on Beale Street. Courtesy Memphis Public Library.

Ryan's Market on Beale Street. Courtesy Memphis Public Library.

Wilson's Drugstore on Beale Street. Courtesy Memphis Public Library.

A partial view of the Memphis Cotton Museum. Courtesy Memphis Cotton Museum.

Cotton Incorporated advertised cotton as glamorous and fashionable. Courtesy NCC.

Advertising cotton as office fashion. Courtesy NCC.

Cotton Incorporated advertised cotton as culturally chic. Courtesy NCC.

Inside a cotton spinning mill. Courtesy NCC.

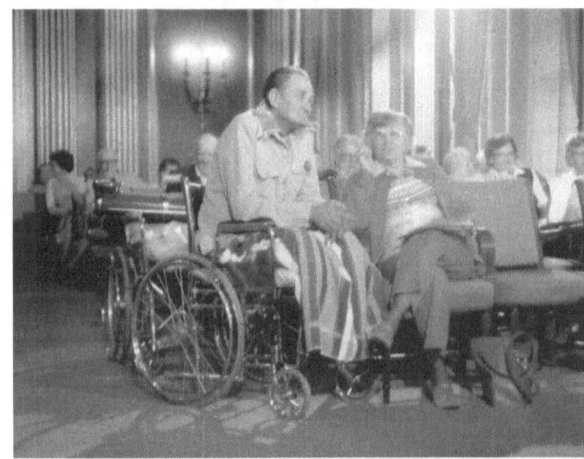

The consequences of byssinosis. Courtesy NCC.

Large-scale mechanization. Courtesy NCC.

A modern rig spraying for insects. Courtesy NCC.

Modules in a field near Brownfield, Texas. Courtesy Kay Brown.

Unloading cotton into module stacker. Courtesy Kay Brown.

An eight-row John Deere 7455 mechanical cotton stripper equipped with a field cleaner in Crosby County on the Texas Plains. Courtesy Randal Boman, Texas AgriLife Extension Service.

CHAPTER 16

RESEARCH
The Key to Viability

"We are in accord as to the necessity for scientific research." When Oscar Johnston made that statement to the founders of the NCC in 1938, he knew that cotton would have to depend on science and technology to remain viable in the U.S. economy. At the end of World War II, USDA research expanded in nearly all areas of agriculture—livestock, fruits and vegetables, grains—and cotton benefited, too. The NCC's division of technical services stepped up its promotion and coordination of scientific projects, and later Cotton Incorporated, created in 1970, opened its research center at Cary, North Carolina. During the last half century, this combination of public and private interests tackled what seemed an endless row of challenges for growers, whether imposed by nature or the vagaries of the world economy.[1]

Agricultural scientists had a definite goal: to advance the productivity of farming. Because cotton had both agricultural and industrial value, it was the chief commercial crop, and because it was a principal export, cotton had to remain economically viable for the sake of public interest. This common purpose drove researchers as they dealt with all manner of threats to cotton—insects, plant diseases, competition from synthetics—as well as improving yields. The breadth and depth of research was extensive because of the wide range of choices as well as threats that growers face each year in making a crop. They must determine the best tillage practices in view of soil conditions; select the variety of cotton likely to provide the highest yield in view of weather, insects, and plant diseases; grow plants suitable for mechanical picking; regulate irrigation to avoid overwatering; apply fertilizers for maximum yield but not allow overgrowth of plants; and choose equipment best suited for the scale of operations. And given that unknown developments arise, such as the appearance of soil diseases or damaging insects, challenges renew themselves regularly and make research imperative.

SEED DEVELOPMENT

Seeds, or the development of new varieties of cotton, demonstrated the importance of research, and the pursuit of better varieties received new energy at the end of World War II. A major breakthrough came in 1945 when Early Ewing stumbled on a plant at D&PL whose leaves glistened in the sunlight because they had no hairs. He anticipated some advantages in this unusual characteristic and crossed the plant with other cultivars that had stronger fiber strength. From this shiny plant, discovered by accident, Ewing introduced Deltapine Smooth Leaf in 1957, which greatly aided growers who were switching to mechanical pickers. Smooth Leaf produced less trash in the picker and made ginning easier for mechanically harvested cotton. It also offered a stronger resistance to fusarium wilt and verticillium wilt, soil diseases that were almost as devastating as the boll weevil.[2] In 1954 Anderson, Clayton and Company released Paymaster 54B and followed it with Paymaster Storm Rider and Paymaster 101. To further its development of superior cottons, the company opened a fiber-testing laboratory in Houston in 1949.

California is an intriguing example of seed development and shows the importance of USDA leadership in cotton research. Cotton farming had always been unique in California, having features and characteristics that separated it from the southern cotton culture. Its one-variety law passed in 1925 and the success of the USDA scientist George Harrison soon after World War II in developing Acala 4-42 had made cotton into white gold there. In 1952 Harrison left the USDA for a position with the California Planting Cotton Seed Distributors (CPCSD), and the USDA asked John Turner of the Georgia USDA Experiment Station to take his place. Turner described how he had a dream in 1952 in which he saw a valley of plush cotton, a "triple hybrid" that he developed, which produced "two bales or more per acre," and on the same day he received a letter from USDA headquarters about the job in California. John and his wife Georgia moved to California despite her misgivings about earthquakes. He faced powerful growers who had come to believe that almost any problem related to farming could be resolved with the development of new varieties, whether it was resistance to disease, the textiles' stiffer demands for quality fiber, or just the need for higher-yielding cotton to combat rising costs. Harrison's miraculous success with Acala 4-42 had engendered this perception and led to the assumption the California research station at Shafter could overcome nearly any obstacle to cotton farming.

California growers faced a hurdle that was not new but had reached intolerable levels: the spread of verticillium wilt. They felt that the USDA ignored this problem, and so they "put political pressure on the federal agency," reported one account, and some frustrated growers had illegally experimented with other varieties.[3] To help Turner with wilt, the USDA sent another researcher to Shafter, Hubert Cooper. But progress was too slow to suit the growers, and they established an independent research program solely for the San Joaquin Valley in cooperation with the University of California. Minor friction arose between the federal and state research programs, but both nonetheless proceeded. Turner and Cooper incorporated several breeding varieties produced by the New Mexico Experiment Station and managed to develop a cultivar able to reasonably resist the wilt, Acala SJ-1, which he released in 1967. In 1973 they released SJ-2 with stronger wilt resistance. Further improvements came with SJ-4, and SJ-5 that provided higher yields and strong resistance. But the lucrative nature of cotton farming in the San Joaquin Valley had caught the attention of private seed companies, and in 1979 the one-variety law was amended so that they could develop varieties for the commercial market. That year the USDA closed down the Shafter station and gave the germplasm to private firms. Growers wanted a greater number of varieties available to them and looked to the CPCSD for new developments. In 1984 it provided growers with GC-510, a slightly coarser cotton, but one that produced much lint and had strong wilt resistance. Infestations of wilt varied in the valley, so that not all growers needed the newer cultivars. In 1990 the CPCSD released Acala Maxxa, which became the most common variety by the mid-1990s owing to its wilt resistance, early maturation, and high yield. Maxxa performed well in areas with great infestations of wilt. By 2000 growers worried less about wilt, and some had begun switching to Pima cotton, the extra-long staple that brings top price.

NEW MEXICO SEED DEVELOPMENT

"Cotton has a long and rich history in New Mexico." That statement by noted cotton agronomists expressed the state's prestige in cultivar development, though New Mexico never grew cotton in large volume. The state had only two cotton-growing areas, the Pecos Valley and Mesilla Valley, where farmers first raised cotton in 1910 and 1919 respectively. With no fear of boll weevils, they increased acreage, so that by 1922 cotton was a chief crop. Cold winters kept down insects, and hot summers encouraged growth of plants. When in 1926 the USDA opened a cotton laboratory in Las Cruces and put G. N. Stroman in charge of seed development, New Mexico's significance to the industry began. In 1940 Stroman introduced Acala 1517, which performed well in the Mesilla Valley and "would prove to be a rich reservoir for future

1517 germplasm improvement." Stroman developed Acala 1517 from seed he brought from Mexico. It was not the same Acala grown in California, though the Acala there had 1517 parentage.[4]

Verticillium wilt spread and forced breeders to develop wilt-resistant varieties. In 1946 the Las Cruces station released 1517 WR and followed it with 1517B in 1949. Acala 1517C, made available in 1951, had still better resistance to wilt and produced a large boll well-suited for spindle picking. It accounted for the premium cotton still grown in the El Paso area. But wilt persisted, and the New Mexico breeders fought back with 1517 BR in 1954, which was followed by still newer cultivars. In 1960 1517D became available and served as a parent of a variety grown in California, SJ-1. Further developments followed, each new cultivar offering an advantage over the previous one, but often with a sacrifice in fiber strength, wind resistance, or other characteristics. In 1995 the Las Cruces station released 1517-95. The seed research program in New Mexico won high respect for its 1517 varieties and contributed to 45 percent of all varieties in the United States between 1970 and 1990, more than any other publicly funded seed development program.[5]

EXTRA-LONG STAPLE

Research to develop extra-long staple (ELS) cotton went back to the late nineteenth century. The USDA operated a nursery for ELS at Yuma, Arizona, but moved it to the Gila River Pima Indian Reservation at Sacaton, Arizona. Thomas H. Kearney of the USDA released a variety of ELS in 1918 that he named Pima. Robert Peebles took over from Kearney in 1934 and produced Pima-32 in 1948. At the University of Arizona, Walker Bryan produced a superior cultivar, Pima S-1, that by 1955 accounted for nearly all the production of Pima cotton.[6] It had excellent spinning properties. The station at Sacaton closed in 1957, and ELS research moved to the University of Arizona under the direction of Carl Feaster. Edgar Turcotte joined him, and they made an effective team in developing Pima varieties. They released succeeding varieties with improved lint yield, early maturation, heat tolerance, and fiber strength. In 1991 the USDA released Pima-7 but shut down its ELS seed development in 1993. Pima production grew in California, however, more than enough to attract private seed companies. During the 1980s, they became active in Pima development, basing their work on the earlier accomplishments by the USDA. One variety, Conquistador, resulted from the work by James Olvey of Olvey and Associates. It proved popular and accounted for much of the Pima acreage. Because of the growing demand for ELS cotton, interest in Pima remained strong.

COLORED COTTON AND ORGANIC COTTON

An interesting step occurred in seed breeding with the development of naturally colored cotton by Sally Fox and Harvey Campbell. Colored cotton had grown in Latin America before the European settlement, but it never developed commercial value for lack of strength and very short staple length. It appealed to Fox, an environmentalist, because colored cotton required less pesticide and dyeing. When she began, Fox worked alone raising plants in flowerpots on her back porch. Her production increased until by 1985 she was growing seeds in a small field. After eight years, she produced a brown fiber that was uniform in color with staple length of one inch, and began selling it to hand spinners. Campbell had begun experimenting in 1988 after several years spent developing Acala varieties, particularly GC-510, which had strong resistance to wilt and had sold well in California. Both expanded their nursery acreage and sold to smaller textile mills. Fox licensed growers to raise her cotton under the trade name FoxFibre, while Campbell, who was associated with F. E. Birds, operated as B. C. Birds Inc.

Campbell described colored cotton as "one step out of the jungle," meaning the germplasm was volatile and wild and made plants keep growing taller. Hand picking worked better with the cotton, since it kept down trash and the discoloration that comes with mechanical picking. Dyes are not used, so the lint must be kept clean. A few retailers offered clothes made of colored cotton, but hand spinners were the principal market; by 2000 sales surpassed fifty thousand bales. Expansion of the market will depend on whether it can match the range of colors available with commercial dyes, whether it can equal the spinning qualities of premium Acala and Pima varieties, and if the further development of transgenic cotton will negate its advantage in resisting insects. Hand picking will limit acreage, too. But research has overcome many obstacles, and the exemplary work of Fox and Campbell sent a message easily understood, that research can produce cotton of a different color.[7]

Organic cotton has drawn interest among consumers who want to wear clothing produced with a minimum environmental footprint. In other words, they prefer organic fiber that meets the criteria established by the U.S. National Organic Standards Board, which defines "organic" as a system of production that practices minimal use of pesticides, herbicides, and chemical fertilizers and avoids genetically modified varieties. In 2009 world production reached about 115,000 bales, but the United States accounted for fewer than 8,000 bales. Turkey was the major producer, owing to the cheaper labor costs. American growers fear that the lack of chemical inputs to fight insects and weeds would lead to infestation of boll weevils or pink bollworms and further spread weed growth. Some growers report lower per-acre yields without the use of fertilizers. Clothiers that offer some organic cotton apparel include Nike, Patagonia, and Cottonfield USA, and demand for organic clothing is expected to grow, but not enough to offset conventional cotton.[8]

NEW TILLAGE PRACTICES

Tillage practices follow a fundamental routine—plowing, planting, and weeding—before harvesting, but variations or improvements occur with new technologies or environmental conditions. One practice that illustrated how growers adjusted their cultivation practices was "no-till" farming, sometimes labeled conservation farming. Most of the pressure to use no-till began when the farm bill of 1985 required farmers to place land subject to severe erosion into the USDA Conservation Reserve Program or implement conservation practices on it. In a few instances, growers had already experimented with no-till because of water shortages or the need to devise lower production costs. With the congressional mandate, however, agricultural researchers at USDA and state experiment stations, sometimes working in conjunction with Cotton Incorporated, conducted field trials to determine how to make no-till a feasible alternative to conventional tillage. This new approach challenged centuries of practices. Farmers enjoy looking at a field clearly plowed and ready for planting; they enjoy the scent of freshly turned soil. "There is no question that it's a different kind of cotton farming," stated the Tennessee grower Allen King.[9]

No-till offered the advantage of less plowing. Cotton might be planted in rows on top of stubble from a crop of wheat, barley, or vetch. An application of weed killer might be made before planting cotton, but the residue stubble remained on the ground. Once the cotton sprouted, it appeared to be growing in weeds or a leftover crop. A few more applications of herbicides usually followed. Defoliants would be sprayed on the cotton in preparation for the mechanical picker. No-till reduced fuel costs, since there were fewer trips in the field with tractors. The savings might be offset, however, with lower yields, so growers approached no-till cautiously. The new method has not been wholly adopted, not because of ingrained orthodoxy, but the economics did not always justify the conversion. No-till carried risk and required an assessment of each farm's conditions. In 2003 *Farm Press* reported that the NCC put the amount of no-till at 59 percent of the total cotton acreage in the United States, but growers were dubious in the Far West, where only 17 percent of the acres went into no-till. Ron Rayner of Goodyear, Arizona, was optimistic about no-till cultivation but recommended that growers use caution in adopting the new method.[10]

Another alternative appeared in the 1970s known as skip-row farming. Congressional legislation encouraged conservation by providing for a "set-aside" of 28 percent of each allotment. There were variations in the skips, perhaps a 2:1 or a 4:2 ratio in the pattern of rows, but an open strip of unplanted land ran between the cotton rows. This practice arose because it allowed farmers to spread their allotment over more acres, and with intense cultivation, irrigation, maximum use of fertilizer, high-yield seed varieties, and weed control, they could get the same yield per acre as with conventional planting. If the yield fell, however, the lower production costs might offset the loss of revenue. But skip-row did not work for everyone because of varying soil conditions, weather, insects, and similar constraints. It worked best in dry areas that could not sustain intense planting. In some areas without irrigation, skip-row farming remains.[11]

THE MODULE

Like other industries, cotton relied on research and innovation to produce and handle its product at the lowest cost. One of the best examples was the module, the compressed stack of cotton, resembling a giant loaf of bread, commonly seen at the edge of the field. In 2002 the American Society of Agricultural Engineers recognized the module as one of the top advancements in mechanized cotton farming. The module broke the bottleneck that came when farmers loaded their machine-harvested cotton into wire-cage trailers and pulled them to the gin by tractor or truck. Drivers had to wait in line to place their trailer under the suction tube that fed the raw fiber into the gin. Here a delay occurred, and drivers sometimes waited for several hours. Since growers had only a handful of trailers, the mechanical picker had to wait in the field. Loss of time meant loss of money as workers still had to be paid. And unpicked cotton deteriorated with exposure to wind and rain or sun and soil. For nearly twenty-five years after the introduction of mechanical pickers, the bottleneck had frustrated growers. "It was common to find long lines of trailers full of cotton at the gin," stated Professor Calvin Parnell of Texas A&M University.[12]

Growers on the Texas High Plains had tried to get around the bottleneck by experimenting with "rickers," a three-sided box-shaped container into which they dumped the seeded cotton out of the picker. Once the ricker filled, field-workers simply stomped the cotton into a compact stack. When the ricker would take no more cotton, a tractor pulled it away and left a stack of seeded cotton on the ground. Workers covered it with a tarp. This early form of a module eased the traffic jam at the gin, but rickers had drawbacks. The cotton was difficult to load onto trailers, sometimes collapsing due to poor compaction, and the trailers still lost time at the gins. In severe wind, common in west Texas, the ricked stack might blow apart, so rickers were only a stopgap measure. The answer to this dilemma began at the University of Arkansas.

Zinn McNeil, an agricultural engineer at the University, devised his "Arkansas Cotton Caddy," a box on a truck in which cotton would be stomped and then hauled to the gin. His caddy eliminated the step of loading the stack but required its own motorized transport. At this point, J. K. "Farmer" Jones of Cotton Incorporated entered the picture. Jones recruited agricultural engineer Lambert H. Wilkes of Texas A&M to join him in searching for a solution. With funding support from Cotton Incorporated, Wilkes devised a module builder equipped with a hydraulic ram to compress the raw cotton. The ram had enough power to tightly compress the cotton so that it could stand alone. Cotton Incorporated licensed G. A. Huskey in Lubbock to manufacture the module builder. Professor Parnell began field demonstrations of the new device, mostly in the western areas, where rain presented the least chance of deterioration to modules standing in the fields. Despite the progress, there remained the troublesome matter of loading the giant bread loaf onto a truck for transport to the gin.

In a chance encounter, the solution appeared. Implement manufacturer Barry Reynolds of Lubbock noticed hay being loaded on a truck. The truck bed tilted and rolled under the hay, and Reynolds saw this technique as the answer to loading cotton modules. After experimenting with several prototypes, he designed a truck that rolled under the cotton like the hay loader. Now a full module of ten to fifteen bales of cotton could be taken to the gin yard and left on the ground. As the ginner needed more cotton, he used a module-loading truck to reload the stack and bring it forward for processing. Gins had to replace the suction tubes with rotating cylinders that tore the compressed stack apart and sent the fiber into the cleaning and deseeding machines. Use of the new system spread across the Cotton Belt, and "the module builder quickly took its place," asserted a team of writers, "as a central instrument in sustaining the profitability of cotton agriculture and is a striking example of cost effectiveness in agricultural research."[13]

Modules are not foolproof. Rain threatens them though growers cover them with a tarp. If rain becomes excessive, or if flooding occurs, water will wick up the loaf and cause the cotton to generate heat and rot. Modules stay in the fields for only a few days or perhaps weeks, depending on the pace of ginning, and under normal circumstances they sustain no damage, but there is an irony connected with them. To protect the modules waiting in the fields, growers learned that tarps made of polyethylene provide better protection than a cotton canvas cover. But modules "revolutionized and helped save the cotton industry nationwide," wrote one author.[14]

THE BREAKTHROUGH OF HVI

The history of assigning a grade to cotton went back to the mid-1770s, when English buyers relied on area descriptions, meaning only a designation of the location where the cotton grew, for example, the American colonies, the West Indies, or India. Grading cotton advanced during the nineteenth century as the industry established a nomenclature and some standardized criteria for staple length and grade, but classing left much to be desired. The vastness of weather, soil, and growing conditions in the Cotton Belt caused cotton to vary in color and staple length, necessitating the need for classing.[15] Cotton classers, men trained to judge samples from each bale, became a vital part of the industry. By eye they set the grade based on color and foreign matter, and by hand they set the staple length. The classers' decisions applied unless the farmer requested arbitration. Merchandisers and textile mills employed most of the classers because grade and staple length determined the price for each transaction. Speedy and accurate classing, especially a system on which seller and buyer would accept, had always been a challenge.[16] It was this set of circumstances that prompted the research into instrument grading by the PCCA in Lubbock.

Before World War II, the colorimeter for determining the shade of white and fibrograph for measuring fiber length were developed by the USDA and university researchers. In 1942 E. H. Pressley at the University of Arizona invented a fiber-strength tester. There remained the need for an objective and reliable test of cotton, however, so that textile manufacturers could anticipate its spinning characteristics. Synthetic manufacturers had a definite advantage in that respect, since their product remained consistent and uniform. This competition drove the search for a way to overcome the human touch, to class cotton more accurately and consistently.[17]

After the progress made by PCCA's Emerson Tucker and the research firm Motion Control, testing of a prototype high-volume instrument (HVI) occurred through 1973 at USDA classing stations in Memphis, Lubbock, and Hayti, Missouri. Another company, Spinlab, developed HVI technology, though Motion Control came first in 1969 and the latter in 1985. Some merchandisers resisted the advancement, fearing it would jeopardize accurate grading or drive them out of business by enabling growers to sell directly to mills. But when in 1987 the NCC endorsed HVI testing, the new technology received a boost, and when in 1991 the CCC announced that cotton going into the loan would require the new testing, HVI became the standard. That year all the USDA classing stations converted to instrument testing. Jane Byers in the Phoenix station was the last government classer to determine staple length by hand. In 2000 engineers made a further refinement when they added color, or the shade of white, to HVI testing.[18]

Like modules, HVI testing was a major step forward. It removed the subjectivity of human classing. It gave textile mills a better grip on the spinning quality of a bale, and mills could now avoid the unintentional mixing of bales with varying characteristics and produce fabric with more consistent quality. Accurate testing made U.S. cotton attractive to foreign mills for the same reason, but the advantage lasted only about five years until other cotton-growing countries began to acquire the technology in the late 1990s. This breakthrough ended a major activity of marketing centers, the presence of classers wearing long white overcoats and whose numbers in merchandising houses had added much to the aura of the cotton markets. Now USDA classing stations, housed in out-of-sight industrial areas, handle all grading. A combined effort led to this advancement, particularly the work at Texas Tech University, Cotton Incorporated, and Motion Control, but the cooperation of the USDA and textile mills was important, too. Each bale began carrying a bar-coded tag showing grade, staple length, and micronaire, and that development enabled growers to sell directly to mills. It was this advancement that gave rise to electronic trading.[19]

NEMATODES

The complexity of farming cotton becomes apparent when considering the impact of nematodes, the microscopic parasite that attacks plant roots. It penetrates the root system and draws nutrients from the plant, causing the leaves to turn yellow and delay setting the bolls. In the most severe infestations, nematodes can reduce yields by 50 percent, but generally losses run closer to 10 to 25 percent. A parasitic worm invisible to the eye, the nematode may be visible as clusters of nodes on plant roots. Two varieties of nematodes thrive in the Cotton Belt, *R. reniformis* and *M. incognita*, but *reniformis* confines itself to the southern states. In some instances, nematodes will encourage the appearance of verticillium wilt, but agricultural scientists recommend applications of nematicide when losses exceed 5 percent. Crop rotation serves as an effective resistance, particularly with corn in the South. The Mississippi grower Bill Pearson encountered an infestation and had to rotate with corn. Progress in fighting the parasite occurred through seed breeding when the station at Shafter, California, developed Nem-X; other varieties using D&PL strains developed in the South provided some protection.

The NCC began keeping records on nematode damage in 1952, which showed that the average yield loss for the Cotton Belt was slightly over 2 percent, but a dramatic increase began to occur in 1987 when the loss was 1 percent, but jumped to 4.39 percent in 2000. In 1995 the Cotton Board inaugurated a cooperative program with Bayer CropScience to promote a nematode and research education program. Nematodes cannot be eliminated, so control is the objective. Turning stalks underground exposes them to drying and temperature extremes, and crops can be rotated, but no single step will accomplish the desired goal in all states. Injecting fumigants into the soil also helps. More than anything, a healthy stand of plants provides the best prevention, so good soil with nutrients and water is recommended. Nematodes remain a nuisance that requires vigilance and further research in overcoming them because though the losses do not match the damage wrought by other pests, this underground pathogen can erode a grower's narrow margin of profit.

COTTONSEED OIL

An overlooked dimension of cotton pertains to cottonseed and the products made from it. Cottonseed oil had little value in the nineteenth century, even being sent to Italy, where it was bottled and resold in the United States as olive oil. Described in one case as a "Cinderella," cottonseed became an integral and valuable part of the cotton kingdom thanks to research.[20] Agricultural and food scientists developed a variety of uses for cottonseed, which used to be cast away as waste. Specifically, the oil from crushing seeds became the basis for oleomargarine, cooking oils, and salad dressings; but nonfood items included paint, cosmetics, soap, plastics, and lacquer. From the small fibers that cling to the seed after ginning, called linters, comes nitrocellulose, an ingredient used in the manufacture of gunpowder, dynamite, and later rocket propellants. Cottonseed has high protein content, and when processed into cottonseed meal or hulls, it makes a feed for cattle. Growers raise cotton for the lint, but the approximately 750 pounds of seed ginned from each bale will account for 10 to 15 percent of their profit. Often they used the seeds to pay for the ginning. Anderson, Clayton and Company built its first oil-crushing plant in 1906 and established itself in the food industry because of its readily available supply of oil. After World War II, the company operated crushing mills in Texas, California, and Arizona. ACCO sold cottonseed oil food brands under the trade names of Mrs. Tucker's shortening, Meadowlake and Chiffon margarine, and Seven Seas salad dressing. The noncotton side of ACCO, meaning mostly foodstuffs, accounted for a large portion of its profit.[21]

Cottonseed oil was popular for frying potato chips because it gave them a nutty flavor. It also had better stability under cooking conditions than soybean or canola oil. For margarines and shortenings, cottonseed oil has desirable crystallization properties. Food processors often mix it with other oils to impart these qualities into their product. Cottonseed oil may be used as a substitute for cocoa butter, and its creaming properties make it a popular ingredient in commercial confectionary items. In 2006, however, Frito-Lay announced that it would no longer use cottonseed oil for cooking potato chips and would switch to sunflower oil because it is 50 percent lower in saturated fats. The company had responded to the growing consumer insistence for healthier foods, although the number of calories in the chips would hardly change. This move demonstrated the fluidity of the cotton market, and how growers and processors have to keep abreast of changing market conditions to be viable.

NCC RESEARCH

The NCC's division of technical services has the responsibility for promoting research and gathering technical information about cotton farming. It has no authority for funding to approve projects, but the political arm of the organization regularly lobbies Congress for funding of omnibus bills for general agricultural research or projects specific to cotton. In 1955 when the NCC made its special study of cotton's future prospects, Rhea Blake referred to a point made by Roger M. Blough, the new chairman of the U.S. Steel Corporation, that American industry had spent more money on research after 1945 than since the founding of the country. "Only through research and the application of research findings," Blake liked to say, could cotton hold its market share.[22] That same year the NCC established the Cotton Foundation in cooperation with several private firms that had a stake in a healthy cotton industry. They wanted to further the practice of scientific farming, since their products would be widely used. But agribusinesses were not eligible for membership in the NCC, so through the Cotton Foundation, implement manufacturers, seed companies, financial organizations, and chemical companies could funnel money into research projects. The foundation supported inquiries into a variety of studies and educational programs on the grounds that viable farming enhanced their profitability. It operated out of the Memphis headquarters of the NCC, which supplied the personnel. In this respect, the NCC tapped into the resources of agribusinesses willing to fund research.

A reliable source of funds for research to supplement or replace the loss of federal dollars began in 1966 with the voluntary assessment on growers through the Cotton Research and Promotion Act. When Congress amended the measure in 1991, making assessments mandatory, it stipulated that 7.5 percent of the funds be divided among the cotton-growing states for research. Known as the State Support Program and administered through Cotton Incorporated, the funds go to the states based on their acreage. Most of the support goes to land-grant universities as grants for cotton-specific projects. Areas of research include tillage practices, weed and insect control, diseases, seed breeding, marketing procedures, and related topics. In 2005 the first multistate program got under way when six states in the South initiated a cooperative effort to control "stinkbugs." Additional funding by the State Support Program partly filled the research gap created by the cutback in public funding.[23]

In the 1980s, the Cotton Foundation began its reference book series on cotton-related subjects. Each book became a comprehensive look at a single subject, written by noted agricultural scientists and specialists. In 1986 the foundation published the first volume, *Cotton Physiology*. Other books followed: *Weeds of Cotton* (1992), *Vegetable Oils and Agrichemicals* (1994), *Cotton Insects and Mites* (1996), *Boll Weevil Eradication* (1999), and *Cotton Harvest Management* (2001). In 1997 the foundation began the *Journal of Cotton Science*, a quarterly publication of refereed scientific papers. Authors generally come from university faculties or agricultural research stations, and in the case of faculty, articles published in the journal are acceptable as progress toward tenure appointments.

As the NCC coursed through the years promoting research along with other activities, there evolved a series of conferences, some initiated by the organization and some outside it, eventually culminating in the annual Beltwide Cotton Conference. After its first mechanization conference in 1947, the NCC held annual meetings on mechanization and added workshops in agricultural engineering. Other meetings pertaining to cotton were the Cotton Weed Science Research Conference and the Cotton Economic and Marketing Conference that had their first gathering in 1977 when they met in conjunction with the NCC's mechanization and production conference. In 1984 the Cotton Soil Science meeting and the following year the Cotton Ginning Research Conference meeting were held with the NCC. In 1985 the NCC combined these conferences into a single gathering and named it the Beltwide Cotton Conference.[24]

At that point, all segments of the industry participated, including USDA researchers, scientists from universities, and researchers and engineers from commercial seed companies, equipment firms, chemical companies and marketing specialists. Foreign visitors attended, and the news media covered the event. In 1988 the Beltwide Conference had about fifteen hundred attendees, but the popularity of the meeting grew, and the extent of scientific and technical reporting increased, so that it became the overpowering cotton conference. By 2000 it had approximately five thousand registered participants and visitors, making it the largest cotton conference in the world.

The Beltwide always comes in early January for four days. There is a program of workshops, scientific reports, equipment demonstrations, and market studies. In the more recent years, Billy Dunavant made an annual presentation about the future market. Growers want information and tips and particularly like panel discussions conducted by other growers on topics such as tillage, insecticides, seed selection, and other farming practices. Faculty from land-grant schools speak to their comrades in science about the physiology, morphology, and anatomy of the cotton plant. Textile researchers discuss cotton from the perspective of finishing cloth. Speakers from other countries share and observe.

But not all is rosy at the Beltwide. Growers may bring their complaints and want answers for mishaps and broken promises from agribusinesses. In January 2001 growers along with textile representatives and merchants took issue with Bt cotton, claiming it was inferior and produced lower-quality lint and brought a loss of $30 million in the Cotton Belt. Dunavant thought genetically modified seeds had brought a "major, major problem," and told Reuters News that "a lot of people agree with that."[25] Seed companies blamed poor weather in 2000 for the lower quality.

Exhibits are provided by agribusinesses, and they furnish personnel to answer questions. There is a Cotton Crafts store that offers gifts such as T-shirts, caps, watercolor paintings, and the ubiquitous miniature bales of cotton. The National Cotton Women's Committee hosts a style show of cotton fashions. For spouses who find the sessions too boring, there are organized tours of shops, museums, historic homes, and restaurants. Evening receptions usually hosted by agribusinesses are plentiful. Other cotton organizations like the American Cotton Shippers Association, the National Cotton Ginners Association, and the National Cottonseed Products Association hold separate annual meetings, but the Beltwide Conference brings the individual interests together under one research umbrella.

CONCLUSION

Research improved quality and yield in nearly all endeavors of agriculture, enabling the United States to enjoy abundant food and fiber. For cotton growers, research improved their livelihoods and kept the United States as a viable producer in the world market. Funding often came as a result of the lobbying by producer organizations and led to breakthroughs in weed control, cultivation practices, and seed improvement. Much credit should go to the disciplined work of extension service and university researchers.

Growers saw scientific inquiry as critical to progress and organized to raise funds among themselves, owing to their belief that nothing should be accepted as untouchable because of tradition or old practices. This change of attitude represented a leap forward from the pre–World War II era, when intellectual sterility characterized much of the cotton community. However mundane and unglamorous cotton research may seem, it kept the United States assured of an independent supply of the fiber of life.

CHAPTER 17

CHALLENGES ANEW

Although the cotton industry achieved many objectives, there always seemed to be another hurdle to overcome. As soon as one challenge had been resolved or made manageable, another took its place, some of which were refashioned versions of older difficulties or unanticipated new perils. Whatever the origin and nature of the threats, the collective power of cotton interests sought to defuse them or accommodate and incorporate them into their operations. In the 1970s, a challenge arose unlike anything encountered before, one that directly touched only one segment of the industry but had threatening ramifications for all. It was byssinosis, or brown lung disease.

The disease affects workers in cotton textile mills and resembles black lung, which occurs in coal miners. Shortness of breath, coughing, and wheezing characterize byssinosis, and these conditions are thought to result from an allergic reaction to inhaling dust from cotton and hemp fibers. In its more advanced stages, the ailment turns to bronchitis or emphysema and can be reversed if treated soon enough, but respiratory or heart failure may come in untreated patients. In 1731 the first record of byssinosis occurred when the Italian physician Bernardino Ramazzini wrote about illnesses associated with workers in flax and hemp shops and described a condition of the lungs that induced coughing and asthmatic difficulties. For the next two hundred years, medical scientists knew that brown lung occurred in cotton mill workers, but otherwise knew little for certain. Occupational illnesses received little attention until around the mid-twentieth century because medical researchers had focused on infectious diseases, and inorganic agents such as cotton dust did not get much notice. As late as 1975, a team of medical clinicians reported that "byssinosis is still poorly recognized in the United States and is seldom listed as cause of either death or absenteeism," but they continued, "there is little doubt that byssinosis remains a significant cause of premature disability and death today." They lamented that few pathological studies had been made, but acknowledged that cotton dust set off the illness.[1]

The earliest efforts to understand the lung ailment occurred in Britain. Clinical observations were made there, but no definite knowledge was established. Because byssinosis affected only a small portion of the population, cotton mill workers, it was thought to be peculiar to textile manufacturing areas. The same symptoms shown by British workers appeared in American textile workers, but most of the concern, which was limited, remained in Britain. In 1936 a hygienist for the North Carolina Department of Health, M. F. Trice, published an article in *Textile World* that recommended better ventilation and vacuuming dust in mills for combating the ailment. James Hammond of the Division of Industrial Hygiene in South Carolina conducted a limited amount of research in the mid-1940s and made similar recommendations, but these studies had no impact on working conditions, nor did they lead to any scientific certainty about the causes of byssinosis. In 1945 Harvard professor Philip Drinker conducted a study in which he observed that dustiness in mills should be addressed, but concluded that cotton dust was not connected with poor health. His colleague Leslie Silverman made another study and claimed that byssinosis was a product of cotton dust and made recommendations for cleaner air in the mill workplace with a warning that mechanical cotton picking might increase the amount of dust in textile work areas. In 1946 a review of Drinker's report appeared in *Textile Labor*, the organ of the Textile Workers Union of America. The union felt the study clearly established cotton dust as the cause of lung ailments among mill workers but made only token efforts to get working conditions improved. In 1950 Silverman published the last piece connected with the Harvard study and asserted that "there is a definite relationship between total concentration of contaminants in the atmosphere and the incidence of the acute disease."[2] Yet a review of studies by the U.S. Public Health Department purported there was no solid evidence to connect cotton dust with byssinosis. Reports and studies were thus too indefinite and conditions for workers changed little, though some mill owners began to install air-conditioning in new plants to improve air quality.[3]

The recognition and understanding of occupational diseases belonged in the era after World War II. In 1945 Oscar Johnston expressed some concern with byssinosis and studied literature on the subject researched by the Institute of Textile Technology in Charlottesville, Virginia. He shared the literature with the Fine Cotton Spinners and Doublers' Association in Britain, but nothing came from this exchange.[4] Public health and medical researchers continued to neglect the ailments of the mill workers, often attributing their shortness of breath and the other symptoms to smoking. Textile management did not think employment in the mills was responsible for the illness. Workers often experienced their worst coughing and wheezing on Monday after being absent from the mill for the weekend, which gave rise to the term "Monday fever" and encouraged the suspicion that lifestyle among the workers accounted for their sickness.[5]

Although medical scientists saw a connection between cotton dust and brown lung, they still acted with no certainty beyond making a link between the two, and the question continued to be a matter of scientific puzzlement. Further work occurred in the 1950s by Dr. Joe Bosworth, medical director for the Liberty Mutual Insurance Company, which insured the mills, but he produced little assurance that cotton dust was the causal agent of byssinosis. But a more militant and sympathetic general attitude toward the plight of the downtrodden began emerging in the 1950s, set off by the civil rights movement and the rising concern over public safety, so when researchers in the Georgia Department of Public Health published articles on brown lung, the suspicion grew that lung disease was synonymous with mill work. In 1967 Dr. Arend Bouhuys, a Swedish scientist interested in byssinosis who had joined the faculty at Yale, published an influential article in the *New England Journal of Medicine*. He made a convincing case that byssinosis was not solely a European ailment but also existed in the United States.

Events began moving faster in the late 1960s and early 1970s and led to the eruption of the byssinosis crisis for the cotton industry. In 1968 Dr. Peter Schrage of the United States Public Health Service (USPHS) finished a study he made of workers in a textile mill that had let him examine its employees. He wanted to publish the report, but the mill refused to grant permission. Schrage complained to the USPHS, which published an abstract of the report, but Schrage wanted more done with the paper. Still the USPHS refused to cooperate. In 1971 Schrage took the report to Ralph Nader, who published an article in *The Nation* titled "Brown Lung: The Cotton Mill Killer." Nader described byssinosis as a little-known disease that affected thousands of textile workers, leaving them ultimately with lung degeneration resembling the effects of emphysema or bronchitis and making them unable to undertake physical activities. He saw brown lung as "an occupational disease of serious proportions that potentially threatens nearly half of the industry's labor force."[6] Nader accused textile managers and the NCC of ignoring the disease, claiming that the managers refused to let public health workers inside their mills and that the NCC would not allocate funds to research the illness. He also indicted the USDA for not undertaking studies of the ailment. Too many state governments were "in the grip" of textile mills to expect improvements at the state level, he continued, so only federal action could address the plight of the workers.

OSHA

In 1970 when Congress created the Occupational Safety and Health Administration (OSHA), byssinosis acquired new meaning because the agency provided a pathway through which the disease could be studied and safety regulations imposed in the mills. In 1968 the secretary of labor, acting under the Walsh Healey Act, had set forth a standard for cotton dust in the workplace that had been adopted earlier by the American Conference of Governmental Industrial Hygienists, and OSHA subsequently established that standard in 1970. In 1972 OSHA added cotton dust to its list of hazardous workplace substances, and the agency's research arm, the National Institute for Occupational Safety and Health, began researching byssinosis.[7] Some workers now filed for disability compensation, which escalated the seriousness of the disease for the mills because they could conceivably be forced to pay settlements for thousands of workers. In the meantime, Burlington Mills had agreed in 1970 for Duke University to study employees and the levels of cotton dust inside the company's mills. Burlington added physicians, nurses, and a hygienist to the payroll to assist in the study. This study, among the first examinations inside a mill, reaffirmed the earlier work by scientists in their efforts to establish the connection between cotton dust and byssinosis. Knowledge within the health profession about the cause of brown lung thus made a step forward, but some reservations remained, and for that reason the workers' general awareness of the disease was not widespread.

Within the cotton community, however, byssinosis received special attention. In 1973 the NCC declared the disease to be the industry's foremost problem and established the Industrywide Cotton Dust Committee, comprising members from each of the segment interests. This meant that growers had joined with textiles, ginners, warehousemen, cooperatives, merchants, and cottonseed crushers in seeking a resolution. Cotton Incorporated had a seat and conducted research in its laboratories.[8] A technical subcommittee consisting of physicians, public health specialists, and economists supported them. Two objectives drove the committee: to raise funds for a crash research program aimed at finding the cause of byssinosis, and to find a way to remove the causative agent from raw bales before sending them to textile mills. It hoped to raise $9 million to add to the $4 million already spent on research by federal agencies and industry interests. All this effort was undertaken to reduce the health hazard to mill workers without jeopardizing the economic solvency of the mills.[9]

A special delegation of the NCC took the plan to Secretary of Agriculture Bob Bergland to plead for federal funding. He praised the plan, but warned that President Nixon's austerity program would likely prevent additional funding from the administration to be awarded for the Industrywide Committee's effort.

A similar disease, black lung, had caused coal miners to strike. Activists on behalf of miners moved to brown lung as they learned about it, and with funding from the Church of Christ, the Campaign for Human Development, a Catholic charity, the AFL-CIO, and the J. P. Whitney Foundation, they set up the Brown Lung Association (BLA) in Columbia, South Carolina, in 1975. The BLA offered breathing clinics, raised funds, promoted and publicized the organization, and filed workers' compensation claims on behalf of workers. Much of the pressure on behalf of workers came from the Amalgamated Clothing and Textile Workers Union, which wanted to step up its membership drive. General awareness of byssinosis and its link with cotton dust began to improve among mill workers and the general public, but scientists still could make no definite connection.

For the cotton farmer, the threat of byssinosis might seem remote. It affected primarily workers in the mills, which were far "downstream" in the flow of cotton from field to fabric. There appeared to be no consequential effects for growers, so though the disease was a health hazard for mill employees, it might appear to pose no danger to cotton farming. But the NCC and related interests in the Cotton Belt saw byssinosis as a threat to the livelihood of the industry when in 1976 OSHA proposed a new and stiffer standard of 0.2 milligrams of dust per cubic feet of air for textiles and all cotton processing and handling plants. The old standard in place since 1971 had been more lenient at 1.0 milligram. OSHA held hearings across the Cotton Belt to get reactions before implementing the standards—and it heard plenty.

To begin with, the Delta Council thought the requirement was unrealistic for gins because it found that when gins had been closed for several months, they still had dust levels beyond the proposed 0.2 milligram. If enforced, the organization feared the new rule would close down all gins in the United States. It further questioned whether reducing gin dust to the OSHA level was possible, and without changes in the standard, it continued, the result would be "the collapse of the entire cotton industry . . . and the denial of essential raw materials to the U.S. consumers."[10] Grover C. Wrenn, deputy director of OSHA, spoke to the PCG in Lubbock a short time later and assured them that though his agency would not weaken the health standards solely because of the cost of installing air-cleaning equipment, every effort would be made to incorporate the economic impact of the new standard into the final decision.[11] But Texas congressman George Mahon planned to state his opposition to the proposals on the grounds they would paralyze the industry. A Lamesa grower expressed the common feeling on the High Plains:

"It would be a shame to . . . lose everything by the passing of some standards that have no relevancy to our area."[12] The *Lubbock Avalanche Journal* reported fierce opposition across the plains from growers who felt the proposed standard "threatens the very future of the fiber crop as an important U.S. commodity."[13] And the Memphis Cotton Carnival King for 1977, Hal Boyd Jr., thought the cotton industry was in danger of being driven out of existence.[14] But the *New York Times* took the other argument, reporting the ill health of a seventy-five-year-old woman who had worked in the mills for most of her life and had been diagnosed with byssinosis. With help from the BLA, she hoped to get workers' compensation.[15] Owners of the mills anticipated, however, that if compensatory payments mounted and the more stringent standard prevailed, the total expenditures for compliance would force small operators out of business and push the larger firms into foreign countries.

SQUABBLING IN THE WHITE HOUSE

Quarreling over dust standards had a political dimension from the beginning. As early as 1972, the subject had stirred enough resentment to prompt OSHA chief George C. Guenther to assure President Nixon that "no highly controversial standards (i.e., cotton dust, etc.) will be proposed by OSHA before the presidential election."[16] When the Labor Department, headed by F. Ray Marshall, took over under President Jimmy Carter in 1977 and OSHA got a new chief, Eula Bingham, the climate of enforcement changed. Now the issue moved forward with the likelihood that the stiffer specifications for clean air would be implemented; and in view of the strong dislike expressed by cotton organizations at the hearings, a fight broke out in the Carter administration. Marshall and Bingham saw the matter in terms of public safety, while the textile industry felt threatened by the costs of installing clean-air equipment estimated to be a staggering $2.7 billion. Carter's Council of Economic Advisors, led by Charles Schultz, worried about the inflationary effects of the standards because the costs incurred by the mills would be passed on to consumers. Marshall and Schultz agreed that OSHA should further review the standards, but Murray H. Finley, head of the Amalgamated Clothing and Textile Workers Union, promised to file suit with the U.S. Court of Appeals to have the standards fully restored. The ATMI considered them inflationary and unrealistic, while the NCC estimated the cost to the mills closer to $2 billion.[17]

Schultz and Marshall had only forestalled a showdown. In April 1978, Mississippi senator Eastland let Carter's senior aide for domestic policy, Stuart E. Eizenstat, know that the OSHA proposals had stirred much resentment, and he wanted an opportunity for cotton representatives to present their case. Eizenstat arranged a meeting of NCC spokesmen with Marshall and Bingham on May 4, but OSHA would not budge, leaving the new standards to go into effect May 31. A few days before the deadline, Eizenstat, Schultz, and Robert S. Strauss, who led the president's anti-inflation program, met. They agreed the new standards would be inflationary, so Eizenstat persuaded Marshall to delay the implementation. Lon Mann, president of the NCC, wrote Carter and commended the delay until the inflationary effects could be more thoroughly studied.[18]

The struggle over the standards had renewed the old fight between cotton and organized labor. In this case, however, each side directed its barbs through federal officials, hoping to win its way behind the closed doors of bureaucracy. Feelings remained strong within the administration, however, and Carter, caught between the interests of labor and his general anti-inflationary program, tried to placate both sides. After being pressured by Schultz and Vice President Walter Mondale, Carter ordered Marshall to issue a "soft" position on the issue, but the move angered Ralph Nader, the AFL-CIO, and sixteen congressmen who claimed the president had abandoned the sick and dying.[19] The uproar led to another meeting with the president that included Mondale, Schultz, Marshall, and Eula Bingham, who threatened to resign if a soft standard went into effect. Carter nonetheless ordered a compromise that would delay the tougher requirement for the textile mills, and ordered OSHA to search for alternative solutions. Carter's chief of staff Hamilton Jordon, miffed by the incident, worried the delay might foster an image of an indecisive president indifferent to the welfare of workers.[20]

The issue had further political ramifications. What had started as a struggle between a relatively weak union and the cotton industry, neither of which usually attracted national attention, became an issue widely reported in the press. And the internal squabbling within the executive branch spilled over into other matters. On one side were the inflation fighters, led by the Council of Economic Advisors (CEA), who wanted to minimize costly expenditures on industry, while the Department of Labor, OSHA in this case, sought to represent the interests of its constituents, the unions. Marshall and OSHA chief Bingham publicly claimed victory over the CEA after the confrontation in the Oval Office, when they actually went along with a compromise more favorable to the textile and cotton-related industries. *Business Week* reported that "OSHA overreacted to the CEA's suggestion that rigid prescriptive standards may not be the only way to respond effectively to health problems."[21] Yet CEA blundered, too. It seemingly chose the cotton dust standards to make the fight for restricting regulations in the name of holding down inflation. Spokesmen on the political left thought CEA should pick on someone its own size, that the Amalgamated Textile Workers Union was too small and weak to be a worthy opponent. "I'd like to see them try this on the steelworkers," *Business Week* quoted a public interest representative.[22] This rock throwing among Washington bureaucrats had stirred groups with other agendas. Common Cause, a public advocacy organization, foresaw an opportunity to push for reductions in maritime subsidies and farm supports for tobacco farmers and dairymen. Even broader ramifications came out of the incident because it generated publicity over regulation of industry, which became a major issue for Republican presidential candidate Ronald Reagan in 1980.

Once the White House infighting had ended, Marshall announced the final OSHA standards in June 1978. For manufacturing yarn, mills had to maintain 0.2 milligrams of dust per cubic meter of air over eight hours; for slashing and weaving operations, the standard was 0.75 milligrams for the same air space; and for nontextile operations such as cottonseed mills and mattress and bedding factories, the standard was 0.5 milligrams. Mills had to post warning signs and provide respirators for workers in areas with high concentrations of cotton dust. No standards would be applied to cotton harvesting, merchant handling of cotton, or dust generated from handling woven or knitted materials and washed cotton. Gins would face no air standards, but gins workers had to submit periodically to medical examinations, and operators of gins had to warn employees of the hazard of cotton dust.[23]

But the matter was not resolved. As many as sixteen lawsuits against the Labor Department were filed, and the NCC filed suit in the Sixth U.S. Circuit Court of Appeals in August, requesting a review of the standards. "They are impossible standards," stated NCC executive director Earl Sears, "or the industry would not be acting in this way."[24] Because of the pressure, OSHA administrator Bingham temporarily suspended the new regulations, but only for cotton-waste-processing industries and their customers that used cotton batting, such as manufacturers of furniture, bedding, and automobiles. Her action prompted George Meany of the AFL-CIO to warn that unions would declare war against the opponents of the OSHA specifications. "We will not surrender the lives and well-being of workers to corporate greed."[25] In September governors from several southern states, meeting at Hilton Head, South Carolina, shot back with a resolution calling on the president to delay the standards.[26] Among the greatest fears of cotton interests were threats from textile mills that if the regulations remained, they would stop handling cotton and switch solely to synthetic fibers. "If our customers, the textile mills, turn to synthetics, then the impact will be severe on farmers," warned B. F. Smith, director of the Delta Council.[27]

The tougher standards owed much to the president's secretary of labor. F. Ray Marshall was a University of Texas academic, a Louisianan trained in labor economics at the University of California, Berkeley. His work focused on the downtrodden: rural workers, black laborers, and union members. As to be expected, his sympathy lay with labor, and he saw unionization as the avenue of progress for southern workers. Marshall was not noisy and inflammatory; he displayed no sense indignation or self-righteousness but acted with the calmness of a professor who methodically pursued an objective and measured his words before speaking. In July 1978, the month after OSHA announced the stiffer cotton dust standards, Marshall agreed to participate in the program of the American Enterprise Institute for Public Policy, a nonprofit and nonpartisan educational organization that promoted better understanding of public policy through discussion of ideas. The "Conversation with Secretary Ray Marshall," a question-and-answer session on current matters of concern, occurred in the midst of the fight over the dust standards, and the session dealt, though not entirely, with the issue.

Marshall's tone and responses were reserved and moderate. He explained the OSHA standards as an effort to protect the health of workers, but recognized the question of cost for the mills and pointed out that OSHA had modified the regulations to lower the cost. He cited the case of cotton ginners. When they "presented their evidence, for example, we decided not to apply the standard to the cotton gin."[28] Marshall further recognized the possibility that if regulations proved too costly for textiles, they might move their operations "to other countries to avoid the standards imposed here."[29] However, he made clear that "the bottom line always will be that we will do everything we can to protect the workers."[30]

THE MYSTERY OF CAUSATION

Throughout the power plays and struggles to protect self-interests, whether by textiles, unions, cotton interests, or for the benefit of workers' health, there remained the mystery: the cause of byssinosis. Evidence showed that only a portion of the employees working in mills developed the ailment. Where bales were opened, or broken apart, and fed into a blending machine, the first step in the mill, and the carding room where the fibers received another cleaning and shaped into a "sliver," the frequency of workers developing brown lung appeared greatest. Agricultural researchers had studied cotton dust, and some research had been under way in medical circles. At the annual Beltwide Conference, agricultural scientists reported their work, but none found a precise link between cotton and byssinosis. The most likely culprit to emerge out of the studies was brachts, the leaves attached to the base of the cotton boll that were pulled into the mechanical picker. By harvest season, brachts would become brittle and break into microscopic particles during ginning and be left in the bale that went to the textile mill. There the brachts became airborne and were inhaled by workers. These tiny bits of leaf reacted with lung tissue over a long period of time and were thought to lead to brown lung among some workers. Yet the scientific community could not conclusively agree that brachts caused byssinosis, and cigarette-smoking employees had a weak case when they filed for workers' compensation.

In 1980 the *Journal of the American Medical Association* (*JAMA*) published a study conducted by Duke University's departments of family medicine and pathology that showed little correlation between cotton dust and brown lung among nonsmokers. Based on autopsies and other evidence, the researchers thought emphysema came from cigarette smoking, not cotton dust. Byssinosis advanced among smokers, according to the study, but not among nonsmokers. The study reported that "the lay term brown lung disease is misleading and should be abandoned." They attributed the assumption that cotton dust caused byssinosis to studies based on patient answers to a questionnaire from an earlier study, and pointed out that the proponents of a relationship admit there is no conclusive identification of a substance that causes the disease, and "there are no characteristic findings on physical examination, chest x-ray, pulmonary function studies or even lung biopsy which are specific for the diagnosis of byssinosis."[31] In summation, the Duke University medical scientists held smoking responsible for the lung ailments of cotton mill workers.

The press gave coverage to the article, which weakened the general effort to get cotton dust recognized as a cause of lung disease. But the article prompted seventy-five mill workers, retired or disabled, to go to Duke University as a group to offer themselves as evidence of the debilitating effects of working in a mill. Among them were thirty-six nonsmokers with byssinosis. To help make their point, they carried a banner that read, "We Offer Ourselves as Evidence," which the author Bennett Judkins used as the title of his book on brown lung.[32]

THE MUSCLE OF COTTON

With the election in 1980 of President Ronald Reagan, who made deregulation a major campaign issue, the question of brown lung served as an example of the lobbying power of cotton. In February 1981, shortly after his inauguration, President Reagan invited a group of agricultural leaders to the White House for a briefing on his economic policy. Among the dozen farm leaders was NCC president Frank Mitchener of Sumner, Mississippi.

When President Reagan asked for the agriculturalists' recommendations for bolstering the economy, nearly all had a negative or gloomy outlook. But Mitchener commended Reagan for his tax cut proposals and promises to cut regulatory reform. Mitchener then described the standards imposed by OSHA and warned that the textile industry faced an estimated cost of $2 billion to comply with the regulations. Mitchener caught the president's attention, and Reagan turned to Murray Widenbaum, the new head of the CEA, and instructed him to look into the matter. Widenbaum replied, "We know about that one, Mr. President. It's an extremely burdensome standard we want to take a hard look at."[33] Reagan ordered Widenbaum to arrange a meeting of cotton representatives with Vice President George H. W. Bush, who headed Reagan's Task Force on Regulatory Relief. Mississippi senator Thad Cochran assisted with the arrangements, and they met with Bush in February. With the vice president's support, they took their case to his general counsel Boyden Gray.[34]

They discussed the specifics of the issue. However, the NCC and ATMI had filed suit against the Labor Department in an effort to get the standards modified, and the case was pending before the Supreme Court. They did not wish to be seen as trying to preempt the Court, but explained that their greatest concern dealt with the clean-air regulations for textiles, where most of the enormous cost for compliance would occur. They further explained that if the Supreme Court ruled against them, textiles would need some administrative relief to reduce the impact of the OSHA regulations. Gray noted that the administration's principal weapon for providing regulator relief was Executive Order 12291, which called for cost-benefit analyses for new and existing standards. Gray worried, however, about the possible effect the Court's ruling would have on the executive order and did not want to jeopardize it. Gray wanted legal research on the matter, and Mitchener offered the services of the NCC. Gray accepted, and the NCC arranged for legal consultation at its expense. To raise funds for the legal work, Earl Sears canvassed various organizations with membership in the NCC.[35]

In June 1981 the Supreme Court upheld the OSHA standards. The ATMI with the NCC acting as a petitioner on its behalf had contended that the standards had been promulgated without a cost-benefit analysis, which, they pleaded, meant the implementation of air-cleaning equipment would threaten the existence of the mills. Robert Bork argued on behalf of the cotton interests, while the deputy solicitor general for the Labor Department responded. In a 5–3 decision, the Court ruled that Congress had incorporated cost-benefit analysis into the legislation creating OSHA and that the purpose of the measure was to eliminate a significant hazard for workers. The Court ruled that OSHA had acted reasonably, and the Justices quoted the lower court: "We are talking about people's lives, not the indifference of some cost accounts."[36] For the mills, this outcome meant they would have to install air-cleaning equipment to comply with OSHA regulations. It strengthened the claims of workers for compensation.

Earlier in February, however, President Reagan had ordered that all regulatory standards costing $100 million or more be sent to the Office of Management and Budget (OMB) for cost-benefit analysis, which made OSHA delay the regulations until the OMB study was finished. In October 1981, the North Carolina Supreme Court ruled in a case involving Burlington Mills that affected compensation claims. It upheld a decision made earlier by the North Carolina Industrial Commission that a longtime worker in a mill was entitled to only 55 percent of her original claim to the commission because she was a cigarette smoker. Of the original $35,931 that her attorneys sought from Burlington Mills, she would receive $20,761. This 5–2 vote by the state court established the precedent that disability due to byssinosis had to be established independently of cigarette smoking. While all parties agreed that the worker suffered total disability, the dispute centered on the extent to which cotton dust caused her illness. Dr. Herbert O. Sieker, chief of chest and allergy studies at Duke University Medical Center, was quoted as assigning about "50 percent to 60 percent for the cotton dust exposure and 40 percent to 50 percent for cigarette smoking." He warned, however, that "at the present time there is no laboratory type of test that would do this."[37] In 1982 the National Research Council, an arm of the National Academy of Sciences, concluded in a study that too little evidence was available to "support an unequivocal relationship" between cotton dust and byssinosis. Not all members of the investigating team agreed because, according to the dissenting researcher, "there is no question from clinical studies in this country's textile belt, that there are numbers of people who have impaired breathing, measurable pulmonary impairment, and disability for work from cotton dust exposure."[38]

In February 1982, OSHA announced it would no longer automatically respond to complaints by workers, and later in June it announced a further study of the regulations. Courts had reinforced the standards, but the zeal of the agency had gone, owing to the new spirit of deregulation in the White House. The BLA had become less effective, since it depended on donations from disabled mill workers. The OSHA standards in the mills had nonetheless withstood challenges and went into effect in 1984, by which time the mills had to install air-cleaning equipment in areas where cotton dust was hazardous. Burlington Mills had spent $49 million by 1984 on ventilation equipment.[39] Mills continued to pay compensation claims, but if workers had a history of smoking cigarettes, their cases were weakened and awards reduced. In

1980 the Supreme Court had ruled in the "benzene case" that OSHA had to establish health risk before issuing standards, a ruling that strengthened the textile defense on individual compensation claims.[40] For other cotton-related interests such as gins, the regulations had been softened or removed. By the 1990s, the brown lung issue had dwindled, mostly because mills had installed automatic equipment to handle the particularly dusty jobs.

The NCC had taken an active part because textiles were a segment member of the organization. But growers faced a direct threat if mills switched wholly to synthetics, not an unreasonable assumption, because artificial fibers had already captured about half the market. Across the Cotton Belt there was widespread anxiety that OSHA's regulation would remake farming, which explained the NCC's persistence. Considerable research went into the disease by land-grant university faculties that hoped to identify the specific agent responsible for byssinosis, but they produced no definite answer. In the fight over the disease, however, the defenders of cotton faced a severe disadvantage: they appeared to oppose public health and to have no regard for the disabled mill workers struggling to breath or finish their lives with the assistance of compensation. When the Supreme Court quoted the Court of Appeals that the question at hand involved the lives of people and not cost accounts, it struck a nerve.

The solution lay outside the realm of cotton. Medical scientists needed to determine the precise cause of the lung disease for the sake of both disabled workers and the economic interests fighting for protection. Agricultural science could not pinpoint the causal factor, though cotton brachts received some attention. To remove all doubt that exposure to cotton dust led to byssinosis required research in the medical realm, with medical surveillance, laboratory research, and even autopsies. Only in the field of medicine could this kind of research be undertaken, but medical researchers, so thoroughly schooled in specificity, could not conclusively isolate cotton dust as the responsible agent. They disagreed among themselves, though there was little disagreement over the association of mill work with the ailment. One Yale University researcher noted, however, that working in a mill for thirty-five years "is the same as smoking one pack of cigarettes a day for fifty years."[41] Complicating the scientific and medical inquiries was the growing recognition that cigarette smoking contributed to lung disease, for just as byssinosis began emerging as an issue in the 1960s, the specter of tobacco overshadowed other considerations about lung ailments. The center of the textile industry lay in the heart of the tobacco culture, where smoking was popular. Despite the role of tobacco, byssinosis declined when manufacturers of spinning machinery began incorporating improved cleaning devices into the textile equipment and developed automated machines to replace human labor. It was apparent the servants of old King Cotton still had muscle, but the byssinosis crisis proved to be a case of accommodating a new challenge.

REGULATIONS

Farming of all types throughout the United States entered a new era with the rise of environmentalism. As the general public began to question the use of insecticides, herbicides, and hazardous air pollutants associated with agriculture, and Congress passed legislation in response to public demand, farmers faced regulations on an unprecedented level. Growers could not overlook or give scant attention to state and federal mandates for protecting the environment, and for cotton growers who relied heavily on new technologies, this development forced them to make adjustments in their operations.

The Cotton Foundation examined this question in its reference book series with the volume *Cotton Harvest Management: Use and Influences of Harvest Aids* (2001). Harvest aids involve the use of defoliants to remove leaves from cotton stalks before mechanical picking to hold down trash and staining of the lint. Hand-picking had avoided this difficulty, but the advent of mechanical harvesting required some way to strip away the leaves, so agricultural chemists looked for a solvent that could readily be applied to the plants. After 1938 when calcium cyanamide was introduced, it remained the most widely used defoliant until 1956, when arsenic acid came into use. Arsenic proved popular with growers, but questions arose over the residual accumulation of the acid in the soil and the risk of long-term human exposure. In 1986 the EPA declared arsenic acid a hazardous air pollutant in several industries, though cotton gins were exempt because the agency considered the health risk small to the workers. In 1993, however, the EPA took arsenic off the market when textile companies expressed concern over the accumulated levels of the substance in the workplaces. Concerns over lawsuits from residents living close to cotton gins and workers' compensation claims also prompted the removal. The Cotton Foundation regarded the acid as a safe product in the fields when used properly, but acknowledged that the anxiety over the safety of mill workers justified its ban.[42] Defoliants became an object of attention because of the residue left on the lint and seed, posing a threat to processors and handlers of cotton downstream from the fields.

The Clean Air Act of 1970, modified and stiffened in 1990, brought new compliance standards for growers and ginners to meet. When in 1987 and again in 1997 the EPA issued clean-air standards for gins, it required some ginners to obtain permits for operating, and some had to install equipment to reduce the emission rates of particulate matter that went into the air. If residents in nearby neighborhoods complained about ginning dust, the likelihood of gins being forced to install equipment became greater. Installation of the air-cleaning machinery reduced the profit margin for ginners, and some had to choose whether to comply or shut down. The number of gins had already fallen, owing partly to the consolidation of gins into larger units, but the stiffer regulations played a role. Ginning did not come to a halt, but the new standards nonetheless required ginners to obtain permits and if necessary to install air-cleaning equipment.

The EPA's announcement in 1997 that it would stiffen the regulation for clean air in workplaces, lowering the standard from 10 to 2.5 milligrams per cubic meter, caused cotton and other agricultural interests to object. EPA had begun reviewing the standards because of a suit launched against it by the American Lung Association.[43] Farmers and ranchers alike, perhaps with no affiliation with cotton, were alarmed that field plowing would be under attack. Philip J. Wakelyn of the NCC testified before the Senate Agriculture, Nutrition, and Forestry Committee that agriculture accounted for a small portion of the particles in the atmosphere at 2.5 milligrams. He wanted the EPA to develop better monitoring technology that would more accurately gauge dust induced by farming, or "otherwise the new rules could result in unnecessary and costly control measures for agriculture."[44] But these fears were unfounded because Carol Browner, administrator of the EPA, assured the committee that the "EPA is not going to regulate farmers as a means of reducing fine particles in the air. We will not restrict tilling."[45] She made it clear that the alarm among agriculturalists was based on "misconceptions and misinformation." Her agency intended to target emissions from electrical plants and heavy industries. Agriculture escaped the requirement because dust emitted into the air from farming operations was much larger than 2.5 milligrams and beyond the interest of the EPA. But the public concern over clean air did not disappear.

In June 2003 the *San Francisco Chronicle* ran an editorial titled "Cows and Cotton versus Clean Air." It took the position that "life is out of whack from Modesto to Bakersfield," because of the emission of air pollutants that originate from farming. It attributed one-quarter of the smog in the region to agriculture and wanted the California legislature to pass bills pending in the state house for making farmers in the San Joaquin Valley comply with air quality standards. The editorial claimed that smog threatened lifestyle in San Francisco and the valley and that remedies had to be found "even at the risk of imposing major changes on farming."[46]

The Agriculture and Resources Center at the University of California agreed that the dust in the valley originated from farming practices such as plowing and harvesting. It mentioned the mandatory tilling of fields for control of pink bollworms as an example of activity that stirred dust, and recommended that growers reduce the number of passes through the fields and practice integrated pest management to control pink bollworm. A similar recommendation came from the University of Arizona's College of Agriculture, where studies made between 1991 and 1995 in cotton fields in the Yuma region showed that plowing contributed to air dust, but nonagricultural activities also contributed. In 1998 the Arizona legislature created a task force to study and promote management practices for growers to reduce dust in the air. In 2005 an implement manufacturer in Bakersfield advertised a new cotton shredder-bedder that would allegedly reduce the number of field passes required for cultivating cotton. The manufacturer pointed out that "concern for the quality of life in the San Joaquin Valley is a high priority as the population continues to increase rapidly in this agricultural area."[47]

From the perspective of growers the regulation of pesticides proved more complex. In 1972 Congress passed the Federal Insecticide, Fungicide, and Rodenticide Act (FIFRA), which superseded a similar act of 1947 that had little or no teeth. In 1996 Congress modified the newer measure through the Food Quality Protection Act. The FIFRA stipulated that manufacturers of pesticides must register and label each chemical with the EPA before making it available for commercial use. Included in the application must be a description of the tests conducted by the manufacturer, directions for its use in easily understood terms, and the formula for the pesticide, which would remain confidential. The EPA must consider the new product in terms of economic, social, and environmental impact, as well as the benefits for its use. Each product successfully registered must undergo review every five years in light of any new evidence that might affect its impact.

Use of pesticides remains strong, though Bt cotton has significantly reduced the need for chemical applications. Environmentalists and public health interests nonetheless charge that cotton growers are a major user of health-threatening pesticides, accounting for up to 25 percent of the total use in U.S. agriculture. They are referring to the use of organophosphates to stop harmful insects not affected by transgenic cotton. Critics argue that cotton growers regularly apply chemicals banned for food crops, and that while pesticides may be safe for use on cotton, exposure to them endangers fieldworkers and residents living in nearby areas because the residues from the cotton fields contaminate water supplies and drift onto food crops and orchards. Questions arose over clothing made of conventionally grown cotton because it received pesticide treatment in the field and again in the dyeing and finishing process. This concern was responsible for the cancellation of arsenic acid as a harvest aid and restrictions on the use of others.[48] If pesticide residue in cottonseed, cottonseed hulls, and meal reached unacceptable levels, it would endanger the market for livestock feed.

The growing perception of cotton farming as a danger to public health presented a new challenge. For the NCC, it required close cooperation with the EPA in protecting the interests of growers. Generally the organization complained that EPA removed pesticides from the market based on political instincts without making a thorough study of the health risks.[49] Members of the NCC serve on EPA advisory committees, and the organization furnishes benefits studies and supplies data to the agency from the perspective of cotton growers. The NCC often pleads with USDA to take steps to rebut an EPA ruling, but generally the rulings stand.[50] Actions by the EPA have required better safeguards for workers such as enclosed tractor cabs and plane cockpits for drivers and pilots when applying the powerful chemicals, or stronger and clearer warning labels for workers who mix and load the chemicals. Executive director of the NCC Gaylon Booker announced in 2002 that Bidrin, the trade name for an insecticide used against plant bugs and stink bugs, could no longer be applied by crop duster planes after two more years.[51] There appeared to be a growing perception that organically grown cotton would remove threats to public health and the environment, but the total production of this cotton accounted for less than 1 percent of the U.S. total in 2006. But Roy Cantrell of Cotton Incorporated pointed out that conventionally grown cotton is sustainable despite the objections associated with cotton farming.[52] He saw a need to improve the message that cotton can be produced in harmony with public interest.

WATER

"Water is a precious and necessary resource for agriculture, and cotton producers are out of business if they don't have access to it." Tommy Horton, editor of *Cotton Farming*, made that comment as recently as 2006 when drought prevailed over portions of the Cotton Belt. He recalled that agricultural researchers had predicted many years ago that water would become a critical issue for farmers and that their prediction had come true. Water had been generally available at low cost in the West until the latter 1970s, when growers began noticing a drop in the underground water table of the "endless" Ogallala Aquifer. That development coincided with the rise of charges from urban interests and environmentalists that farmers wasted water while benefiting from federal water programs at the taxpayers' expense. The growth of the general population and the rising popularity of the Sun Belt presented a new threat to water supplies, and this combination—less water and more competition for it—challenged growers in the western states.[53]

Even if many plains growers saw no danger in the supply of water, some Texans understood the importance of conservation and the wise use of water. Warnings about overuse of the underground reservoir occurred in the late 1930s from hydrologists like Walter N. White and the Texas Board of Water Engineers, and some farmers in the Texas Plains reported a drop in the water table during the expansionist era immediately after World War II. In 1949 the Texas legislature passed a bill that allowed the creation of water districts, and two years later, landowners in thirteen counties around Lubbock voted to establish the first underground water management district in the state, the Texas High Plains Underground Water Conservation District No. 1, with offices in Lubbock.[54] Although the new district sought to regulate the use of water, the underground table continued to drop, since regulation relied on voluntary action by growers.[55] From 1951 to 1958, the table dropped over twenty-seven feet. Overuse of the Ogallala Aquifer could not be stopped because of the firm belief among farmers that they were entitled to use the water beneath their land with no restriction. The normal low rainfall, plus droughts, particularly the second dust bowl of 1950 to 1956, made use of irrigation compulsive and forced down the water table. By 1960, however, growers began to recognize the importance of conservation as they saw the lower water pressure from their wells. "The discouraging sight of an eight-inch discharge pipe spewing out only half its capacity of water," one author effectively stated, "forced them to acknowledge the depleting supply of water."[56]

Cotton nonetheless remained the principal crop on the Texas Plains, though the planting restrictions imposed by Congress made growers add other crops such as grain sorghum, wheat, and alfalfa. To produce any crop successfully required irrigation, so water usage continued, despite the drop in cotton acreage because "irrigation was the lifeblood of the High Plains."[57] Farmers of all stripes relied on irrigation, but the regional economy benefited, too, so that towns and cities expanded in population. Cotton remained king on the Texas Plains, since the area accounted for almost one-third of the U.S. crop.

Usage of the Ogallala Aquifer technically fell under the jurisdiction of Water Conservation District No. 1, but landowners continued to pump water as needed. When the water table fell, they drilled deeper. When fuel prices jumped in the 1970s, however, they began to feel the pinch of unrestrained watering; and as farming costs, including land, also began climbing, the margin of profit became narrower, and pumping expenses had to be minimized. When in 1978 the Texas Supreme Court ruled that usage of groundwater could be declared wasteful and negligent, careless irrigating began to stop.

These developments prompted the district to begin urging farmers to abandon open-ditch irrigation, a wasteful system, and move to the more-efficient center-pivot sprinklers invented in 1950 by a Colorado farmer. By the mid-1980s, efficiency of water usage became the new ethic. When a particularly severe drought hit the plains in 1988, and subsequent dry years came in the 1990s, farming on the Texas Plains became more risky. Growers faced the challenge of a deeper water table with little room to maneuver irrigation costs. Unfettered watering on the High Plains had ended.[58]

The questions over water in Texas may seem mild compared with the complexity of water usage in California. Nowhere did Mark Twain's adage that "whiskey is for drinking, but water's for fighting" seem more apparent. Water wars are part of the Golden State as feuds are to the Bluegrass State. Development of water for agricultural irrigation in the two cotton-growing areas of California paralleled the Texas Plains, except that irrigation in the San Joaquin Valley depended significantly on the snow pack in the Sierras, while the Imperial Valley relied on the Colorado River bringing water from the snowy slopes of the western Rockies. Growers in the San Joaquin Valley had the added advantage of underground water, so wells had been part of the infrastructure of agriculture there, but California's underground supplies were no match for the Ogallala Aquifer. California contained major cities such as Los Angeles, San Diego, Sacramento, San Jose, Oakland, Santa Barbara, and San Francisco; intermediate cities such as Bakersfield, Modesto, Merced, and Fresno dotted the Central Valley. Both snow pack and underground water fed the urban centers. The Texas Plains had Lubbock and Amarillo, but larger cities such as Dallas, Fort Worth, San Antonio, and Houston did not rely on the Ogallala for water. And total population for California had reached about thirty-five million by 2000 compared with eighteen million in Texas.

Agricultural interests have long battled with cites, and more recently with environmentalists, over water. Certainly the best-known case involved the Owens River Valley, when in 1913 the city of Los Angeles finished an aqueduct to divert water from the Valley over two hundred miles for its use. It turned the Owens Lake into a dry bed and retarded development of farming in the valley. In 1970 the city built a second aqueduct that took more water, including pumping underground water to feed it. Agriculturalists and environmentalists refer to the Owens Valley case as an example of theft of water rights by urban centers, and the legacy of this drama, which served as the basis for the movie *Chinatown* (1974), left a strong determination among rural landowners to protect their water rights. With the completion of the Central Valley Project (CVP) and the State Water Project after World War II, agriculture was poised to take off, so for the next generation, farming interests enjoyed little interference.[59]

When a six-year drought hit California in 1987, it set off new concern. Cities enviously eyed agricultural water supplies, which prompted a round of assertions that farming used too much of the state's water supply, some 70 percent, at the expense of other users. Farmers, so it was charged, wasted their liquid gold through needless irrigation to maintain their water allotments. Critics of agriculture singled out cotton farmers on the grounds they received a water subsidy from the federal government to raise a crop in surplus. "We have a situation, as taxpayers, where we are providing extremely large subsidies to provide very cheap water to grow surplus crops for which we've paid billions in subsidies," asserted Ralph Abascal of the California Rural Legal Assistance Foundation.[60] He wanted growers to repay past underpayments in water and urged that a new price structure for water be implemented to "reflect the full cost of water." As cities cast a covetous eye on agricultural water, the concept of water marketing grew, meaning that landowners could sell their water. In the Central Valley of California, landowners paid a subsidized rate of $10 for one acre-foot of water when cities were willing to pay $200.[61] One Democratic assemblyman considered letting farmers market their water to cities without endangering crop production. But landowners believed that once sold, their water would be lost forever, and they often referred to the Owens River Valley. Business interests in the farming areas oppose water transfers on the grounds that agriculture is the lifeblood of the economy, and environmentalists in some cases favor irrigation because it partially restores underground supplies. One example of successful water marketing occurred in the lower Rio Grande Valley, where agricultural interests sold rights to surface water to municipalities. Proponents of free enterprise such as the Cato Institute nonetheless advocate water marketing, and the Reagan administration embraced the idea.[62]

Transferring or marketing water has practical drawbacks, such as the means of transporting it, whether by canal or pipelines, and whether landowners may suspend flow to irrigate crops. In 1985 the irrigation district in the Imperial Valley agreed to transfer water to the Los Angeles Metropolitan Water District, but growers in the valley stopped the plan when they feared they would lose their water rights. A plan instigated by businessmen in Denver to dam a tributary of the Colorado River and sell the water to San Diego fell apart when Colorado officials feared they would lose control of water to an entity outside the state.[63] In 1993 the water crisis eased with the end of the drought in California, but the issue of access to the state's precious water, held in large part by farm interests, had reached a new level, prompting the remark by one writer that "the agricultural wing of the water industry finds itself today increasingly on the defensive."[64] In the particular round of accusations that accompanied the drought, cotton growers received the most criticism.[65]

But the fortunes of cotton farmers are related to the most unpredictable factor of all—the weather. In 1992 a drought hit the Texas Plains and lasted for four years, bringing a renewed sense of the importance of water conservation. Steps taken by farmers had reduced the losses of water, so that by 1985 the drop in the water table had begun to stabilize, but the new drought and the growing urban population caused the table to drop again. The High Plains Water District encouraged growers to install modern equipment for the sake of efficient irrigation, and farmers, anxious to preserve their underground supplies, switched from a philosophy of production, which had little emphasis on conservation, to mitigation and efficiency. As reported by the author John Opie, mitigation involves conserving water at the well head, and efficiency means striving to obtain more yield with less water.[66] Plains farmers had adopted the "Gospel of Efficiency," Opie continued, hoping to preserve their livelihoods and lifestyles.

When California's drought ended in 1993, the concern over water did not stop. Pressure for a reevaluation of the state's supplies continued, with agriculture being the main target, since it consumed the largest part of the water. Environmentalists claimed that if farmers forfeited 10 percent of their water, there would be enough for decades of growth in the state. Landowners questioned such claims and viewed any promises about protection of their water rights with a skeptical eye. By 2000 water marketing nonetheless gained strength as a feasible solution. Through the California Agricultural Efficient Water Management Practices Act, irrigation districts may mandate to follow management practices aimed at less usage of water: the installation of meters on irrigation wells and the construction of tailwater reuse systems. Agricultural use of water in California had stabilized before the last drought, but with the pressure of a mounting urban population, landowners anticipated encroachment on their water rights. Some speculated that if severe shortages occur, irrigation might only be allowed on the most lucrative crops such as pistachio nuts, grapes, and almonds. And there was fear of a "regulatory drought," meaning that water for farmland would be siphoned to maintain habitats for fish and wildlife.[67]

For good reason, the editors of *Cotton Farming* reported that "in these two regions of cotton production in the Southwest and West, water supplies are indeed being threatened."[68] Cotton interests urged conservation of water to stabilize water supplies, but there appeared little expectation that reduced use of water would sustain farming in the arid areas. Indeed, the concept of sustainable agriculture began to grow near the end of the 1990s, which meant that use of water and soil should be managed in a manner not to threaten use of resources by future generations, whether rural or urban. To achieve this objective, the proponents of sustainability believed that cooperation would have to replace quarreling over water, and that all interests—consumers, farmers, industry, government, environmentalists—would have to respect each other's right to water and distribute it fairly. Threats from the specter of global warming, the reshuffling of world economies that affect cotton and textiles, and the relentless growth of world population that strains resources put agriculture's dependence on unlimited supplies of water at peril. Such global pressures might make sustainability only a wish. In California and on the Texas Plains, farmers know that "shutting down the pumps would tear the social fabric."[69]

URBAN SPRAWL

"They paved paradise and put up a parking lot." *Cotton Farming* used that lyric from a Joni Mitchell song to express the frustration of growers whose land had been swallowed by urban expansion.[70] Different terms are used to refer to the topic—development, sprawl, or fragmentation—but the result is the same: the loss of fields to housing tracts and shopping malls. Rural landowners have mixed feelings as they see their farms paved with streets and covered with cookie-cutter houses. They resent the unstoppable intrusion into their lives, while they acknowledge the legitimate need of space for people to live. Growers embrace progress and know that population growth is normal, that if fate puts their land into the path of development, they should accept it. Besides, there is the relief of selling the land, perhaps at $8,000 to $10,000 per acre compared with the profit of $200 per acre each year from producing cotton. "It makes my land worth more," related Tennessean John Sullivan, "but at the same time I don't want to sell it."[71]

Loss of land, or the conversion from farming to housing, generally falls into two categories: development and sprawl. Development refers to urban expansion and landowners accept it as normal growth. Sprawl, however, causes resentment because that often means fragmentation, the practice of developers "leapfrogging" across farm land and leaving valuable fields between sites of housing tracts. Farming becomes more difficult in these cases if the new neighbors complain about dust from tilling or the grower's application of insecticides and herbicides. Residents in subdivisions might regard a field as a dumping ground for all manner of trash and discarded materials. In some cases, ginners faced the consequences of sprawl when new residents complained of dust from gins.[72] A particular annoyance for growers comes from the practice of new homeowners buying plots of two or three acres on the outskirts of cities and erecting extraordinary large houses, often called McMansions, on them. And when new homeowners jog or walk dogs in cultivated fields, they worsen the annoyance.

California again had the most obvious clash over farmland protection. To preserve rural land, the state assembly passed the Land Conservation Act in 1965, which would protect land from development for ten years, but only if the owner requested it. In 1996 the state extended the time to twenty-five years through the Agricultural Land Stewardship Program. Both measures required participation to be voluntary, so the sale of farmland for nonagricultural purposes continued with little change. The availability of water that accompanied the land drove up the value and made lucrative offers hard to resist. Between 1982 and 1992, California lost three million acres of farmland, and the state's Department of Conservation reported a loss of 65,827 acres from 1994 to 1996 alone. Based on records compiled by American Farmland Trust, a nonprofit organization, the southern half of the San Joaquin Valley lost the most. Loss of cotton land alone did not account for the change, but in view of the large acreage devoted to the crop, it was included in the loss. As reported by the USDA Natural Resources Conservation Service (NRCS), the total loss of farmland in the United States between 1992 and 1997 was over fifteen million acres.[73]

Two areas of the rural-urban conflict are Memphis and Phoenix-Tucson. Cotton farmers in Shelby County, the location of Memphis, and nearby Fayette County have experienced urban growth. The Fayette County Chamber of Commerce welcomed the growth but recognized the importance of keeping open spaces and not damaging the rural economy. By Tennessee state law, counties must have a policy for regulating growth, the annexation of farm land, and the incorporation of land into municipalities. Along the I-10 corridor of Tucson and Phoenix, agricultural land is being lost to urban development at a rapid pace. The town of Cotton Lane sits in the path of development. "The speed at which growth comes can be tremendous," stated the Arizona cotton producer Steve Sossaman.[74] To avoid confrontation with new residents over air quality, he switched to minimum-till cultivation and planted ground cover. He must carry out aerial and ground spraying cautiously and under the right climatic conditions.

Loss of farmland could legitimately be interpreted as the expected outcome of population growth, but growers nonetheless face adjustments that affect their livelihoods. The aggregate loss of land in the United States began to intensify in the 1980s, according to the USDA.[75] Two areas in Texas particularly affected were the blackland prairie and the lower Rio Grande Valley. In both instances, mushrooming growth gulped fields and pasture. Landowners in the Rio Grande Valley were overwhelmed by the impact of winter tourists, known as snowbirds, and the border development induced by NAFTA.

CONCLUSION

Socioeconomic change imperils cotton farming. The dynamics of change are too powerful to stop, and a halt would be detrimental to the general welfare, but the exploitation of land that arises from the ethic of exaggerated consumption, as seen in the popularity of McMansions on mini-estates, should be reexamined. Farm Bureau member and Tennessean farmer Harris Armour thought land should be treated wisely regardless of its use. As farmland converts to other uses, however, he felt that "we need some open land and we need production agriculture."[76]

General agriculture faced new challenges. Cotton in particular received criticism, since it was not a foodstuff and could be imported, which made it expendable in the eyes of nonfarm interests. The increased costs of cotton farming rising from regulations, coupled with threats to federal subsidies, could conceivably leave only the extraordinarily large operators in business. Foodstuffs are increasingly imported, mostly fruits and vegetables, but consumers see cotton as an industrial item and are quicker to question the supports and protections for it. The old cultural image that associates cotton with exploitation and greed weakens its ability to withstand attacks, so that the production of the soft lint popular with consumers could nonetheless undergo a significant drop in importance for the general economy. For the next generation of growers, however, water supplies will likely be the most pressing threat.

CHAPTER 18

THE GLOBALIZATION OF COTTON

A feature of American life since World War II, driving economic, military, and cultural behavior, was the expanding and interconnected world economy commonly called globalization. The links of international markets, the emergence of transnational corporations, and the flow of capital across borderless economies, all recognized as powerful forces in the lives of people, made the United States the leading world power. Cotton had been the first item of trade to advance globalization when after the American Civil War cotton wove "together the far-flung threads to create the warp and woof of a new global political economy," according to one account.[1] Foreign markets had declined after World War I, which had motivated cotton interests in 1945 to recapture them. Their early successes, as seen in the Marshall Plan, were the beginning of efforts undertaken for the rest of the century to maintain cotton's place in the international market. In one respect, this meant meeting the competition from developing countries that saw cotton as a path to their economic betterment, but in a broader context cotton had value beyond its use as a fabric, because trade fostered peace.

In 1945 cotton shippers understood the European market, but they had little trading experience in developing countries. Even Anderson, Clayton and Company had few or no contacts in the smaller cotton-producing nations. Little information could be obtained from the Department of Agriculture because it, too, had little knowledge of these countries' potential to grow cotton. It was known that countries in Asia and the Far East were purchasing new textile equipment in an effort to take advantage of the war damage in Western Europe, but the extent of these purchases remained a mystery, and the likelihood of any new mills there becoming customers of the United States was only speculation. The NCC proceeded to make in-depth studies of the "current and potential cotton production" in nine countries and regions: China, Mexico, Africa, Pakistan and the Indian Union, the Middle East, South America, Central America, and the Soviet Union.[2]

READ DUNN: AMBASSADOR OF TRADE

Dunn's trip to Europe in 1946 on behalf of the NCC, combined with the similar tour by Bob Jackson in 1945, had served as the foundation for the federally sponsored assistance programs for war-torn Europe. Dunn's trip with E. D. White to Japan in 1947 had the same purpose. Since these programs were temporary, the NCC realized that trade had to be restored on a permanent basis and conducted through private channels. Dunn went to China in 1947 after completing the arrangements for Japan to receive U.S. cotton with a loan from the Export-Import Bank. China was a vast and undeveloped market that had sustained much damage during the war but now found itself caught in political and military turmoil as the communist forces under Mao Zedong resumed their civil war against the Chinese Nationalists. At best the spindle capacity of the country was 80 percent of its level in 1939, but shortages of raw cotton hampered the restoration of Chinese mills, and farming methods remained primitive. "If a farmer has a caribou or an ox, he is fortunate," Dunn reported to the NCC.[3] Chinese cotton grew on small plots, one to eight acres, and yielded about two-thirds the per-acre level of the United States. Total production for 1947 reached 1.9 million bales. Opportunities for trade looked dismal because communists held Manchuria and some northern provinces. Outside investors that could best restore the textile industry and create a market for U.S. exports had begun to avoid China.[4]

Dunn made further studies during 1952, but then Frank Barlow and Leonard Mobley took over the studies and finished with the last country, the Soviet Union, in 1959. These reports demonstrated that most of the cotton-growing countries needed food and were not likely to increase fiber production soon. Dunn felt he had misjudged the growth of production in Pakistan, however, because construction of dams and irrigation developed faster than he anticipated. Acreage also expanded faster in China than he predicted.[5]

In 1949 the NCC conducted a special study of Europe under Dunn's direction. Clothing and other goods were in short supply in both European Cooperation Administration (ECA) and non-ECA countries owing to a shortage of raw lint. Europe's dollar shortage made textile mills there dependent on ECA cotton, but the shortage had stimulated production of synthetics. Dollar shortages in Europe were expected to hold down imports of American cotton, which prompted the NCC to see extending U.S. credit and bartering "as interim means of furthering cotton exports."[6] Dunn and Caffey Robertson, adviser to the NCC, had accompanied E. D. White, now chief of the Cotton, Wool, and Fiber Branch of the ECA, through ten countries that included Britain, France, and Germany. The American Cotton Shippers Association had assisted with the organization and arrangements for their touring study. The NCC became the supplier of data and information on the European cotton trade to the USDA and New York Cotton Exchange because of its network of contacts there. Only when the USDA expanded the Agricultural Research Service (ARS) did it forgo this activity.[7]

Despite the proficiency of the NCC in gathering information and analyzing world trade, the competition with foreign production continued to grow. Overseas production went from twelve million bales in 1947–48 to 16.9 million in 1950 and 19.2 million in 1951. The temporary restraints on U.S. exports during the Korean War had encouraged expansion of foreign grown cotton, and when the price of U.S. cotton rose above world prices temporarily during the conflict, it sent overseas buyers to competitors, particularly if they had dollar shortages. Foreign synthetic production began to bite, too. Losses to artificial fibers manufactured overseas seemed irreversible because "once factories . . . have been built and markets for the product developed," wrote the research director for the Giannini Foundation of the University of California, "this much of the market for cotton has virtually been lost forever."[8] The United States thus faced the double threat of greater competition in cotton and a world increase in the output of synthetic fabrics. With the stimulus of sales wrought by the Marshall Plan scheduled to end in 1952, the export trade looked discouraging.

It became obvious that cotton exports could not continue to rely on recovery programs for countries damaged during the war. In late 1949 Dunn made a poignant analysis of agricultural exports in the full context of the world economy and political tensions. The NCC printed the report and circulated copies among its seven segment member groups, as well as various cotton organizations. Titled "After the Marshall Plan—What?" the report repeated the assertion made within cotton circles as far back as the mid-1930s that the United States must buy from other countries for them to acquire dollars to purchase our goods, whether bales of cotton or industrial machines. Dunn pointed out that losses or severe reductions of agricultural exports would threaten the U.S. economy, which he thought was not an unreasonable likelihood, since food production and general recovery were making progress around the globe. The strong dollar, based on victory in war, the trade surplus and sound economy, and the hoards of gold in U.S. vaults, further encouraged foreign buyers of cotton to shop in soft-currency countries, the competitors of the Cotton Belt.

These were not Dunn's only worries. He anticipated cutbacks in cotton allotments to coincide with the expiration of the Marshall Plan in 1952, and since dollar shortages would likely persist among U.S. trading partners, the Cotton Belt could experience a serious drop in income, enough to "be chaotic if not dangerous." Unrest and political instability in parts of the world also threatened markets. Dunn acknowledged that cotton, though important, was only one commodity among many that traded; he ranked it in importance with iron and steel, electrical machinery, nonferrous metals, chemicals, and other industrial items and saw a thriving general trade as the best chance for cotton. "For its own security," he wrote, "and for the peace of the world, the United States must work toward maintaining the flow of goods to the world."[9]

He used his report to advocate the Atlantic Union, a concept for a trading organization of Atlantic nations that was popular among cotton leaders. Will Clayton, now in retirement, was cofounder of the concept, and Rhea Blake saw merit in the proposal, occasionally speaking on its behalf. To establish conditions in Europe and North America for a sustainable and prosperous flow of trade, Dunn urged an economic union of Europe, the United States, and Canada. Such a body would control vast natural resources and industrial capacity and presumably reduce the likelihood of war. In the late 1940s and early 1950s, the notion of an Atlantic Union had appeal owing to the general apprehension about American trade: what would happen at the end of the Marshall Plan? Truman appointed a special committee to study the matter, and Senator Estes Kefauver offered a resolution urging congressional discussion of the subject. This proposal could be construed as a forerunner of the European Union (EU) and World Trade Organization (WTO), but it went nowhere at the time, likely because of the outbreak of the Korean War. Dunn did not mention the Atlantic Union in his memoir, and Blake made no reference to it in his oral history interview conducted by the University of Southern Mississippi after his retirement.[10]

THE BUXTON CONFERENCE

An example of the Cotton Belt fomenting its philosophy of consumption rather than restriction at the international level occurred when in 1952 leaders of British textiles invited representatives of prominent world textile enterprises to a conference at Buxton, England. The British had begun to experience rising global competition from Japan and hoped to impose restrictions or quotas on the textile trade. This would be a meeting of private interests known as the International Cotton Textile Conference, but they had "in mind restricting production and distribution of textiles to limit the competition from Japan and India," Dunn wrote one correspondent.[11] Delegates from most countries of Western Europe, as well as the United States, India, and Japan, received invitations, and American raw cotton became involved when the ACMI requested input from the NCC. Because the countries invited to the conference depended on the United States for almost 90 percent of their raw lint, the meeting had obvious importance for agricultural interests. Most of the American delegation consisted of textile executives, but Robert Jackson and Claudius Murchinson of the ACMI came with them, and from raw cotton there was Mississippi planter Howard Stovall and the NCC's Ed Lipscomb, who along with Dunn had responsibility for organizing the U.S. delegation to the conference.

Restrictions on trade had no appeal to the Americans, who preferred to think in terms of improving consumption. In 1952 U.S. textiles felt no threat from the Japanese, but Jackson knew that American trade laws would not allow them to enter into an agreement that established a cartel.[12] Hence the delegation had a different perspective from that of its British hosts. To prepare for the conference, which was scheduled for September, Dunn had arranged a daylong meeting of the American representatives on August 18 at the Peabody Hotel and prepared an analysis of the background, problems, and proposed remedies for discussion. W. M. Garrard, head of Staplcotn, and Lamar Fleming of Anderson, Clayton and Company joined them.

At the Peabody meeting, raw-cotton interests considered the British proposal a mistake. To begin with, since Japan depended on exports to fund its heavy importation of foodstuffs, it was in the interest of American and European security not to encumber Japan's textile exports, because Japan occupied a strategic position in the Pacific from both an economic and a military perspective. In view of the war in Korea, this assertion had significance. Adding to the instability of world politics was the fall of European colonialism. The rise of self-governments and nationalism, particularly in East Asia, meant that "a new system must be devised," insisted Stovall, a planter from Mississippi, "for supplying capital and know-how for the development of these areas."[13] Colonial powers had "worn out their welcome," and it behooved Western security not to be seen imposing restrictions on development in Asia, where vast resources in timber, rubber, copra, coal, and oil were available. The Hawley-Smoot fiasco had demonstrated the futility and danger of restrictive controls, and Japan's future had vital links to the Far East, a land of potential wealth that could benefit western countries. Better to encourage the free exchange of goods and promote prosperity among Japan and former colonial provinces than risk alienation in view of the struggle against communism. Ambitious and competent people cannot be blocked for long, Stovall believed, and "I don't think the cotton farmers," he continued, "want to block them."[14] For Stovall, the United States would be wise to devise policies that would encourage economic growth for the Japanese and the Pacific Rim. "Cartels restricting production, limiting competition and artificially holding prices will result in lower consumption and more trouble," he concluded.[15]

Upon reaching Buxton in September, the American delegation, which was a combination of textile and grower representatives, decided to stress cooperation for increasing consumption. A persuasive and compelling presentation for market expansion came from Ed Lipscomb, whose appeal along with discussion from other delegates had an effect. The conferees approved a resolution recommending that countries participating in the Buxton meeting initiate programs for increasing consumption. No principle or statement suggesting restrictions came out of the meeting. For the Cotton Belt interests, this outcome was heartening, because they hoped new trade policies, expected to come soon after the postwar assistance programs ended, would be integrated into overall foreign policy. For Dunn, the outcome of the Buxton Conference encouraged the formation of international programs to promote consumption.[16]

PUBLIC LAW 480

The pressure to find outlets for cotton quickly mounted with the buildup of surplus stocks and accompanying stagnant prices starting at the end of the Korean War. In 1954 Congress passed the Agricultural Trade Development Act, or Public Law 480 (PL 480), which provided for the export sale of surplus commodities for foreign currencies rather than dollars, up to a maximum of $700 million for the next three years. Another $300 million of surplus commodities could be granted to countries for humanitarian considerations. This unorthodox plan originated with Gwynn Garnett of the Farm Bureau, whose military experience in the U.S. occupation of Germany gave him the idea. He had seen bartering and exchanging of currencies there. According to Dunn, "many considered the proposal to be an economic fluke," and he described it as a gimmick, but "it had appeal" because it offered a way to dispose of surplus stocks other than cotton. Garnett included a provision stipulating that a portion of the currencies be used to develop overseas markets. Only countries with balance of payments difficulties would be allowed to buy U.S. cotton with foreign currencies. Not all customers of American cotton would qualify, but Germany, Japan, and Italy, as well as some Mediterranean countries, were expected to be eligible. For the sake of overseas sales, the concept appealed to cotton interests, especially shippers, and "the Cotton Council got on the bandwagon," Dunn remembered.[17]

Dunn represented the welfare of cotton, but other commodity interests informed Congress of their fondness for the proposal. The USDA endorsed the concept, and it next went to the House where Congressmen Poage of Texas and Clifford Hope of Kansas sponsored it. Hope's position as chair of the House Agriculture Committee gave the measure extra strength. The NCC lobbied Congress and met with other commodity organizations. Wheat, tobacco, and soybeans were enthusiastic supporters, but cotton led them all. Mississippi Senator James Eastland introduced the Senate version of the proposal that became PL 480. It passed Congress easily.

This measure extended the global reach of cotton's domain because it developed markets that previously were small or practically nonexistent, such as Turkey, Greece, Taiwan, and South Korea. Much of California's Acala already had overseas outlets, but PL 480 opened sales for the traditional varieties. By enabling developing countries to buy cotton even if they had no items to sell for dollars, the United States expanded its exports and created outlets for humanitarian assistance and foreign aid. Sales of cotton went well under the program; Dunn reported that eight million bales went to previously untapped markets from 1955 to 1963.[18]

THE COTTON COUNCIL INTERNATIONAL

Cotton interests encountered in the global market the same problem that faced them at home: the growing popularity of artificial fibers. The NCC did not intend to let the growing sales of synthetics in the international market go unchallenged. Western Europe consumed about three million bales of synthetic fibers and ten million bales of cotton, but consumption of artificial fibers was expected to grow. An international program for advertising and promoting demand for cotton had to be developed, but Dunn, Blake, and others realized that such an undertaking would benefit foreign producers as well as American farmers. The NCC decided to stick with its founding principle that consumption must be improved, which would in this case benefit all producers. "Bake a bigger pie," Dunn wrote, "so that the slices would be bigger."[19] To carry out its objectives, the NCC established the Cotton Council International (CCI) in 1956 as an independent body and empowered it to make contractual agreements with the USDA-FAS and foreign governments for marketing development. Blake would serve as the executive director, and Dunn oversaw day-to-day operations. In other words, the CCI acted as an overseas arm of the NCC, but with its own identity as a legal entity.

Germany quickly asked to participate, but the CCI had no operating funds, so along with officials of the FAS, Dunn arranged an agreement for Germany to buy frozen poultry in the United States, and those funds kicked off CCI operations. For each country making agreements to promote cotton, Dunn negotiated the arrangements and learned to incorporate at least one cotton industry leader from each country, usually from textiles, in the negotiations. With such support, the program would go forward. Dunn leaned on his contacts made at the Buxton Conference to establish working relationships with European cotton leaders. France and Japan signed agreements along with Germany, Italy, Belgium, and Austria. Britain took longer; retailers provided the impetus for the CCI there, particularly Marks and Spencer. Within one year, Dunn had brought Japan and all western European countries into the fold, except Portugal. Each country furnished half the funds spent within its borders, with the FAS picking up the balance. Though the synthetic manufacturers still spent more money on promotion, the CCI now had some funds to fight back. Other countries signed agreements with the CCI: Canada, Mexico, India, and Columbia.

For all practical purposes, the CCI operations in foreign lands followed the general approach of the NCC in the United States. The new organization had a publicity program that provided information to the media about events and sales pertaining to cotton merchandise, and it kept textile and apparel leaders informed of advances in fabricating cloth. Public relations and promotion of sales took great importance, which came through fashion campaigns, special advertising of children's clothing and men's wear, sales training programs for retailers, and educational seminars for retailers about shrinkage and laundering of cotton. As was the case in the United States, the CCI engaged in market research for comparisons of cotton and synthetic uses, pricing, quality studies of cotton goods, and consumer preferences. No better connection could be found between the NCC and these overseas programs than the foreign tours made by the Maids of Cotton. They traveled with CCI personnel and a business secretary to handle their activities. "The programs were modeled," Dunn wrote, "on those in operation in the United States by the National Cotton Council."[20]

The CCI advertised cotton, primarily as a generic item, widely in the participating countries. In 1965 alone it had over 800 pages of magazine advertisements, almost 300 radio and television spots, and over 240 all-cotton fashion shows. Dunn reported that 14,000 retail outlets participated in special "cotton weeks." Market research was new in Europe and Japan, but the CCI stirred much interest in this area, which stimulated some companies to start their own research programs. It arranged meetings of the Maids of Cotton with various royal families, which encouraged newspaper coverage and lent credence to cotton as prestigious clothing. Positive results came: per capita consumption of cotton increased about 12 to 15 percent in the European participating countries. This increase paralleled the improved standard of living there, but synthetics' share of the market gained nonetheless on cotton, though on a per capita basis cotton consumption still outranked artificial fibers. The popularity of artificial fibers continued to rise until by 1987 they equaled cotton in per capita consumption. Oscar Johnston's fear that synthetics furnished the greatest threat to cotton had come true.[21]

Since the CCI relied on public funds from the USDA-FAS, raised through the PL 480 program, government auditing of the organization occurred. In 1963 an evaluation team reported the CCI program had more impact with overseas industry leaders than with consumers and urged that funding be increased and more advertising directed to consumers rather than generic promotions. But the organization came under fire when the chair of the House Subcommittee on Executive and Legislative Reorganization requested an examination of public funding of market development activities conducted overseas. The auditor from the General Accounting Office (GAO) considered the Maid of Cotton program to be an unnecessary use of government funds, and speeches against the Maid of Cotton promotions appeared in the *Congressional Record*. The final report from the GAO, though critical, treated cotton more kindly than other commodities with market development programs in Western Europe.

THE INTERNATIONAL INSTITUTE FOR COTTON

Despite its success, the CCI again made apparent the Achilles' heel of cotton promotion: lack of funding. However much money was raised to stimulate consumption, the amount never matched the millions of dollars that synthetic interests invested for the same purpose. It was an arms race for market share in which cotton started out ahead but slowly lost ground to its competitor at home and abroad. By the mid-1960s, the conviction grew that foreign producers should contribute directly to funding their own interests. Under the CCI, countries that produced no cotton, such as Britain, Germany, and France, had been making the largest contributions after the United States, but more money was simply needed. Dunn had stated on record in 1959 that another $25 million should be available for overseas promotion.

The NCC went to work organizing countries with a stake in cotton exports to establish a new international body for cotton promotion. Dunn did the legwork. Textile leaders in Japan and Europe provided assistance, thanks to the leadership of W. T. Kroese of Holland. Through the International

Cotton Advisory Committee (ICAC), Dunn discussed funding with cotton-producing countries, and they agreed to collect one dollar per bale from the grower for each bale exported to Western Europe or Japan. In 1966 the participating countries signed a formal agreement, and the International Institute for Cotton (IIC) went into operation. Under Secretary of Agriculture John Schnittker expressed the rationale for the new body: "This is not just a matter of overcoming sentiment or tradition, it is a matter of necessity to maintain income for literally millions of persons who depend on cotton for a livelihood."[22] Participants other than the United States were Spain, India, Egypt, Mexico, and Sudan, all of which accounted for 60 percent of the world's production. A few months later, Greece, Brazil, Uganda, and Tanzania joined the organization.

The CCI promoted only American cotton, but the new organization had an international foundation. And it had a larger budget. Never before had competing producers joined together for a common cause. It remained unclear if farmers around the world felt a sense of camaraderie with their counterparts in the Cotton Belt, but they were nonetheless linked together for a common cause. Dunn served as the IIC director, and the headquarters were located in Washington; he hired non-Americans for his staff. Overseas offices were located in Brussels, London, Manchester, Paris, Frankfurt, Oslo, Milan, and Osaka. "The U.S. cotton industry had fought the rising tide of man-made fibers since World War II," stated one account, "and it was now fully engaged in combat with its nemesis on an international scale."[23]

In Europe synthetic manufacturers had acquired some spinning mills, which prevented the IIC from entering into agreements with textile associations because they included cotton's competitors. This forced the IIC to concentrate on the end products of cotton: clothing, sheets and towels, children's apparel, and designer fashions, which focused the advertising on brand name products. In this respect, American interests were pleased, but the use of brand names varied from country to country. Marks and Spencer bought cotton goods manufactured according to its specifications and sold them under its brand names in Britain and Europe. Indeed, the IIC liked making agreements with nationally known brands so that it could advertise the soothing feel of a natural fiber. Herein lay a major difference with the CCI: it could only promote cotton as a generic product, but the IIC could advertise popular brands. Dunn arranged IIC tours with the Maid of Cotton program, since it operated independently of the General Accounting Office.[24]

Like other programs to promote consumption, the IIC always ran short of money. Sudan and Egypt had never made their initial membership payments and withdrew soon after the organization started. The Egyptians suspected the IIC favored the United States. In 1984 they offered to rejoin on condition that 30 percent of Egypt's own assessment be spent on research in Egypt and that the IIC start promoting extra-long staple. Peter Pereira, who took over when Dunn retired in 1975, refused on the grounds that special considerations could not be granted on an individual basis. Sudan was riddled with internal strife, and Pakistan did not join because it saw "nothing to lose by staying out," according to one IIC delegate.[25] Australia discussed membership, but growers there questioned the assessment fee. China's production of cotton had been growing fast, and there was a general expectation that it would become a principal exporter of both raw cotton and textile goods. Due to its communist ideology, however, China would not engage in retail advertising, which kept it out of the IIC. The Soviet Union showed no interest. At the IIC's peak, member organizations accounted for 60 percent of the world production.

The international cotton trade had been undergoing change, a "geographic shift," according to Dunn. Brazil and Mexico, longtime exporters and supporters of the IIC, developed their textile industries to the point that they became net importers of raw lint and manufactured a sizable amount of the cotton goods that went into American and European stores. When Brazil, outranked only by the United States in supporting the IIC with funds, dropped out in 1989, owing the organization over one million dollars, the IIC's future looked dismal.[26]

Membership fell even more, until by the mid-1990s the IIC faced its end. The dropout nations saw the value of the organization but could not pay their assessment fees. Pleas for funding to the United Nations Advisory Committee on the Application of Science and Technology to Development (UNACSTD) and the International Textile Manufacturers Federation (ITMF) in Zurich went unrewarded. Support for the IIC eroded as nonmembers realized they would benefit at the expense of paying members. The United States withdrew in 1991, and three small African nations, the last members, disbanded the IIC in 1996.

But the CCI continued and remained the principal promotional organization for American cotton overseas. Dunn attributed increases in per capita consumption to the CCI, reporting that during the first five years of its operation, consumption jumped 12 to 15 percent. In Holland the shirt market went up owing to new fabric finishes. But consumption of all textiles rose throughout the world, with cotton "barely holding its own during the latter 1960s and 1970s."[27] It started to climb again in the 1980s, but not as fast as artificial fibers. Precise measurement of the impact of the promotional programs is impossible to determine, but in 1993 two economists at Alabama's Tuskegee University and Auburn University made the general conclusion that with countries in the Pacific Rim there was "a significant relationship between promotion expenditures and U.S. market share in four of the six countries examined."[28] No comparable study had been made of trends in Europe, but Dunn noted that with the decline of textile manufacturing, mills lost their interest in promoting cotton and focused on multifiber spinning, which meant that cotton could no longer command the enthusiasm and devotion it once had.[29]

In 1985 Congress authorized the Targeted Export Assistance (TEA) to help commodities whose exports had been harmed by unfair practices in other nations, at least for short-term promotions. Advertorials funded by the TEA featured a U.S. cotton logo in both Asia and Europe, and the NCC produced advertisements that incorporated the logo while brandishing the finished goods of private companies. In 1990 the Market Promotion Program (MPP) replaced the TEA, and the total expenditure of funds for the sake of cotton consumption reached over $50 million from 1985 to 1992. This represented nonprice support expenditures. Economists have been reluctant to make anything other than general claims about the impact of these promotional campaigns, though they saw a generally positive result. All acknowledged, including Dunn, that while the per capita consumption of cotton increased over the years, owing likely to these efforts, it did not rise as fast as consumption of artificial fibers.

In 1994 the CCI expanded its campaign to include the export of yarn, fabric, and finished goods made of U.S. cotton. The value of these exports from 1990 to 1994 averaged $2.3 billion, but the average jumped to $4.6 billion after the new program started.[30] Had the cotton used in these finished products remained as raw lint, the gain would have been less, amounting to 18 percent of the value of the exported finished products, so claimed the CCI. Since the intent of the CCI was the export of American cotton, it started a licensing agreement for overseas advertising and promotional programs, using the logo "COTTON USA." Each item of clothing has a card attached with the logo attached, which certifies that the piece contains a minimum of 50 percent U.S. cotton. From the perspective of CCI, this assures consumers that high-quality cotton has been used. Estimates made by the CCI show that COTTON USA accounted for an increase of 16,000 bales of American cotton consumed in Belgium, German, and Italy in 1996. In Asia, the CCI reported, the results were more rewarding, with the licensing program generating over 66,000 bales in 1994 to 1996.[31]

Cotton was not alone in promoting trade. Globalization depended on no commodity and belonged to no country. Other commodity interests such as rice used funds generated by the PL 480 arrangements to increase exports. The Big Three automakers, Ford, Chrysler, and General Motors, expanded their European markets, and even tourists increased the flow of capital and culture across the borders as they traipsed through ancient ruins or absorbed the ambience of boutiques and salons. All these activities made trade grow, so cotton, for all its importance among trading partners, must not be viewed singularly in the rising global economy.

SHIFTS IN WORLD COTTON TRADE

By the 1990s, cotton faced a new set of circumstances unlike anything it had known before. Only about a dozen countries grew cotton for the world market before 1945, with the preponderance coming from the United States, Brazil, Egypt, and India. Cotton had been grown in many areas of the world solely for local markets and provided no competition for American growers, but as developing countries sought to strengthen and expand their economies, they turned to cotton production for export as a means to raise funds for foreign exchange. With the addition of former Soviet satellites, the number of producers increased until by 2000 about one hundred countries grew cotton, and world supplies in stock reached over forty-nine million bales that year compared with over twenty-one million bales in 1945.[32] In 1997 China produced over twenty million bales. New and better varieties of cotton spread across the world, too, though American Upland continued to lead the way. Among some major foreign producers, the native varieties, usually short staple, disappeared. This development reflected the widespread adoption of American techniques in farming cotton engendered by the large number of foreign agricultural scientists who trained in the universities in the Cotton Belt. "In fact," wrote Dunn, "the United States was the only country where practical, hands-on training in cotton breeding and production was available."[33] Educated by American standards, the foreign-born agriculturalists embraced the science of farming practiced in the cotton kingdom and used it in their homelands.

Textile manufacturing shifted from one side of the globe to another. Before World War II, Britain had been the principal fabricator of cloth, which had fueled its industrial revolution. But the British warnings at the Buxton Conference had come true because textile manufacturing shifted to Asia, which changed the destination of much cotton grown in the South. In 1958 about 65 percent of the world's textile spindles were in Western Europe, but the percentage fell to 35 percent by the mid-1980s.[34] Britain became a net importer of textile goods by the early 1990s. Growers in the western United States had long depended on the Asian market, so the adjustment to new markets did not bother them, and brokers and merchants in the United States had no particular difficulty in establishing these new outlets, owing to

Table 5. United States and World Production, 480-Pound Bales (Thousands), 1945–2005		
Year	United States	World
1945	9,015	21,410
1955	4,698	44,329
1965	14,938	56,900
1975	8,302	54,000
1985	13,432	80,100
1995	17,900	93,900
2005	23,890	116,600

Source: USDA Economic Research Service, *Statistics on Cotton and Related Data, 1920–73*, Bulletin No. 535 (Washington, D.C.: Government Printing Office, 1974), 40; Leslie Meyer, Stephen MacDonald, and James Kiawu, *Cotton and Wool Situation and Outlook Yearbook* (USDA Economic Research Service, November 2008), 35.

their experiences on behalf of western cotton and the activities of the CCI. In 1950 the world-renowned Liverpool Cotton Exchange closed, but this owed much to the presence of the Labour government in Britain, and the Bombay Cotton Exchange had no further activity after the Indian government froze cotton prices. American traders gained more business in world markets with these losses. Over time cooperatives such as Staplcotn in Mississippi, the Plains Growers Cooperative Association in Lubbock, and Calcot in California picked up much of the trade.

TRADING BLOCS

A new avenue for maintaining or enlarging the export market emerged toward the end of the twentieth century: trading blocs. Partly in response to the creation of the European Union, the conviction grew that the United States needed to establish a similar bloc before it was closed out by the rest of the world. This conviction gained strength as the difficulty worsened in reaching trade agreements during the various rounds of talks conducted through GATT. Because it championed free trade, the Cotton Belt had endorsed the creation and operation of GATT and supported Clayton, who negotiated the original talks establishing it in 1947. Free-trade interests, led by Clayton, had also sought to get an international trade organization established that would have gone beyond GATT. An international trade organization had been the dream of Cordell Hull, but when in 1950 Congress refused to approve U.S. participation, Hull's dream vanished, and GATT remained the world's only multilateral body governing trade. For agriculture the important provision in GATT was Section 22 of the Agricultural Adjustment Act of 1938, which protected the domestic and export subsidies for cotton. Reduction of tariffs remained the principal goal of GATT, and it succeeded in significantly lowering them during its life span, and membership grew. But the organization could not resolve issues pertaining to farm trade, so the general practice during its rounds of negotiations was to exempt agriculture and permit countries to subsidize and otherwise provide special support for farm products. Europe and the United States preferred this method of operation, as each wished to protect its agricultural interests. Textile products received considerable attention in the GATT rounds of discussions starting in the 1960s because developing countries began shipping low-value cotton goods to the United States, but raw cotton did not come under scrutiny, since many of the members in GATT imported raw fiber for manufacturing. These circumstances kept cotton out of the spotlight of foreign affairs.

A major development in global trade with particular consequences for cotton did occur with the creation of the World Trade Organization (WTO) in 1994, originating with the Uruguay Round of GATT in 1986. Trade distortions and protectionism in agriculture had grown since World War II, while industrial tariffs had dropped from an average of 40 percent to about 4 percent. Trade reform in agriculture had been purposely neglected in the GATT rounds of the past, but in the Uruguay Round, the developing countries resolved not to agree to reform in areas vital to the rich nations—intellectual property, services, and investments—unless progress came in farm trade. The European Community (EC) had been extremely protective of its agricultural subsidies, which far outstripped U.S. subsidies, about "50 percent to 3 times higher than in the U.S.," according to USDA press releases.[35] Europe had no cotton to protect, but the United States had prevented the importation of raw cotton, except in minor amounts, through Section 22 of the 1938 Agricultural Adjustment Act. The U.S. delegation expressed a willingness to make concessions on agriculture, but the EC refused to negotiate its farm subsidies, so progress went slowly at the Uruguay Round. The NCC monitored the talks and kept its membership updated on the issues.[36]

World pressure, including the self-interest of the United States, kept the Uruguay Round under way. It became clear that only a broader and more authoritarian trade body could resolve disputes, and when the GATT ministers proposed the creation of the WTO, they included concessions by the American delegation on agriculture. Congress had to ratify the agreement, and President Clinton made membership a priority because it was expected to increase exports and create new jobs. The NCC was cautious, holding back an endorsement because the proposed agreement sacrificed Section 22.

There would be a new system of "tariffication" that set a duty of 16 cents per pound on bales imported over a quota of 238,000 for the first year (1995), falling to 14 cents by 2000, with an additional 159,000 bales added to the quota. Loss of Section 22 had been a matter of concern in the Cotton Belt.[37] Textile quotas under the Multi Fibre Arrangement would be phased out over ten years, and duties on textile imports would be reduced by an average of 12 percent. China would not benefit, since it was not a member of GATT, but U.S. textile manufacturers nonetheless remained nervous about China's strength. Aside from the WTO issues, the White House proposed in the 1995 budget to eliminate all funding for the Cottonseed Oil Assistance Program (COAP), which promoted exports for cottonseed oil. The proposal alarmed cotton interests, who saw a threat to overseas market programs such as the CCI. To fund the implementation of the WTO, the Clinton administration further wanted funding taken from agricultural appropriations, reported to be as high as $5 billion.[38]

Objections quickly erupted. Bruce Brumfield, a Mississippi grower and president of the NCC, testified before the Agriculture Committee of the House of Representatives. He accepted that import quotas could be changed, but wanted the United States to retain the authority to invoke Section 22 if needed in the future. He believed the domestic subsidies were not affected, a point highly important to growers. The NCC understood that the U.S. support program qualified by the rules of the GATT, so Brumfield stated: "We don't want it changed."[39] He wanted congressional funding maintained for "greenbox" programs, referring to export assistance and market development programs. For the cuts in agricultural funding, Brumfield thought such a reduction would undermine the ability of farmers to compete in world markets, but the White House wanted ratification and had the support of powerful nonagricultural interests, such as pharmaceuticals, that wanted protection of intellectual property. Cotton interests could not afford to be obstinate, and the NCC looked for a settlement.

In September the NCC's Phil Burnett reported a compromise between Cotton Belt congressional leaders and the Clinton administration. Section 22 would be replaced by the tariffication schedule, but the White House committed to funding export enhancement programs with an interagency review process to ensure that sales would be carried out promptly. Spending on "greenbox" programs would reach $600 million over the next five years. Agriculture would be expected to sacrifice $1.7 billion from the federal budget instead of the reputed $5 billion. For textiles, the administration imposed a "rule of origin" that meant a tag must be attached to fabric showing the country of manufacture rather than the location of apparel sewing. This last provision was meant to prevent China from circumventing the new WTO. In regard to subsidies, the president wrote Patrick J. Leahy, chair of the Senate Committee on Agriculture, Nutrition, and Forestry, that "my next two budget requests will safeguard spending for agricultural programs."[40] For both raw cotton and textile manufacturers, a new era of competition was expected, but domestic subsidies for growers were expected to remain intact. Textile leaders predicted that mills would move into foreign countries and only large companies could survive. Roger Milliken, well-known textile leader, worried about the loss of jobs.[41]

COTTON AND THE WTO

Cotton became the object of international attention when it became the focal point of trade disputes in 2002. Brazil, a cotton exporter, filed through the WTO a complaint against the U.S. cotton subsidy programs, claiming they held down the world price of raw cotton and thus brought injury to Brazil's growers and general economy. The complaint targeted export subsidies: deficiency payments, Step 2 payments, countercyclical payments, and marketing loan assistance payments. Suddenly the spotlight shone on American interests as more foreign producers accused the United States of distorting the world market and causing poverty among the farmers of West Africa. This charge received widespread attention and led to charges of greed and selfishness against growers in the Cotton Belt. The United States had erected a support system for cotton farmers "bizarrely reminiscent of Soviet state planning principles," and taxpayers in 2001 furnished "the cotton barons of Texas and elsewhere" with support greater than the real value of their crop, according to Oxfam, a charity and relief advocacy organization based in Britain.[42]

Behind Brazil's dissatisfaction lay the new developments in the global cotton trade. The United States, now the second-largest producer, remained the world's largest exporter, accounting for 40 percent of the international trade from 2001 to 2003.[43] The growing U.S. importation of finished cotton goods had brought a drop in demand from domestic mills, which had caused raw-cotton interests to maximize exports and compete more vigorously in foreign markets. But world commodity prices had been falling since the mid-1990s, with cotton in decline since 1997, spiraling downward 55 percent by June 2002, and the fall of the so-called tiger economies of Southeast Asia in 1997 also contributed to the decline. As Brazil's complaint and the U.S. response sifted through the WTO resolution process, the issue became embroiled in the general negotiations over trade.

In 2001 the Doha Round of WTO discussions, conducted in Doha, Qatar, set a major objective for negotiations at future meetings. Ministers agreed to improve trading opportunities for undeveloped countries as a means to promote prosperity and thwart the rise of terrorism. The next meeting of the Doha Round came at Cancun, Mexico, in 2003, but U.S. negotiators found the air poisonous. The Brazilian challenge had gotten under way, and four cotton-growing countries in West Africa—Benin, Burkina Faso, Chad, and Mali—joined the protest over American subsidies, insisting the supports caused poverty by forcing their farmers to sell at depressed prices. Two days before the opening discussions at Cancun, officials from many of the 146 member states and some nongovernmental organizations sympathetic to the African countries, staged an event they called World Cotton Day. Among the most outspoken critics of U.S. policy was German minister Heidemarie Wieczorek-Zeul. Joining her were Pascal Lamy and Franz Fischler, trade commissioner and farm commissioner respectively for the European Union (EU). Cotton had suddenly become the hot item of world trade and was pushed to the top of the agenda for the first day of the Cancun discussions. To woo friends and protect its gigantic agricultural subsidies, the EU sympathized with the African countries, an easy move, since Europe grew little cotton. This posturing reflected the shifting alignments in world trade, and with the creation of the G20, consisting of both industrialized and developing countries that held about 65 percent of the world population and nearly 75 percent of its farmers, four members held much influence with the developing countries: Brazil, China, India, and South Africa. Known as the "Cancun Collapse," the meeting broke down after four days because of disagreements principally over cotton, but the collapse had implications for industrial goods. Later negotiations in Geneva and Paris brought promises over lowering agricultural subsidies, but the discussions needed new energy to expect real progress. France proved to be particularly recalcitrant over its own farm supports.

In the course of the WTO discussions in the Doha Round, opinion makers on the home front joined the chorus of critics asserting that U.S. cotton interests were responsible for the poverty of West African farmers. In 2002 the *Wall Street Journal* featured a lengthy comparison of farming in the Mississippi Delta and Mali. It contrasted the fully mechanized operations in the Delta, including the use of tractors equipped with GPS, with a single-plow oxen team used near Korokora. There African farmers walked barefoot while Mississippi growers sat in air-conditioned tractor cabs with a cushioned seat. American producers, so the article continued, relied on subsidies to float their investments and return a profit, but Africa's poor slipped deeper into poverty. The Africans were becoming alienated and angry, and since they were Muslims, the newspaper continued, the United States risked the growth of anti-Americanism.[44]

From the *New York Times* came a similar condemnation, but it held Europe and Japan equally responsible owing to their farm subsidies. The *Times* editorialized that U.S. subsidies perpetuated hardship and ill will toward the United States because American growers could afford to keep overproducing and sell in the depressed world market while being assured of a profit. They glutted the world market, said the *Times*, at the expense of poor African farmers who had no safety net. The "club of rich nations" had rigged the game and was "harvesting poverty" and "plenty of resentment as well." The *Times* urged a revision in agricultural policies for the sake of the African poor. "Such madness is no longer sustainable."[45] Similar statements came from the *Washington Post* and several national television commentators.[46] But Oxfam remained the most severe critic, alleging that subsidies benefiting the Cotton Belt were "at the heart of a deep crisis in world cotton markets."[47]

The Brazilian challenge kept the issue of U.S. subsidies alive and in the forefront of trade news. In March 2005 the WTO Appellate Body upheld the original finding made in 2002. It specified that the "Peace Clause," a proviso in the WTO that exempted agricultural subsidies, did not apply in this case because the payments in question, export guarantees and Step 2 marketing payments, amounted to export subsidies prohibited by the WTO general agreement. The Appellate Body affirmed such payments as trade distortions. The four African countries had endorsed the complaint and now urged the United States to modify and adjust its policies by the next round of negotiations scheduled for December in Hong Kong. Oxfam, cotton's fiercest critic, warned that a delay by the United States in complying with the ruling would "cost poor African farmers the chance to trade their way out of poverty and perpetuate an unfair system of rules rigged for the rich."[48] And the Environmental Working Group (EWG), based in Washington, released a statement: "Brazil's victory proves that the WTO isn't a plaything of subsidized U.S. agriculture."[49] Disappointment and frustration came from the United States. A representative for U.S. trade minister Peter Allgeier stated that "negotiation, not litigation, is the most effective way to address distortions in global agriculture."[50] Texas grower Joe Alspaugh worried that he would have to abandon farming.[51]

American cotton interests, led by the NCC, had always insisted that the subsidies complied with the WTO General Agreement, but indicated that trade should be conducted in a "rules based international system."[52] The NCC intended to work with Congress and the administration of President George W. Bush to "formulate an appropriate response to the decision."[53] The organization had regularly pointed out other cotton-growing countries, including the four African countries involved in the dispute, furnished subsidies in different forms: taxation, setting the price of raw cotton, loans, and help with seed and fertilizers. The NCC suggested to Oxfam that it rely on more accurate data when analyzing the world cotton market; that it recognize the use of agricultural supports in other countries; and that it incorporate the impact of the monumental growth of synthetics into its assessment of trading practices. But the WTO case had killed the Step 2 payments; Congress ended them effective August 2006.[54] The WTO ruling also called for the end of Export Credit Guarantees, a program that committed the United States to guarantee loans furnished by U.S. creditors to foreign banks for purchases of cotton and other agricultural products grown in the United States. Remedying these differences would require congressional action.

When the Hong Kong round of WTO discussions occurred in December 2005, cotton was on the defensive. Opposition to U.S. cotton supports had subsided to an extent, owing to the approval of the Dominican Republic–Caribbean Area Free Trade Agreement (DR-CAFTA), which put Brazil in a weaker bargaining position.[55] The dependence of smaller countries on currency and capital flows from the industrialized nations, whether through loans or normal business, also weakened their position. Consequently there was no failure as in the Cancun Collapse in 2003. At the Hong Kong meeting, however, U.S. trade negotiators promised to reduce domestic subsidies, which drew criticism from the Cotton Belt.[56] No offer to reduce subsidies came from the EU, but it attacked cotton subsidies, which demonstrated, stated NCC executive director Mark Lange, "that the priority of most of the members of WTO in Hong Kong was cotton specifically, not any kind of overall agricultural agreement."[57] Singling out cotton irritated U.S. growers, and Woods Eastland, head of Staplcotn, felt that the NCC should cooperate with Congress and examine the implications.[58] William Gillon, outside counsel for the NCC, saw a potential erosion of support for multilateral agreements by U.S. agricultural interests.[59]

NAFTA-CAFTA

In 1994 the North American Free Trade Agreement (NAFTA) went into operation. The treaty had three members: the United States, Mexico, and Canada. The United States and Canada had operated a bilateral agreement starting in 1989, but the United States increasingly needed leverage to deal with the growing economic power of the EU. By bringing Mexico into the agreement, the United States expected to gain access to markets and resources, particularly oil, but the opportunity to expand exports of agricultural commodities also appealed. The agreement provided that nontariff barriers affecting farm trade would be lifted immediately, and that tariffs would be phased out through stages, with all tariffs eliminated in fifteen years. NAFTA incorporated the "rule of origin" that meant only items imported from a member would receive preferential tariff treatment. In other words, goods imported from other countries would be assessed a tariff. In 1991 the United States exported almost two hundred thousand bales per year to Mexico, but rising income levels there increased the likelihood that imports would increase. Mexico grew cotton but needed to import additional quantities to supply its textile mills. For this reason, farming interests in the Cotton Belt supported the proposal to join NAFTA. When the new trade bloc went into effect in 1994, Phil Burnett of the NCC felt that cotton exports to Mexico "could conceivably grow by around one million bales over the next ten years."[60] In 1998 both Mexico and the United States ended their cotton tariffs with each other. No major dispute erupted over the cotton trade within the framework of NAFTA.

The pursuit of free trade continued, and when in January 2002 President Bush made a trade agreement with countries in the Caribbean, completion drew nearer for the Caribbean Area Free Trade Agreement. Negotiations were conducted with Costa Rica, El Salvador, Guatemala, Honduras, and Nicaragua. When the Dominican Republic joined the talks, the proposal became known as the DR-CAFTA. In the United States, stiff opposition came from antiglobalization forces such as Ralph Nader's organization Public Citizen and the Sierra Club. Opponents believed the trade bloc would lead to exploitation of labor in the Caribbean countries and harm the environment. Workers in the textile industry expressed strong dislike for the proposal.

Cotton growers, particularly in the southeastern states, favored it. For one thing, they thought it would lead to increased exports of cotton yarn and fabric into the Caribbean market. The DR-CAFTA region already accounted for the second-largest number of bales exported in 2005, with about 200,000 bales of raw cotton and the equivalent of 2.5 million bales in yarn and fabric made in the United States. A formal trade agreement was expected to increase these figures and thus benefit growers in the southeastern states. Mark Lange estimated that the content of U.S. cotton in apparel made in the Caribbean countries would increase, since the agreement would eliminate tariffs on their importation of American cotton. The proposal had strong appeal, furthermore, as "a solid platform to compete against the surging Chinese cotton product imports to our country."[61] The NCC lobbied Congress on behalf of the proposal, which passed by a bare margin in the House of Representatives, 217 to 215, but the Senate had a wider margin, 61 to 38. Only when the White House agreed to pressure China to restrain shipping apparel goods to the United States did DR-CAFTA pass Congress.

CONCLUSION

Since 1945 the U.S. cotton industry faced a series of challenges in world trade. By 2000 it had lost its dominance of the international market in production of raw fiber and the manufacture of textiles. It initiated overseas marketing organizations, the CCI and IIC, to promote foreign sales, but the steady growth of popularity of synthetics around the world kept cotton interests on the defensive. The rise of trade blocs, meant to enhance trade, led to a stinging indictment by the WTO of the federal support program for overseas sales, reminding growers they had to produce quality fiber at low cost to survive in a competitive market. Along with recent charges about U.S. growers acting as a special interest at the expense of poverty-stricken African farmers, the industry was thrust into the media spotlight. These developments kept cotton in our current affairs, if negatively, and meant that cotton farming remained more than a matter of agricultural production.

CHAPTER 19

THE NEW COTTON CULTURE

In 1988 when the Cotton Council celebrated its fiftieth anniversary, Albert Russell felt that the old connection of cotton with slavery, exploitation of sharecroppers, and general poverty had been eradicated. "Cotton was linked historically with slavery . . . and was associated in many people's minds with keeping blacks in poverty," but Russell felt that cotton had achieved a perception as a "contemporary and progressive industry."[1] Nearly all the symbols and trappings of cotton farming so common a half century earlier—mules, walking plows, sharecroppers, shacks—had been replaced with machines, computerized systems, mega-gins, and slick advertising. No longer did the image of the cotton farmer evoke wretched poverty or entrapment in a life as a peasant. The caricature of the hapless southerner meagerly existing on a small patch had been relegated to history. If the old cotton fraternity had been insulated from the forces of progress and innovation, the new generation eagerly sought to move forward with no regard for past ideologies or farming by custom and habit. Modernization and the business ethic had become the common denominators across the Cotton Belt. As stated by a president of the Agricultural History Society, "The old structure—small cotton farms worked by people and mules—had been changed to a highly capital-intensive, diversified, mechanized, and labor-efficient agriculture."[2]

THE NEW CULTURE

Certainly the cotton culture of the past—and all that it meant—no longer exists. Gone are the social injustices and poverty that prevailed before and during World War II. Mechanization, the advancement of civil rights, and the expansion of the national economy, all of which gave cotton's toiling class better opportunities, were chiefly responsible for this achievement, but for the growers who remain and are responsible for supplying the United States with its primary fiber, the cultivating, harvesting, and marketing of a crop continues to be their life's work. For them life on a cotton farm is rewarding and satisfying despite its endless frustrations and challenges.

To earn a livelihood, they have to sacrifice their ease and comfort to the royal plant from mid-April through December. But the sensitive green stalks that blossom white demand more than time and labor; they require management skills and diligence on a level commensurate with industry and commerce. Growers have much to understand: government programs, marketing on a global scale, EPA regulations, new technologies, and international trade discussions, all while staying abreast of political developments that might affect operations. For the cotton grower, staying organized politically is essential, while the potential loss of a crop in a storm, flood, or hurricane never disappears. If insects can be controlled reasonably well, weather cannot. Even the irrigated areas of the West have lower yields during drought. There is little room for mismanagement in an enterprise of such risks, and whether southerner or westerner, a successful producer must be entrepreneurial, well informed, and highly diligent. The old notion of slow-paced figures clad in overalls and spitting tobacco has been replaced by business-minded farmers whose management style belongs on main street rather than a town square.

Growers know that many of the pitfalls of producing cotton have been overcome. Nothing satisfies better than the conquest of the boll weevil and not having sleepless nights wondering if the old scourge of the South would ruin a year's work. Research by agricultural scientists had led to the breakthrough, but much organized lobbying lay behind the research. The discipline of monitoring the weevils exemplifies the new cotton culture in which guardsmen stay posted not only for marauding worms and weevils but also for swings in world prices. The appearance of a hurricane in the Gulf of Mexico with the potential to hit the mid-South and wreak havoc on the Mississippi Delta affects world cotton prices. International trade policies may affect supply and demand, so representatives of cotton organizations accompany U.S. trade ministers at world trade conferences, if only to advise. Success through precision farming makes yields of fifty years ago seem like relics of bygone centuries.

Certainly the smaller population currently involved in cotton farming stands out as the most abrupt difference. Slightly fewer than twenty-five thousand farms grew cotton in 2002, and only 60 percent of them devoted half or more of their acreage to the fiber. Two of the major growing areas are now in the West: the Texas Plains and the San Joaquin Valley. The Mississippi Delta still retains its importance, and the Southeast, particularly Georgia, has regained stature, surpassing California in 2003 in total production of bales. With the lower density of population in cotton-growing areas, the physical landscape has changed. Gone are the remote sharecropper shacks that dotted the countryside, sometimes lined up in rows or scattered through the fields. Small-town street corners no longer bustle with activity on Saturday afternoons, since tenants and field-workers long ago moved away. Migrant laborers do not descend on the West during picking season; no longer does the rush of the season occur in western towns. Old cotton towns have lost much of their vibrancy.

Gone, too, are the gangs of cotton hoers moving through fields, chanting in rhythm, and clumps of hand-pickers dragging long white sacks with shoulder straps. Mule teams followed by drivers faded away, which brought the end to mule trading centers and the large corrals and barns housing them on plantations. Country schools stopped setting their academic calendar according to the cycle of the cotton plant. Indeed, the migration off farms encouraged small country schools to consolidate, which enabled local school districts to concentrate resources better and improve instructions. If only by the drop in farming population, the socioeconomic life of the Cotton Belt changed.

Another development of significance occurred when textile manufacturing fell dramatically in the United States and forced over one million textile and apparel employees out of work. Here a parallel drop in population went side by side with cotton farming and had similar ramifications on textile towns in the Southeast. Textiles had accounted for much of the aura of the Cotton Belt, so as mills closed, a part of the old culture of the South died with them. King Cotton had become only a figure of history, and no more does a crowned sovereign hold sway over a kingdom.

Industrialization displaced agriculture as the principal way of life, and cotton become just another crop by the mid-1970s. Its penetrating influence disappeared. Planting and picking cotton now affects few people; money does not ebb and flow with the fortunes of the weather or the onslaught of insects. Cotton only rarely catches the attention of journalists and essayists because the suffering and oppression of tenants are no longer an issue, and novelists can only refer to a time past if they wish to conjure up the drama and pathos of cotton. Though poverty is visible in cotton-growing areas, the pain that went with sharecropping and migrant labor has fortunately gone, but so has the romanticism. Machine spindles replaced the touch of fingers plucking lint from bolls, and diesel tractors replaced long-eared mules.

THE NEW IMAGE

From the perspective of the consumer, cotton has a new meaning. It has the cool, sleek look of fashion and sophistication; cotton had become a fabric of lifestyle, be it work, recreation, or couture. Advertisers portray young urban professionals wearing stylish, upscale clothes with the soothing feel of cotton. The contemporary fondness for casualness and comfort in the office or country club blends well with natural fiber, but a dressier appearance is available with cotton, too. In the home, the smart look goes with cotton sheets and towels, large pillows, and comforters. Snazzy full-page magazine ads equate bed sheeting with luxury and success, and upscale hotels feature bed sheets and pillow cases of 400 to 800 thread count. The consumer fondness for softer and more luxuriant fabrics stimulated the production of Pima cotton in California, which ended the one-variety requirement that had once allowed only Acala strains there.

The stylish outfits on fashion's corner are a great leap from the flour sack prints of the past. Specialty advertising created the new image, with Supima, the promotional organization for extra-long staple cotton, promoting the "world's finest cottons." Pima had constituted a small portion of the total U.S. production, accounting for only four thousand bales in 1945 and rising only to eighty-one thousand bales in 1957. When Supima went into operation in 1954, it concentrated on improving farming practices and handling government commodity regulations for members, but in the 1970s the organization put more effort into promotion. It effectively promoted Pima, taking advantage of the rising standard of living and demand for quality, and concentrated its advertisements in fashion magazines like the *New York Times Style Magazine*, *Vogue*, and *Glamour*. Slick photos now associate cotton with youth and glamour, beauty and style.

POLITICAL ORGANIZATION

Certainly the organization of growers, political and economic, stands out as an important feature of the new cotton culture. Only a few new organizations appeared after World War II, mostly in the West, but membership and participation in all organizations zoomed. Unity had ramifications for interests other than growers because they saw the advantage of being identified with farmers. When textiles faced the crisis of byssinosis, the industry drew on the resources of the NCC and its enviable lobbying record in Congress. The effectiveness of cotton lobbyists comes from thorough preparation and the esprit de corps of the producer element to fight for a common cause. This sense of purpose survived into the new century, admittedly kept fresh by economic necessity, but also by growers' belief in the worthiness of their way of life and the national interest in the fruit of their labor.

Maintaining unity requires vigilance. The decline of the textile industry requires adjusting to the severe drop in the domestic market and concentrating on exports, just as EPA regulations pertaining to the use of insecticides and herbicides keep growers alert. They accept environmentalism but want to control insects that threaten their livelihood or control weeds that reduce the value of a crop. The responsibility for keeping the expectations of farmers before Congress in view of environmental concerns rests heavily on the NCC.

Some changes developed in the power structure of the cotton industry. Merchant interests no longer have the muscle of the post–World War II era. Many old-line merchandisers disappeared, with Anderson, Clayton and Company being the best known. It had been a force in the development of cotton farming in the West, and the leadership of its founder, Will Clayton, followed by Lamar Fleming, had been invaluable to the industry. When the firm withdrew from the cotton business, arguably being the strongest player, the ability of a few large companies, usually merchandisers, to affect the operations of the industry had already begun to end. Well-known firms like Paul Reinhart Inc. in Texas and Weil Brothers and Stern in Alabama disappeared; others were acquired by large agricultural conglomerates, as in the case of grain-based Cargill taking over Hohenberg Brothers of Memphis in 1976. With the rise of Dunavant Enterprises, a merchandiser solely devoted to cotton, merchandisers continued to have influence in the Cotton Belt, but not to the extent of former years. The loss of power over grading cotton to HVI removed one of merchandisers' principal functions and thereby gave a degree of independence to growers.

Cooperatives now have a strong presence and wield great influence. Woods Eastland, head of Staplcotn, the second-largest employer in Mississippi, served as chairman of the NCC in 2006, from whence he worked on behalf of all cotton interests. On the West Coast, Tom Smith helped manage and lead Calcot through tumultuous growth. The power of cooperatives like PCCA of Lubbock, which developed HVI testing and the first electronic cotton trading system, TELCOT, came after World War II. Roger Haldenby and Steve Verett of PCG in Lubbock help make the political fight on behalf of cotton on the Texas High Plains.

WATER AND URBAN ENCROACHMENT

The rising competition for water and urban encroachment cause unease. In Texas the PCG deals with the threat of dwindling supplies of water by encouraging conservation and promoting irrigation efficiency and new varieties. Drought remains common on the Texas Plains, and when rain is inadequate and the fuel costs of pumping water rise, cotton yields go down. "Drought is part of business," the *Texas Observer* quoted a plains cotton grower, "but it sure wears on people's hearts, souls, and bank accounts."[3] PCG campaigns for disaster relief from Congress when dire weather strikes, but growers in Texas have a new competitor for water from speculators trying to buy water rights and resell them to cities willing to pay lucrative prices. The falling water table in the Ogallala Aquifer had begun to endanger farming and brought a sense of unease to the Cotton Patch.

In Arizona and California, a tussle over land as well as water had broken out by 2000. A new and formidable foe, more powerful than any previous threats, challenged cotton for water: state governments, cities, and land developers laden with cash. In the Tucson-Phoenix corridor, population growth caused housing subdivisions to replace fields of cotton and alfalfa, either through the pressure of irresistible offers for land or through the right of cities to exercise eminent domain. "Farmland has become housing tracts in central Arizona," stated Tom Smith.[4] The two cotton organizations for the Grand Canyon State, the Arizona Cotton Growers Association and Supima, work with political representatives at all levels, but the pressure for space and water had already caused a drop in cotton production, from a ten-year average of 866,275 bales from 1990 to 1999 to 711,167 bales from 2000 to 2002. A similar trend appeared in California for the same years: an average production of 2,625,080 bales from 1990 to 1999 compared with the average of 2,342,867 for 2000 to 2003.[5]

REVIVAL OF THE SOUTHEAST

If cotton farming retrenched in one area, it expanded in the Southeast. "The hoofs of the cotton bull are thundering and perhaps in no region are they louder than in the Southeast." That statement by the editor of *Cotton Farming* in 1988 accurately forecast the return of cotton to its oldest growing region. In the 1930s, acreage planted with cotton in the Southeast began shrinking; Georgia acreage began dropping from 3,414,000 acres in 1930 until by 1967 it had fallen to 335,000. Five states, Georgia, Alabama, Florida, and the

Carolinas, make up the Southeast, which planted only 689,000 acres in 1980. The boll weevil, which arrived in 1915, had eaten so deeply into profits that cotton lost its attraction, but the overworked soil had also made farming a miserable existence. Gins had gone idle or were torn down. Poultry farming expanded, but in the late 1980s, cotton started to resurge in the Southeast until Georgia ranked second in total production by 2003. The boll weevil eradication program accounted for the new interest in Georgia, bringing a surge of production in 1994 with a steadily rising number of bales each year until 2005, when it hit 2,140,000 bales. North Carolina ranked next in the Southeast with 1,437,000 bales the same year compared with its 263,000 in 1990. Alabama and South Carolina more than doubled output for the same period, though their total production was about one-third of Georgia's.[6]

Most of Georgia's production occurred in its southwestern counties, but the renewal did not regenerate the old adulation for the natural fiber. Urbanization and economic diversification had modernized the Southeast, so the small proportion of farmers in the general population recently converting acreage to cotton had little impact on the general economy and culture. The refreshed activity of the Southeast nonetheless brought the reconversion or construction of gins, an upswing in sales of mechanical pickers and module builders, further sales of planting seed, fertilizer, insecticides, herbicides, and other production inputs. Revival of the original growing region in the Cotton Belt demonstrated the continuing viability of the crop, but in the more diversified economy, cotton's resurgence received little recognition.

TECHNOLOGY

Technology drives cotton farming. It is not unique in this respect, but modern farmers' adaptation of technology and the willingness—almost the rush—to experiment with advances contrast with cotton farmers of a half century earlier, who were known to stick with old habits. In some cases, the use of technology might seem extreme as some growers, particularly large-scale operators, incorporate satellite imagery and GPS into their operations, but the pressure to hold down production costs makes these technologies economically feasible for commercial farms. With the use of GPS, a tractor driver can plant only the amount of seeds that the soil will bear in various spots. A similar illustration would be the application of plant-growth stimulants because blanket applications would be wasteful, since the growth rate of cotton stalks in fields tends to vary from spot to spot. Known as precision agriculture or smart farming, such steps foster the most efficient use of equipment, fertilizer, seeds, and chemicals, often termed "inputs." Technology encourages sustainability, a method of farming embraced by environmentalists that involves the use of natural resources without exploiting or destroying their value and saving them for future generations.[7]

Advances in technology may, however, be only temporarily beneficial. Certainly the new no-till method of farming saves water, fuel, and money, though it requires the applications of weed retardants. By 2000, however, weed resistance had increased across the Cotton Belt, meaning that weeds would grow in fields despite the application of retardants. The ability of weeds to develop resistance demonstrates how technological breakthroughs may be short-lived. Farmers are again resorting to old-fashioned methods of controlling weeds through rotation of crops and increased cultivation. "It's a good thing I didn't sell off my conventional farming equipment," one landowner stated.[8]

But the rising weed resistance did not mean the end of high-tech farming, nor did it endanger the faith in technology. It is clear that the repeated use of the same herbicides year after year without following crop rotation or applying different retardants, known as mode of action, encourages resistance. Researchers thus remain vigilant to stay ahead of weeds just as they strive not to let insects return en mass. For the grower, each new technology must be profitable, which may vary from area to area depending on soil and weather conditions or the farming practices of each grower. In some cases, the advantages might not warrant the investment; landowners with smaller acreages tend not to jump into new technologies until they are proven. But the belief that technology is an asset to be used to the fullest advantage has been accepted.

Although only about twenty-five thousand farms produced cotton by the new century, technology kept total production high, reaching over 17,188,000 bales in 2000 and climbing to 21,567,000 in 2006. In 1945 lint production averaged 245 pounds per acre but had climbed to 725 pounds in 2003. From 1991 to 2002, the average production of bales was 16.8 million bales, about 20 percent of the total world output. Costs of production rose steadily, however, and though profit margins have always been narrow for cotton farming, costs exceeded price between 1999 and 2002, which accounted for the dependence on subsidies. From 1991 to 2003, farmers received an average market price of 57 cents, but with the subsidy supports they obtained an average of 78 cents per pound, or 21 cents in government support. "In the absence of support programs," reported the Congressional Research Service, "the data suggests a sizeable proportion of cotton would not be profitable."[9] The small range of profitability accounts for the emphasis on efficiency and large-scale production, and without government supports, the number of growers would unquestionably shrink. In Arizona, where urban development pushes up land values, landowners sell their holdings when the return on their investment in cotton farming yields too little. Rick Lavis, executive vice president of the Arizona Cotton Growers Association, lists the combination of high land values and low profitability as primary causes of Arizona's falling cotton acreage.[10]

Because technology requires investment in a risky venture, skillful management must be employed, management that includes overseeing financial affairs, smart cultivation practices, a willingness to experiment, political awareness, and shrewd marketing. Experience and the desire to acquire knowledge are essential, since a single misstep can bring disaster. Luck plays a role. For a family to stake its livelihood on fields of cotton demands a level of business acumen and general cultural awareness not associated with farming only a few decades ago.

A COTTON FAMILY OF THE NEW CENTURY

Alison and Keith Deputy operate Deputy Farms on the northern outskirts of El Paso. Keith owns and leases about twenty-two hundred acres and produces an even mixture of Upland and Pima varieties. Most of the acreage lies in New Mexico, but all the fields in both states are close to his headquarters. On the eastern side lie the Hueco Mountains, and the Potrillos that border the west make a backdrop for the flat vistas and valleys within the greater complex of the Mesilla Valley. Sandy loam deposited by the Rio Grande makes for intensive farming if water is available. Trees are sparse, and irrigation canals crisscross the farmland because flood irrigation prevails. Urban sprawl from El Paso creeps near the Deputy Farms, which lie near Interstate 10. The area receives little rainfall, about ten inches per year, so all farming relies on irrigation water from the Elephant Butte Reservoir, upstream on the Rio Grande. Competition for water rights has mounted in recent years as the city acquires them for housing developments. Fields without water grow only weeds.

Deputy Farms became known for high per-acre yields, getting up to five bales per acre in spots, but the average usually falls slightly over two bales. To achieve that level, Keith fully mechanizes with equipment as basic as field levelers to sophisticated GPS guidance systems mounted on tractors. He operates eighteen tractors and three spindle pickers along with trucks, a fueling pump, module builders and the like. At his headquarters he keeps a mechanic on full-time duty in a large maintenance building and equipment yard. Workers either drive equipment or handle irrigation. Soil quality varies, so fertilizer must be applied, but only according to the particular specifications for each field. Pima will grow only in his Texas fields, so he plants Upland, including some Acala, in the New Mexico fields.

Keith regards close management as the difference between success and failure. He enjoys farming and has lived since the age of sixteen on the acreage that his father and grandfather cultivated. As a child he learned to drive tractors and trucks, to move water from field to field, and became acquainted with the working end of a shovel. "I acquired my farming skills from years of watching, talking, and listening to both my father and grandfather." He shows enthusiasm for his work and feels "that the passion I have for farming gets me up every day."[11]

The day begins no later than 5:30 a.m. Daily routines vary with the season, but during spring, when activity jumps with plowing and planting, which must be accomplished quickly after the last frost to get the most days for maturing the plants, Keith starts patrolling the fields and supervising workers. Breakfast may come at 9:00 or 10:00, perhaps at a local taqueria. He meets with the irrigators and ditch riders before they go off shift, and since his acreage extends into another state, dotted about eighteen miles from tip to tip, Keith supervises from his pickup on a cell phone. He feels that his job has grown more intense in the past few years, which he attributes to his cell phone and the complicated nature of high-tech farming. Irrigation mandates a close eye on soil moisture, and water must be released at the right moment. Keith runs two shifts of workers during irrigation, which normally comes at sixteen- to twenty-two-day intervals, so those days may not end until midnight. Such intensity is inescapable for a successful crop; only in May after planting does he sometimes have a slack period in the growing season.

Like many farmwives, Alison helps. She grew up on a farm near the edge of El Paso and likes outdoor life. She understands almost all aspects of the operation and handles payroll and bill payments. Together Keith and Alison discuss business operations, generally on Sundays or slow days. Alison enjoys horseback riding, plays tennis, and keeps several novels under way. A favorite pastime is her volunteer work in their son's farm club, the Future Farmers of America. When the children were younger, she participated in parent organizations and the Junior League of El Paso. Now she often joins Keith in inspecting the fields, which keeps her updated on business operations.

This busy life has moments for pause. Keith travels on occasion, generally to meetings of farm organizations, and Alison frequently accompanies him. Trips to see family are common, but some of their best memories are trips to Washington, D.C. or California with their children. Family reunions are popular and they enjoy taking an occasional sea cruise. For Keith a game of golf comes about once a year, but Alison meets regularly as a member of Therapeutic Horsemanship of El Paso. Their greatest enjoyment was their children, but their two daughters now attend college and their son plans to attend after finishing high school.[12]

This lifestyle of work, travel, and participation in business and community affairs, with children in college, is not uncommon among cotton growers. The Deputys do not have an insular life, they do not avoid change, nor do they seek solitude and security. Cotton farming carries too much risk to allow a slow pace of life. Growers across the Cotton Belt may be more or less involved in business and community organizations, but if they depend on cotton for their livelihood, they have much in common with this couple in El Paso.

THE RISK-DEBT COMPLEX

Probably the most common characteristic among growers regardless of location in the Cotton Belt is their awareness and management of risk. All farming involves risk, and when coupled with the cost of producing a crop on enough acreage for a livelihood, the possibility of losing a crop affects their outlook. Crop yields tend to be cyclical, as a fat year will likely soon lead to a lean one; men and women on the soil expect this development and can usually withstand the usual drought or other occurrence. Drought hit South Texas in 1998, but rain ruined the crop in 2001. Rain was so heavy in Fort Bend County, Texas, in 1998 that cottonseed sprouted in open bolls on the stalk. "This left the farmer with a severe damage situation that they don't know how to handle," stated ginner Jerome Kulcak.[13] Hurricanes that hit the South capture public attention, but the insidious droughts, not limited to the West, come frequently. Droughts are more likely to occur than floods; throughout the twentieth century, some portions of Texas had a drought each decade. Price volatility can also play havoc with the best management. The high price of cotton in 1996 and low price of 2001, which went below production costs, demonstrated the cyclical nature of commodity prices.

Weather accounts for the greatest risk of farming and pushes growers off the land more than world competition or the impact of artificial fibers. Synthetics hold down the price of raw lint, but thanks to the resiliency of farmers, they stumble along with low prices. Prolonged drought is the worst enemy for both dryland and irrigation farmers and nearly always puts them in debt. Many rely on government disaster relief payments until the next planting season, but droughts may persist for several years. This situation hit High Plains growers in Cochran County, Texas, starting in 1996, where landowner Sam Burnett estimated his losses at $50,000 per year. Production costs do not decline during drought, so he was forced to take a nonfarming job as a loan officer with a mortgage bank in Clovis, New Mexico. "I'm not going to starve my family," he told reporters. "If farming is not going to make it, we need to do something that will."[14]

The drought persisted. "People are as despondent as I've seen them," Steve Verett of the PCA in Lubbock told reporters in 2001.[15] But growers in the lower Rio Grande Valley were also caught in nature's grip. That year bankers reported losses from borrowers, with most growers surviving on government assistance and nonfarm employment. Irrigation partially offsets the parched conditions, but the fuel price of pumping water further reduces the narrow margin of profit. Planting other crops provides no safety net during drought; corn uses more water than cotton.

Farm crises tend to go unnoticed as long as food supplies remain plentiful on supermarket shelves, and the hardship of farm families remains out of sight. "I don't think people know there is a farm economy," commented a midwestern landowner.[16] But the ramifications nonetheless come with serious effects in farm communities. One businessman in Brownfield, Texas, in the heart of the Cotton Patch, saw sales drop 25 percent in four months owing to drought in 1999. Tax revenues fell until the school district had to dismiss seventeen teachers. Not since the debt crisis of the 1970s and 1980s had the situation been so severe.[17]

Living with debt runs across the Cotton Belt, and farmers with smaller acreages tend to suffer more. Debt creates a sense of helplessness among families and a general lack of power to control their lives. A rural sociologist at the University of Nebraska reported a drop in self-confidence among farm dwellers when debts rose, and farmers young enough to start another life have an itch to move. "When I look down the road ten years, I don't see it getting better," stated a young grower.[18] Bankers notice that farmwives are more likely to contact them about refinancing or asking for relief, owing presumably to men's shame at not fulfilling loan agreements and reluctance to go to the banker's office.[19] Small towns lose business, schools lose teachers, and young people give up farming. Debt and its consequences are part of the cotton way of life, so a risk-debt complex permeates the culture of cotton farming, and even among those who successfully remain on the land there is the haunting awareness that disaster prowls nearby.

In 1985 when Willie Nelson hosted his first Farm Aid concert in Champaign, Illinois, before eighty thousand fans, he drew attention to the debt crisis. Other performers at the charitable performance were B. B. King, Billy Joel, Roy Orbison, Bob Dylan, Tom Petty, and others. They raised $7 million. In 1986 Farm Aid II played in Austin, Texas, and millions of viewers saw it live on VH1. This performance featured artists such as the Beach Boys, Stevie Ray Vaughan, and Alabama. Each year the Farm Aid concert raises money across the United States, and farmers of all manner and persuasion are highlighted, so cotton growers get no special attention. In 2005 Farm Aid published *Farm Aid: A Song for America*, which recalled the organization's twenty-year history of fund-raising. As a child, Willie Nelson lived on a small farm near Abbott, Texas, in the blackland strip, and picked cotton, hauled hay, and tended livestock. For him farm life has meaning and worth preserving.

To preserve farm life, however, cotton growers must remain highly competitive in the world market, which forces them to obtain high yields and hold production costs to a minimum. In 1982 the United States lost its top-ranked position as a cotton producer to China. The presence of world competitors adds to the stress of farming efficiently and marketing under the most optimal conditions, which means growers must be knowledgeable, skillful, and able to make adjustments quickly. The new cotton culture has no room for the lassitude of the past.

Table 6. World's Top Ten Producers, 1945–2005			
1945	1965	1985	2005
United States	United States	China	China
India	Soviet Union	United States	United States
Soviet Union	China	Soviet Union	India
China	India	India	Pakistan
Brazil	Mexico	Pakistan	Uzbekistan
Pakistan	Brazil	Brazil	Brazil
Egypt	Pakistan	Turkey	Turkey
Mexico	Egypt	Egypt	Australia
Peru	Turkey	Australia	Greece
Uganda	Sudan	Mexico	Syria

Source: USDA Economic Research Service, *Statistics on Cotton and Related Data, 1920–73*, 111–12; National Cotton Council, http://www.cotton.org/econ/cropinfo/cropdata/rankings.cfm.

FUTURE DIRECTIONS

In the current world climate of terrorism, political instability, and post–Cold War trade shifts, cotton is too important to the general welfare and national security not to be safeguarded. Evaluations of federal support must consider the potential impact of subsidy declines on trade agreements and the stress on local economies and public services. Large-scale growers might move their operations overseas, for corporations may establish plantations in foreign lands. Industrialized farming in overseas locations would almost certainly escape tough environmental regulations and obtain lower labor costs. With fewer restrictions and less governmental oversight, growing cotton could proceed in an unfettered manner, making greater use of insecticides and herbicides.

American growers operating in other countries would be subject to the sovereign rights of governments, and the risk of seizure or confiscation could not be ruled out. Changes in governments, not an unlikely development, might reconfigure the conditions under which growers would operate, and jeopardize production and interrupt the reliable stream of fiber to textile mills in the United States or elsewhere. Just as the United States embargoed wheat to the Soviet Union in 1980, the United States could face an embargo on raw cotton. If the United States became dependent on overseas cotton, it would have less to spin and weave into fabric or cottonseed to process into foodstuffs and various products. Such a development borders on the extreme and seems unlikely in the global economy, but it would be unwise to assume that the supply of cotton would remain steady and consistent. Interference of shipments by terrorists does not seem implausible, since labor disputes and inadequate port facilities have occasionally kept container ships from unloading in American ports.

A severe loss in domestic acreage would damage local and regional economies. Cotton farming generates more jobs than other crops because it produces the material used in two sectors of the economy: textiles and apparels. Both employ thousands of workers, despite the closings of many mills. The infrastructure of cotton farming also includes jobs in ginning, irrigation, trucking, and warehousing. Tax revenues would fall and affect schools and other public services if the United States lost its self-reliance in cotton. Nearly all proponents of maintaining federal assistance for agriculture regard the benefits to local businesses and schools as a major consideration.

If the industry dwindled to an inconsequential level, cotton research would decline in the United States. Researchers at Experiment Stations and land-grant universities would focus on other crops or livestock, and the development of new technologies in cotton farming would not receive priority. Should new cotton diseases appear or damaging insects resurface at significant levels, the ability to bring them under control might be weakened. Only the United States has an institutional research structure established on behalf of cotton, and just as new diseases such as SARS or AIDS erupt, so plant pathogens could appear and jeopardize the world crop. Herein lies an important dimension of American agriculture: the well-developed infrastructure of personnel, laboratories, demonstration fields, and other facilities to monitor and stay abreast of threats to the supply of food and fiber. If cotton farming becomes insignificant, research would decline. Indeed, the drive to overcome cotton pests led to the development of integrated pest management (IPM), the concept that led to an increased world food supply.

In some fashion or another, cotton farming will remain. A reduction in subsidies would almost certainly force a portion of the twenty-five thousand growers to convert acreage to another crop, and a period of consolidation in which large commercial operators lease land from smaller farmers would occur, but the production of cotton would remain, though likely with less acreage and smaller annual production. Cotton's future will resemble its past: it will swirl in controversy. In her novel *In the Land of Cotton*, Dorothy Scarborough's character Hog-Eye best expressed the nature of the fabric of life: "There's a romance, a drama, a glory about a cotton field. A story needs a struggle, a conflict, a complication. . . . I'd say cotton has them all."[20]

NOTES

PROLOGUE

1. *Cotton Farming*, November–December 1989, 13.
2. For discussion of the origins of King Cotton, see David C. Boller and Robert W. Twyman, eds., *The Encyclopedia of Southern History* (Baton Rouge: Louisiana State University Press, 1979), 213.
3. Rupert B. Vance, *Human Factors in Cotton Culture: A Study in the Social Geography of the American South* (Chapel Hill: University of North Carolina Press, 1929), vii.
4. B. B. King with David Ritz, *Blues All around Me: The Autobiography of B. B. King* (New York: Avon Books, 1996); for further discussion of the origin of the blues, see William Ferris, *Blues from the Delta* (Garden City, N.Y.: Anchor Press/Doubleday, 1978), 31–36. In 2003 the Center for the Study of Southern Culture at the University of Mississippi began hosting an Annual Blues Symposium.
5. Billy Joe Shaver, *Honky Tonk Hero* (Austin: University of Texas Press, 2005), 85.
6. For a general discussion of subsidies and related topics, see Stephen Yafa, *Big Cotton: How a Humble Fiber Created Fortunes, Wrecked Civilizations, and Put America on the Map* (New York: Viking, 2005); and Gerard Helferich, *High Cotton: Four Seasons in the Mississippi Delta* (New York: Counterpoint, 2007).

1. THE CULTURAL IMAGE OF COTTON, 1945

1. James C. Cobb, *The Most Southern Place on Earth: The Mississippi Delta and the Roots of Regional Identity* (New York: Oxford University Press, 1994), 69–97; James H. Street, *The New Revolution in the Cotton Economy: Mechanization and Its Consequences* (Chapel Hill: University of North Carolina Press, 1957), 25–27.
2. Douglas A. Farnie and David J. Jeremy, "The Role of Cotton as a World Power, 1780–1990," in *The Fiber That Changed the World: The Cotton Industry in International Perspective, 1600–1990s* (Oxford: Oxford University Press, 2004), 4.
3. *The Statistical History of the United States from Colonial Times to the Present* (Stamford, Conn.: Fairfield Publishers, 1965), 301–2; Street, *New Revolution in Cotton*, 25–26.
4. Rupert B. Vance, "The Profile of Southern Culture," in *Culture in the South*, ed. W. T. Couch (Chapel Hill: University of North Carolina Press, 1935), 30.
5. *Statistical History of the United States*, 289–90.

6. Wilbur J. Cash, *The Mind of the South* (New York: Random House, 1941), 285.
7. Fredrick Lewis Allen, *Only Yesterday* (New York: Harper and Brothers, 1959), 113.
8. Ibid.
9. George Soule, *Prosperity Decade: From War to Depression, 1917–1929* (New York: Holt Rinehart and Winston, 1947), 229–251, quote on 316.
10. Allen, *Only Yesterday*, 113.
11. Frank Tannenbaum, *Darker Phases of the South* (New York: G. P. Putnam's Sons, 1924), 116.
12. Ibid., 22.
13. Ibid., 116, 144–45.
14. John Shelton Reed, *The Enduring South: Subcultural Persistence in Mass Society* (Chapel Hill: University of North Carolina Press, 1972), 57.
15. George B. Tindall, *The Emergence of the New South, 1913–1945* (Baton Rouge: Littlefield Fund for Southern History of the University of Texas, 1967), 211.
16. William H. Skaggs, *The Southern Oligarchy* (New York: Devin-Adair, 1924; Negro Universities Press Reprint, 1969), 279.
17. Ibid., 201, 353; the cost estimate appears on p. 352.
18. Tindall, *Emergence of the New South*, 208.
19. Dorothy Scarborough, *In the Land of Cotton* (New York: Macmillan, 1936), 101.
20. Ibid., 141.
21. George B. Tindall, "The Benighted South: Origins of a Modern Image," in *Myth and Southern History*, ed. Patrick Gerster and Nicholas Cords (Chicago: Rand McNally College Publishing, 1974), 246.
22. Bruce Clayton, "Southern Intellectuals," in *Debating Southern History: Ideas and Action in the Twentieth Century*, ed. Bruce Clayton and John Salmond (Lanham, Md.: Rowman and Littlefield, 1999), 6.
23. Claudius T. Murchinson wrote *King Cotton Is Sick* (Chapel Hill: University of North Carolina Press, 1930), a study of the unhealthy textile industry that lent credence to the growing recognition that all was not well in the cotton South.
24. John Faulkner, *Dollar Cotton* (New York: Harcourt, Brace, 1942), 83.
25. Paul Oliver, *Blues Fell This Morning: Meaning in the Blues* (New York: Cambridge University Press, 1960), 3.
26. Tindall, *Emergence of the New South*, 416.
27. Arthur Raper, *Preface to Peasantry: A Tale of Two Black Belt Counties* (Chapel Hill: University of North Carolina Press, 1936), 406.
28. Ibid., ix.
29. Ibid.
30. Dan Dickey, "Corridos y Cauciones de las Pizcas: Ballads and Songs of the 1920s Cotton Harvests," *Western Folklore* (Winter 2006), http://www.findarticles.com.
31. Tindall, *Emergence of the New South*, 588.
32. Howard W. Odum, *Southern Regions of the United States* (Chapel Hill: University of North Carolina Press, 1936), 399.
33. Rupert B. Vance, *Human Factors in Cotton Culture: A Study in the Social Geography of the American South* (Chapel Hill: University of North Carolina Press, 1929), 314.
34. Robert E. Syndor, *Cotton Crisis* (Chapel Hill: University of North Carolina Press, 1984), xvii.
35. Ibid., 129.
36. Tindall, *Emergence of the New South*, 419.

37. Donald Grubbs, *Cry from the Cotton: The Southern Tenant Farmers' Union and the New Deal* (Chapel Hill: University of North Carolina Press, 1971), 125.

38. Sidney Baldwin, *Politics of Poverty: The Rise and Decline of the Farm Security Administration* (Chapel Hill: University of North Carolina Press, 1968).

39. Tindall, *Emergence of the New South*, 426. For further discussion see Lawrence J. Nelson, *King Cotton's Advocate: Oscar G. Johnston and the New Deal* (Knoxville: University of Tennessee Press, 1999), 206–27; Baldwin, *Politics of Poverty*, 405–19. The FSA ceased operation in 1946.

40. Original quote appears in Tindall, *Emergence of the New South*, 421.

41. National Emergency Council, *Report on Economic Conditions of the South* (Washington, D.C.: Government Printing Office, 1938), 10.

42. Ibid., 45.

43. Thomas D. Clark and Albert D. Kirwan, *The South since Appomattox: A Century of Regional Change* (New York: Oxford University Press, 1967), 103.

44. http://newdeal.feri.org/works/wpa06.htm.

45. *Cotton Trade Journal*, International Edition, 19 (1939): 34.

46. Herman C. Nixon, *Possum Trot: Rural Community, South* (Norman: University of Oklahoma Press, 1941), 85.

47. William Alexander Percy, *Lanterns on the Levee: Recollections of a Planter's Son* (New York: Alfred A. Knopf, 1941), 282.

48. Frank E. Smith, *The Yazoo River* (New York: Rinehart, 1954), 182–83.

49. H. L. Mitchell, *Mean Things Happening in This Land: The Life and Times of H. L. Mitchell, Co-founder of the Southern Tenant Farmers Union* (Montclair, N.J.: Allanheld, Osmun, 1979), 127.

50. Smith, *The Yazoo River*, 301–2.

51. David Southern, *Gunnar Myrdal and Black-White Relations: The Use and Abuse of an American Dilemma, 1944–1969* (Baton Rouge: Louisiana State University Press, 1987), xvi.

52. John Hope Franklin, *From Slavery to Freedom: A History of Negro Americans* (New York: Vintage Books, 1967), 493.

53. Thomas D. Clark, *The Emerging South* (New York: Oxford University Press, 1961), 51.

54. H. L. Mitchell, *Mean Things Happening*, 103.

55. Wayne Gard, "The American Peasant," *Current History* 46 (April 1937): 47–52.

56. T. J. Woofter and A. E. Fisher, "The Plantation South Today," Federal Works Agency, Works Progress Administration, Social Problems Series No. 5 (Washington, D.C.: Government Printing Office, 1940), 19–20.

2. THE NEW POLITICS OF COTTON

1. Lawrence Nelson, *King Cotton's Advocate: Oscar G. Johnston and the New Deal* (Knoxville: University of Tennessee Press, 1999), 161–62. I drew from Nelson's work for information in this chapter.

2. *Fortune*, March 1937, 125–132, 158.

3. Albert R. Russell, *U.S. Cotton and the National Cotton Council, 1938–1987* (Memphis, Tenn.: National Cotton Council, 1987), 1.

4. Ibid., 3.

5. William Rhea Blake, Mississippi Oral History Program of the University of Southern Mississippi, CXXXIII (1978), 13. Interviews through the Mississippi Oral History Project are cited hereafter as MOHP.

6. Read Dunn Jr., *Mr. Oscar: A Story of the Early Years in the Life and Times of Oscar Johnston and of His Efforts in Organizing the National Cotton Council* (Memphis, Tenn.: National Cotton Council, 1991), 73.

7. Ibid., 73–74.

8. Quotes in Blake, MOHP, 19–22.

9. Dunn, *Mr. Oscar*, 71.

10. Ibid., 74–75.

11. Ibid., 75; Edward L. Lipscomb, MOHP, CXLV (1979), 26–27.

12. Dunn, *Mr. Oscar*, 76–77.

13. Blake, MOHP, 35–36.

14. Ibid., 26.

15. Norris Blackburn, MOHP, CXVIII (1979), 26.

16. Russell, *U.S. Cotton*, 3.

17. Ibid.

18. Blackburn, MOHP, 26; Nelson, *King Cotton's Advocate*, 192.

19. Lipscomb, MOHP, 34.

20. Dunn, *Mr. Oscar*, 77.

21. Ibid., 80.

22. Nelson, *King Cotton's Advocate*, 195.

23. *Cotton Trade Journal*, November 12, 1938, 1, 5.

24. Nelson, *King Cotton's Advocate*, 199.

25. Aubrey L. Lockett, MOHP, CXXVIII (1978), 4.

26. David L. Cohn, *God Shakes Creation* (New York: Harper and Brothers, 1935), 14.

27. *Commercial Appeal*, November 22, 1938, 10.

28. Ibid.

29. Oscar Johnston, "Address to the Committee on Organization, National Cotton Council," Memphis, November 21, 1938, Executive Vice President files, box "Jan 1–Dec 31, 1945," folder "Gasoline Rationing," 78–79, National Cotton Council Archives, Memphis, Tenn. Hereafter cited as NCC-EVP files.

30. *Commercial Appeal*, November 22, 1938, 6.

31. Ibid.

32. Russell, *U.S. Cotton*, 23.

33. Robert C. Jackson, MOHP, CXLIII (1980), 109.

34. Russell, *U.S. Cotton*, 22.

35. Claude L. Welch to Huntsville, Texas, Selective Service Board, March 31, 1945, NCC-EVP files, box Jan 1–Dec 31, 1945, folder C.

36. Robert C. Jackson, interview by the author, Clemson, S.C., September 27, 2002.

37. Sam Bledsoe, Oral History Collection, Columbia University, 682–83.

38. Ibid.

39. Robert C. Jackson, MOHP, 42.

40. Walter W. Wilcox, *The Farmer in the Second World War* (Ames: Iowa State College Press, 1947), 217.

41. Oscar Johnston to Donald Nelson, August 7, 1943, NCC-EVP files, box "Feb 1943–Jan 1944," folder "Letters to Congressmen and Senators."

42. Circular letter to Senators, August 7, 1943, NCC-EVP files, box "Feb 1943–Jan 1944," folder "Letters to Senators and Congressmen."

43. *Cotton Trade Journal*, international edition, 1945–46, 136.

44. Oscar Johnston to Milton Eisenhower, December 14, 1946, NCC-EVP files, box "Jan 1–Dec 31, 1946," folder "Eisenhower Statement." For a study of German POWs, see Jason

Morgan Ward, "Nazis Hoe Cotton: Planters, POWs, and the Future of Farm Labor in the Deep South," *Agricultural History* 81 (Fall 2007): 471–92.

45. USDA Economic Research Service, *Statistics on Cotton and Related Data, 1930–67*, Statistical Bulletin 417 (Washington, D.C.: Government Printing Office, 1968), 1.

46. Wilcox, *Farmer and the Second World War*, 218.

3. THE COTTON CONFERENCE

1. Congress, House, Subcommittee of the Committee on Agriculture, *Hearing Cotton*, 78th Cong., 2nd sess., December 4–9, 1944, 37. Hereafter cited as *Hearing Cotton*.

2. Ibid., 37.
3. Ibid., 36.
4. Ibid., 37.
5. Ibid., 95.
6. Ibid., 185.
7. *Commercial Appeal*, December 3, 1944, sec. 2, p. 3.
8. Stephen Pace to Will Clayton, October 2, 1944, box 19, folder 44-10-2, William Lockhart Clayton Papers, Woodson Room, Rice University.
9. Will Clayton to Stephen Pace, October 5, 1944, box 19, folder 44-10-5, Clayton Papers, Rice University.
10. *Hearing Cotton*, 41.
11. Ibid., 499, 848.
12. Ibid., 424.
13. Ibid., 196.
14. Ibid., 113–14.
15. Ibid., 812.
16. Harold U. Faulkner, *American Economic History*, 8th ed. (New York: Harper and Brothers, 1960), 693.
17. *Hearing Cotton*, 812.
18. USDA Economic Research Service, *Statistics on Cotton and Related Data, 1930–67*, 64.
19. *Hearing Cotton*, 534.
20. Ibid., 100, 112, 215, 379, 485–500, 538.
21. Ibid., 648.
22. Ibid., 485.
23. Ibid., 121.
24. Ibid., 110.
25. John Turner, *White Gold Comes to California* (Bakersfield: California Planting Cotton Seed Distributors, 1981), 141–45.
26. *Hearing Cotton*, 512–15; USDA Economic Research Service, *Statistics on Cotton*, 84–93.
27. James H. Street, *The New Revolution in the Cotton Economy: Mechanization and Its Consequences* (Chapel Hill: University of North Carolina Press, 1957), 159.
28. Ibid., 157–60; USDA, *Farmers in a Changing World: The Yearbook of Agriculture, 1940* (Washington, D.C.: Government Printing Office, 1940), 512–17.
29. *Hearing Cotton*, 96, 739.
30. Ibid., 536.
31. Ibid., 504.
32. Ibid., 110.
33. Herman C. Nixon, *Possum Trot: Rural Community, South* (Norman: University of Oklahoma Press, 1941), 95.

34. *Commercial Appeal*, December 4, 1944, 1.
35. *Cotton Trade Journal*, international edition, 1945–46, 146.
36. Quotes in Bureau of Agricultural Economics, *The Cotton Situation* (December 1944), 5–7.
37. Frank E. Smith, *The Yazoo River* (New York: Rinehart, 1954), 329.
38. Charles Campbell, interview by author, Memphis, Tenn., February 21, 2005.
39. *Commercial Appeal*, December 10, 1944, sec. 4, p. 5.
40. Stephen Pace to Will Clayton, October 2, 1944, box 19, folder 44-10-2, Clayton Papers, Rice University.
41. *Hearing Cotton*, 213.
42. Ibid., 444.

4. A NEW ERA BEGINS

1. Herman C. Nixon, *Possum Trot: Rural Community, South* (Norman: University of Oklahoma Press, 1941), 88.
2. Deborah Fitzgerald, *Every Farm a Factory: The Industrial Ideal in American Agriculture* (New Haven, Conn.: Yale University Press, 2003).
3. *Commercial Appeal*, May 31, 1945, 1.
4. "Summary Report of Research Conference on the Postwar Agricultural and Economic Problems of the Cotton Belt," Peabody Hotel, May 30–31, 1945, 12, NCC-EVP files, box "Jan 1–Dec 31, 1945," folder "Report of Research Conference."
5. Alfred Steinberg, *Sam Rayburn: A Biography* (New York: Hawthorn Books, 1975), 350.
6. Fred Halliday, *The World at 2000: Perils and Promises* (New York: Palgrave, 2001), 22.
7. Oscar Johnston to Lamar Fleming Jr., May 10, 1945, NCC-EVP files, box "Jan 1–Dec 31, 1945," folder F.
8. *Progress Bulletin*, April 15, 1945, 3.
9. Rhea Blake, MOHP, 101.
10. *Press Scimitar*, May 31, 1945, 1.
11. Rhea Blake to McDonald K. Horne, March 30, 1945, NCC-EVP files, box "Jan 1–Dec 31, 1945," folder H.
12. McDonald K. Horne, "Reflections of a Cotton Economist," Speech to Egyptians Memphis, May 16, 1968, papers of McDonald K. Horne, in author's possession.
13. Rhea Blake to Lt. Read Dunn, August 17, 1945, NCC-EVP files, box "Jan 1–Dec 31, 1945," folder D.
14. Read Dunn Jr., MOHP, CXXXIX, 51.
15. Oscar Johnston to Commanding General Percy Jones, October 30, 1945, NCC-EVP files, box "Jan 1–Dec 31, 1945," folder C; Rhea Blake to Ernest B. Stewart, October 30, 1945, NCC-EVP files, box "Jan 1–Dec 31, 1945," folder S.
16. Rhea Blake to Ernest B. Stewart, October 22, 1945, NCC-EVP files, box "Jan 1–Dec 31, 1945," folder S.
17. Blake, MOHP, 102–3.
18. McDonald K. Horne, interview by author, Memphis, Tenn., June 1, 1998.
19. Tom Miller, interview by author, Memphis, Tenn., May 25, 2004.
20. Lewis T. Barringer, MOHP, CXVI (1979), 3.
21. Lewis T. Barringer, Oral History Interview (April 15, 1969), 3, Harry S. Truman Library.
22. Robert H. Ferrell, ed., *Off the Record: The Private Papers of Harry S. Truman* (New York: Harper and Row, 1980), 297.

23. Lewis T. Barringer, Oral History Collection, Truman Library, 33.

24. Read Dunn, interview by author, Washington, D.C., October 14, 1999; Robert C. Jackson, interview by author, Clemson, S.C., September 27, 2002; John Barringer, interview by author, Nashville, Tenn., June 7, 2005.

25. Lewis T. Barringer, Oral History Collection, Harry S. Truman Library, 33.

26. Earl Sears, interview by author, Memphis, Tenn., February 15, 2005.

27. Charles S. Murphy, Oral History Collection, June 3, 1963, Harry S. Truman Library, www.trumanlibrary.org/oralhist/murphy.htm.

28. "Hop in Again, Tony," typescript, n.d., Papers of John Barringer, Nashville, Tenn. Vaccaro preserved a typescript of the trip to Independence in which he recalled how he and Barringer accompanied Truman to the Potsdam Conference of 1945, that after dinner on board the *Augusta* he and Barringer played cards, watched movies, and shared drinks with Truman and his advisors. The Truman Library had no record of Barringer accompanying the president to Potsdam, and I could not corroborate Barringer's trip to Potsdam.

29. Fredrick J. Dobney, ed., *Selected Papers of Will Clayton* (Baltimore, Md.: John Hopkins University Press, 1971), 3.

30. Ibid., 4–5.

31. Ibid., 3–4.

32. Congress, Senate, Committee on Foreign Relations, Hearings, Nominations–Department of State, 78th Cong., 2nd sess., December 12–13, 1944, 67–68, quote on 73.

33. *Progress Bulletin*, July 15, 1943, 1–2.

34. Edward Lipscomb, MOHP, XXI, 34.

35. Ibid.

36. Ibid.

37. National Cotton Council Headquarters, Storage Files, folder "Fashion Campaign 1945"; quote in Albert Russell, *U.S. Cotton and the National Cotton Council, 1938–1987* (Memphis, Tenn.: National Cotton Council, 1987), 29.

38. *Christian Science Monitor*, March 13, 1945, 11.

39. *Progress Bulletin*, January 15, 1945, 19.

40. Ibid.

41. Peter L. Bernstein, *The Power of Gold: The History of an Obsession* (New York: John Wiley and Sons, 2000), 331.

42. *Cotton Trade Journal*, September 1, 1945, 1.

43. *Progress Bulletin*, February 15, 1945, 3.

44. Rhea Blake to Sam Bledsoe, April 6, 1945, NCC-EVP files, box "Jan 1–Dec 31, 1945," folder "Bretton Woods Proposals."

45. Rhea Blake to Fred C. Crawford, April 18, 1945, NCC-EVP files, box "Jan 1–Dec 31, 1945," folder C.

46. George Schild, *Bretton Woods and Dumbarton Oaks* (New York: St. Martin's Press, 1995), 184.

47. John Van Sickle, *The Stake of the Cotton South in International Trade* (Berkeley: University of California Press, 1945), 3.

48. Murray R. Benedict, *How Much Tariff Protection for Farm Products* (Berkeley: University of California Press, 1945); Oscar B. Jesness, *The Dairy Farmer and World Trade* (Berkeley: University of California Press, 1945).

49. *Christian Science Monitor*, March 1, 1945, 2.

50. Rhea Blake to H. G. Holbrook, March 14, 1945, NCC-EVP files, box "Jan 1–Dec 31, 1945," folder H.

5. AMBASSADORS OF FOREIGN POLICY, 1945-1950

1. *Cotton Trade Journal,* July 18, 1942, 3.
2. *Progress Bulletin,* July 15, 1943, 2.
3. W. D. Anderson to Rhea Blake, February 17, 1945, NCC-EVP files, box "Jan 1–Dec 31, 1945," folder A.
4. Will Clayton to Henry L. Stimson, July 3, 1945, box 23, folder 45-7-3, Clayton Papers, Woodson Room, Rice University.
5. "Report of Trip to Europe of the Committee of the American Cotton Shippers Association," quote on p. 1, NCC-EVP files, box "Jan 1–Dec 31, 1945," folder "Robert Jackson—European Trip."
6. E. D. White to William L. Clayton, July 13, 1945, box 1077, folder "Cotton, Record Group 16," National Archives, College Park, Md.
7. Camillo Livi to Syndor Oden, May 28, 1945, NCC-EVP files, box "Jan 1–Dec 31, 1945," folder F.
8. Lamar Fleming Jr. to C. C. Smith, July 11, 1945, NCC-EVP files, box "Jan 1–Dec 31, 1945," folder F.
9. E. D. White to Will Clayton, July 13, 1945, box 1077, RG 16, National Archives.
10. Oscar Johnston to Gerald Creekmore, New Orleans Cotton Exchange, July 11, 1945, NCC-EVP files, box "Jan 1–Dec 31, 1945," folder C.
11. Robert C. Jackson, MOHP, 69–71.
12. Oscar Johnston to Hugh Comer, July 16, 1945, NCC-EVP files, box "Jan 1–Dec 31, 1945," folder C.
13. Robert C. Jackson to E. D. White, July 30, 1945, box 1077, folder "Cotton, RG 16," National Archives.
14. E. D. White to J. B. Huston, August 2, 1945, box 1077, folder "Cotton 1 of 2," National Archives.
15. Robert C. Jackson, interview by the author, Clemson, S.C., September 27, 2002; Ross Joseph Pritchard, "Will Clayton: A Study of Business-Statesmanship in the Formulation of United States Economic Foreign Policy" (Ph.D. diss., John Hopkins University, 1955), 211–14.
16. James L. Forsythe, "World Cotton Technology since World War II," *Agricultural History* 54 (January 1980): 208–222; Robert C. Jackson, MOHP, 74–75; Robert C. Jackson, interview by author, September 27, 2002.
17. Jackson interview.
18. Robert C. Jackson to Rhea Blake, October 2, 1945, NCC-EVP files, box "Jan 1–Dec 31, 1945," folder "Robert C. Jackson—European Trip."
19. Ibid.
20. Ibid.
21. Ibid., 9.
22. Robert C. Jackson, Joint Intelligence Objectives Agency, Office of Military Government for Germany, Office of Director of Intelligence, "Factors Relating to Prospects for Exporting U.S. Cotton to Germany, Final Report No. 457, Restricted" (November 14, 1945), 22–23, NCC-EVP files, box "Jan 1–Dec 31, 1945," folder "Robert C. Jackson—European Trip."
23. Ibid., 34.
24. E. D. White and C. C. Smith, November 9, 1945, NCC-EVP files, box "Jan 1–Dec 31, 1945," folder S.
25. Will Clayton to Robert P. Patterson, Undersecretary of War, September 10, 1945, box 24, folder 45-9-10, Chronological File, Clayton Papers, Rice University.

26. *Cotton Trade Journal*, International Edition, 1945–46, 126.
27. Ibid., 120.
28. *Cotton Trade Journal*, December 8, 1945, 1.
29. *Cotton Trade Journal*, International Edition, 1945–46, 150.
30. Memorandum of Conversation, October 25, 1945, Office Files of Assistant Secretary of State, box 41, record group 59, Harry S. Truman Library.
31. McDonald K. Horne, interview by author, Memphis, Tenn., June 1, 1998.
32. Read Dunn, MOHP, 50–51; "Personal History of Read Dunn, Jr.," typescript, 25, in author's possession.
33. Read Dunn, *Remembering: An Account of the Organization and Early Days of the Cotton Council International and the International Institute for Cotton; Experiences Traveling throughout the World in the Interest of United States Cotton during 30 Years and Other Stories* (Memphis, Tenn.: National Cotton Council, 1992), 50.
34. Read Dunn, MOHP, 52.
35. Ibid., 53.
36. Read Dunn, "A Preliminary Report on Germany, Received by Air Mail, 1946," NCC-EVP files, general file A-2, box "Jan 1–Dec 31, 1946," folder "Program Committee, Foreign Trade."
37. Ibid.
38. Ibid.
39. Ibid., 1.
40. Ibid.
41. Ibid., 2.
42. Will Clayton to Robert B. Rice, January 1, 1945, box 21, folder 45-1-1, Clayton Papers, Rice University.
43. Will Clayton to Lt. Thomas E. Samules, box 24, folder 45-8-03, Clayton Papers, Rice University.
44. Will Clayton to Fairfax Crow, November 2, 1944, box 20, folder 44-11-2, Clayton Papers, Rice University.
45. Quotes in *Cotton Trade Journal*, International Edition, 1945–46, 18–19; James Ashe to Will Clayton, August 28, 1945; and memorandum, Livingston T. Merchant to Will Clayton, August 30, 1945, Subject Files A-Com, Office Files of the Assistant Secretary of State for Economic Affairs, 1944–46 and Under Secretary for Economic Affairs, 1946–47; both in box 1, RG 59, Clayton Papers, Harry S. Truman Library.
46. Lamar Fleming Jr., "Possibilities of Post War Trade in Cotton," Fourth Annual Cotton Research Cotton Congress, July 9, 1943, 5, reprinted in ACCO Press, copy in NCC-EVP files, box "Jan 1–Dec 31, 1943," folder "Anderson-Clayton"; R. L. Dixon, "Post War Cotton Prospects," mimeograph, NCC-EVP file room 217, folder "Agricultural and Economic Problems of the Cotton Belt, May 31, 1944."
47. W. L. Clayton to Lamar Fleming Jr., October 8, 1944, box 19, folder 44-10-8, Clayton Papers, Rice University.
48. Fleming to O'Daniel, June 7, 1945, box 22, folder 45-6-7, Clayton Papers, Rice University.
49. John W. Snyder, Oral History Collection, January 15, 1969, 977, Harry S. Truman Library, http://www.trumanlibrary.org; *New York Times*, May 11, 1945, nytimes.com archives; "The Reminiscences of Will Clayton," Oral History Collection, Columbia University, 1962, p. 174.
50. Oscar Johnston to Will Clayton, March 22, 1946, NCC-EVP General Files, box "Jan 1–Dec 31, 1946, folder "Program Committee, Foreign Trade."

51. Oscar Johnston to Read Dunn, March 22, 1946, NCC-EVP General Files, box "Jan 1–Dec 31, 1946, folder "Program Committee, Foreign Trade."

52. Congress, Senate, *Nominations Department of State*, 78th Cong., 2nd sess., December 12–13, 1944, 66.

53. Office Files of Assistant Secretary of State, subject file N, box 5, RG 59, Harry S. Truman Library.

54. Claire Wilcox to Rhea Blake, March 20, 1946, NCC-EVP files, box "Jan 1–Dec 31, 1946," folder "Program Committee, Foreign Trade."

55. Richard M. Freeland, *The Truman Doctrine and the Origins of McCarthyism* (New York: Alfred A. Knopf, 1972), 71.

56. Herbert Hoover, "The President's Economic Mission to Germany and Austria," Report No. 1, German Agriculture and Food Requirements, February 28, 1947, 21, http://www.trumanlibrary.org/whistlestop/study_collections/marshall/large/index.

57. Copy of Resolution in NCC-EVP files, box "Jan 1–Dec 31, 1947," General File A–Z, folder A; Oscar Johnston to Robert P. Patterson, February 25, 1947, and Howard C. Peterson to Oscar Johnston, March 17, 1947, both in NCC-EVP files, box "Jan 1–Dec 31, 1947," folder "Program Committee, Foreign Trade."

58. *Bonham Daily Favorite*, January 13, 1947, 1.

59. Will Clayton, "The Reminiscences of Will Clayton," Oral History Collection, Columbia University, 217.

60. http://www.trumanlibrary.org/publicpapers.

61. Edwin McCommon Martin, Oral History Collection (July 6, 1970), http://www.trumanlibrary.org/oralhist.

62. *Cotton Trade Journal*, International Edition, 1947, 155.

63. Ibid., 157; USDA Economic Research Service, *Statistics on Cotton and Related Data, 1930–67*, 1.

64. *Cotton Trade Journal*, International Edition, 1947, 129.

65. *Proceedings of 8th Cotton Research Congress* (Dallas: July 16–18, 1947), 16.

66. Burris Jackson, *Proceedings of the Seventh Cotton Research Congress* (Dallas, July 8–9, 1946), 10.

67. *Cotton Trade Journal* 26 (April 1947): 1; *Progress Bulletin*, May 15, 1947, 3.

68. Charles L. Mee, *The Marshall Plan: The Launching of the Pax Americana* (New York: Simon and Schuster, 1984), 88.

69. Forrest C. Pogue, *George C. Marshall: Statesman* (New York: Viking, 1987), 210.

70. Dean Acheson, *Present at the Creation: My Years in the State Department* (New York: W. W. Norton, 1969), 227–28; Lewis T. Barringer, Oral History Collection, Harry S. Truman Library, 54; Chip Morgan, interview by the author, January 6, 2006, Stoneville, Miss.

71. Acheson, *Present at the Creation*, 227–28.

72. Mee, *The Marshall Plan*, 94.

73. Document 21, Papers of Joseph M. Jones, in Dennis Merrill, ed., *Documentary History of the Truman Presidency: Establishing the Marshall Plan, 1947–1948* (Bethesda, Md.: University Publications of America, 1996), 131–132.

74. Ibid., 141.

75. Lamar Fleming Jr., "Possibilities of Post War Trade in Cotton," *ACCO Press*, July 1943, 5.

76. *Commercial Appeal*, May 9, 1947, 6.

77. *Cotton Trade Journal*, May 2, 1947, 1.

78. *Current Popular Opinion on Foreign Trade Issues*, Division of Public Studies, Office of Public Affairs, Department of State, Office Files of the Assistant Secretary of State of Economic Affairs, 1944–50, and the Under Secretary for Economic Affairs, 1946–47, box 2, RG 59,

Harry S. Truman Library. In January 1945, R. S. Hecht of the Hibernia National Bank of New Orleans urged the South to take the leadership role in promoting trade because southerners were more "internationally minded." See *Cotton Trade Journal*, January 20, 1945, 2. For articles promoting free trade, see *Cotton Trade Journal*, International Edition, 1943–44.

79. Copy in Clayton Papers, box 30, folder 47-5-27, Clayton Papers, Rice University.

80. Acheson, *Present at the Creation*, 231.

81. Mee, *The Marshall Plan*, 89.

82. "Interview with General George C. Marshall," October 30, 1952, 2, Harry B. Price Papers, http://www.trumanlibrary.org. In this interview, Marshall insisted that "speeches and statements" by Acheson or Clayton "did not represent trial balloons."

83. *Kiplinger Newsletter*, February 14, 1948, copy in NCC-EVP files, box "Jan 1–Dec 31, 1948," folder "Marshall Plan."

84. Read Dunn, "Future Prospects for U.S. Cotton Exports," April 1, 1947, 8, copy in NCC-EVP files, box "Jan 1–Dec 31, 1948," folder "Circulars."

85. Copy of speech in NCC-EVP files, box "Jan 1–Dec 312, 1948," General File A–Z, folder "Tenth Annual Meeting Speeches."

86. Editorial reprinted in *ACCO Press*, February 1948, in D&PL Records, box 26, folder "D&PL-IX-History, Biog Information, Oscar Johnston," Special Collections, Mississippi State University Library.

87. NCC-EVP files, General File A–Z, box "Jan 1–Dec 31, 1948," folder "Cotton Belt Senators and Congressmen, General Letters."

88. Interview with George C. Marshall, October 30, 1952, Harry B. Price Papers.

89. NCC-EVP files, box "Jan 1–Dec 31, 1948," folder "Marshall Plan"; Read Dunn, MOHP, 54–55.

90. NCC-EVP files, box "Jan 1–Dec 31, 1948," folder "Marshall Plan."

91. Albert R. Russell to Field Service Personnel, March 1, 1948, NCC-EVP files, box "Jan 1–Dec 31, 1948," folder "Marshall Plan."

92. Will Clayton, "Reminiscences," Oral History Collection, Columbia University, 171.

93. Kenneth D. Hairgrove, "Sam Rayburn: Congressional Leader, 1940–1952" (Ph.D. diss., Texas Tech University, 1974), 207.

94. Will Clayton, "Reminiscences," Oral History Collection, Columbia University, 178.

95. E. D. White, "Foreign Markets for American Cotton," *Proceedings Fourteenth Annual American Cotton Congress* (June 25–27, 1953), 78.

96. Lamar Fleming Jr., "Possibilities of Post War Trade in Europe," *ACCO Press*, July 1943, 2.

97. Lamar Fleming Jr. to Rhea Blake, January 18, 1945, box 1, folder 45-1-18, Clayton Papers, Rice University.

98. Joseph M. Jones, *The Fifteen Weeks* (New York: Viking, 1955), 276.

99. D. Clayton Brown, *Electricity for Rural America: The Fight for the REA* (Westport, Conn.: Greenwood Press, 1980), 112–20.

6. THE DINNER TABLE WAR

1. Frederick Eugene Melder, "State and Local Barriers to Interstate Commerce in the United States: A Study in Economic Sectionalism," *Maine Bulletin*, University of Maine Studies, 2nd ser., no. 43 (November 1937): 104.

2. Ibid.

3. George B. Tindall, *The Emergence of the New South, 1913–1945* (Baton Rouge: Littlefield Fund for Southern History of the University of Texas, 1967), 596.

4. National Emergency Council, *Report on Economic Conditions of the South*, 58.

5. *Dairy Record*, June 18, 1941, clipping in W. R. Poage Papers, box 253, file 3, Special Collections, Baylor University.

6. Oscar Johnston to George M. Grant, September 1, 1943, NCC-EVP files, box "Feb 1943–Jan 1944," folder "Letters to Congressmen and Senators"; *Progress Bulletin*, November 15, 1943, 6.

7. NCC-EVP files, box "Jan 1–Dec 31, 1947," General File A–Z, folder "Trade Barriers."

8. Robert (Bob) Poage Papers, box 253, file 3; NCC-EVP files, box "Jan 1–Dec 31, 1947," General File A–Z, folder "Trade Barriers"; Sam Bledsoe, Oral History Collection, Columbia University, 684–85.

9. *Congressional Record*, 80th Cong., 2nd sess., vol. 94, pt. 1, 195.

10. Ibid., 196.

11. NCC-EVP files, box "Jan 1–Dec 31, 1948," folder "Circular Letters."

12. Sam Bledsoe, Oral History Collection, Columbia University, 685.

13. *The Democratic National Platform on Margarine*, copy in Robert Poage Papers, box 253, file 3.

14. *The President's Economic Report to the Congress*, January 7, 1949, copy in Robert Poage Papers, box 253, file 3.

15. Emanuel Celler, news release, February 10, 1949, Poage Papers, box 253, file 3.

16. *Memorandum on the Position of the American Farm Bureau Federation on Margarine*, Poage Papers, box 253, file 3.

17. Copy in Poage Papers, box 253, file 3.

18. NCC-EVP files, box "Jan 1–Dec 31, 1949," folder "Circular Letters."

19. Ibid.

20. "Editorial Comment," *Cotton Gin and Oil Mill Press*, April 1, 1950, copy in NCC-EVP files, box "Jan 1–Dec 31, 1950," folder "Program Committee, Margarine Legislation."

21. Will Clayton to J. C. Dillinger, December 5, 1951, Alphabetical File, 1950–53, box 95, folder "Domestic Offices, Anderson-Clayton Company, Los Angeles," Clayton Papers, Harry S. Truman Library.

22. *Progressive Bulletin*, April 15, 1952, 7.

23. *Cotton's Week* 1, no. 38 (July 28, 1945).

24. *Cotton's Week* 1, no. 49 (October 13, 1945).

25. *Cotton's Week* 1, no. 51 (October 27, 1945).

26. L. T. Barringer to Mr. President, August 15, 1946, White House Central Files 258, Official Files, box 879, Harry S. Truman Library.

27. Memorandum, the President to Secretary of Agriculture, September 3, 1946, White House Central Files 258, Official Files, box 879, Harry S. Truman Library.

28. Clifford P. Anderson to the President, January 10, 1947; The President to L. T. Barringer, January 14, 1947; both in White House Central Files 258, Official Files, box 879, Harry S. Truman Library.

29. "Statement of National Cotton Council," December 5, 1945, NCC-EVP files, box "Jan 1–Dec 31, 1945," folder "Circulars."

30. Rhea Blake to John Bankhead, December 8, 1945, NCC-EVP files, box "Jan 1–Dec 31, 1945," folder "Circulars."

31. *New York Times*, March 16, 1946, 14.

32. Ibid., 1; *New York Times*, March 17, 1946, 34.

33. *International Teamster* 43, no. 7 (June 1946), copy in NCC-EVP files, box "Jan 1–Dec 31, 1946," folder "Circulars"; *Progress Bulletin*, July 15, 1946, 1.

34. *International Teamster* (June 1946), NCC-EVP files, folder "Circulars."

35. Oscar Johnston to the People of the Cotton Belt, June 14, 1946, NCC-EVP files, box "Jan 1–Dec 31, 1946," folder "Circulars"; *Progress Bulletin*, July 15, 1946, 5.

36. Oscar Johnston to the People of the Cotton Belt, June 14, 1946, NCC-EVP files, box "Jan 1–Dec 31, 1946," folder "Circulars."

37. "The Cotton Monopoly Squeals," *International Teamster*, August 1946, copy in NCC-EVP files, General Files A–Z, box "Jan 1–Dec 31, 1946," folder "Circulars."

38. Rhea Blake to Oscar Johnston, November 22, 1946, NCC-EVP files, box "Jan 1–Dec 31, 1946," folder "Labor."

39. Oscar Johnston to Richard Russell, March 1, 1947, NCC-EVP files, box "Jan 1–Dec 31, 1946," folder "Labor."

40. Oscar Johnston to Members of the Cotton Industry, March 10, 1947, NCC-EVP files, box "Jan 1–Dec 31, 1947," folder "Circulars."

41. Oscar Johnston to Richard Russell, March 13, 1947, NCC-EVP files, box "Jan 1–Dec 31, 1947," folder "Circulars."

42. Oscar Johnston to James B. Murphy, May 5, 1947, General Files A–Z, box "Jan 1–Dec 31, 1947," folder M.

43. USDA Economic Research Service, *Statistics on Cotton and Related Data, 1930–67*, 84.

44. McDonald K. Horne to Rhea Blake, February 24, 1945, NCC-EVP files, box "Jan 1–Dec 31, 1945," folder "Investigational Study—Competitive Price of Cotton."

45. Congress, House, Special Subcommittee on Cotton, Hearings, *Study of Agricultural and Economic Problems of the Cotton Belt*, 80th Cong., 1st sess., July 7–8, 1945.

46. Ibid., 16.

47. James H. Street, *The New Revolution in the Cotton Economy: Mechanization and Its Consequences* (Chapel Hill: University of North Carolina Press, 1957), 77–78.

48. W. M. Davlin to Rhea Blake, August 13, 1947, NCC-EVP files, box "Jan 1–Dec 31, 1947," folder "National Planning Association, Committee of the South."

49. McDonald K. Horne to Rhea Blake, January 13, 1948, NCC-EVP files, box "Jan 1–Dec 31, 1948," folder "Cotton Fact Finding Study."

50. Alan Greenspan, *The Age of Turbulence: Adventures in a New World* (New York: Penguin, 2007), 32–33.

51. Allen J. Matusow, *Farm Policies and Politics in the Truman Years* (Cambridge, Mass.: Harvard University Press, 1967), 171.

52. Ibid., 171, 190.

53. *Arkansas Gazette*, June 1, 1949, 26, copy in box 11, folder "Brannan Plan," in Charles F. Brannan Papers, Harry S. Truman Library.

54. Rhea Blake to W. B. Coberly, May 4, 1949, NCC-EVP files, box "Jan 1–Dec 31, 1949," folder C; Rhea Blake to Lorenzo K. Wood, May 5, 1949, NCC-EVP files, box "Jan 1–Dec 31, 1949," folder W.

55. Read Dunn, interview by author, Washington, D.C., October 14, 1999.

56. W. E. Hendrix, "The Brannan Plan and Farm Adjustment Opportunities in the Cotton South," *Journal of Farm Economics* 31 (August 1949): 495.

57. Matusow, *Farm Policies*, 195.

58. Ibid., 196.

59. Ibid., 199.

60. Ibid.

61. Hendrix, "Brannan Plan," 495.

62. Dunn interview.

63. *Des Moines Register*, February 19, 1950, 1; NCC-EVP files, box "Jan 1–Dec 31, 1950," folder "Policy Statement"; Matusow, *Farm Policies*, 218.

64. *New York Herald Tribune*, September 24, 1950, clipping in NCC-EVP files, box "Jan 1–Dec 31, 1950," folder "Cotton Mobilization Committee, 1950."

65. Lamar Fleming Jr. to Will Clayton, September 1, 1950, Alphabetical File, box 96, folder "Lamar Fleming–Anderson Clayton," Clayton Papers, Harry S. Truman Library.

66. Robert Poage to Eugene Butler, September 20, 1950, box 813, file 4, Poage Papers; USDA Press Release 2407-50-2, October 3, 1950, copy in NCC-EVP files, box "Jan 1–Dec 31, 1950," folder "Cotton Mobilization Committee, 1950"; Stephen Pace to Charles F. Brannan, October 5, 1950, NCC-EVP files, box "Jan 1–Dec 31, 1950," folder B.

67. NCC Press Release, September 18, 1950, NCC-EVP files, box "Jan 1–Dec 31, 1950," folder "Circular Letter"; statement, Walter L. Randolf, American Farm Bureau Federation, October 13, 1950, NCC-EVP files, box "Jan 1–Dec 31, 1950," folder "Circular Letter."

68. Street, *New Revolution*, 86.

69. Matusow, *Farm Policies*, 231–32.

7. THE SOUTH TRANSFORMED

1. Peter F. Drucker, "Exit King Cotton," *Harper's*, May 1946, 473.

2. Donald Holley, *The Second Great Emancipation: The Mechanical Cotton Picker, Black Migration, and How They Shaped the Modern South* (Fayetteville: University of Arkansas Press, 2000).

3. D. J. Pledger and D. J. Pledger Jr., *Cotton Culture on Hardscramble* (Shelby, MS: D. J. Pledger and D. J. Pledger Jr., 1951), 10.

4. Charles R. Sayre, MOHP, CXXXV (1979), 22.

5. James H. Street, *The New Revolution in the Cotton Economy: Mechanization and Its Consequences* (Chapel Hill: University of North Carolina Press, 1957), 51; D. Clayton Brown, "Health of Farm Children in the South, 1900–1950," *Agricultural History* 53 (January 1979).

6. Gilbert Fite, *Cotton Fields No More: Southern Agriculture, 1865–1980* (Lexington: University Press of Kentucky, 1984), 225.

7. "Mr. Little Ol' Rust," *Fortune*, December 1952, 200.

8. David Halberstam, *The Fifties* (New York: Villard, 1993), 446.

9. Ibid., 447.

10. "The Origin and Development of the Cotton Picker," typescript, box 1, folder 1, 22, John Rust Papers, Special Collections, University of Memphis; *Reader's Digest*, October 1936, 43–47; *Literary Digest*, September 5, 1936, 45–46.

11. Quote appears in Street, *New Revolution*, 126.

12. John Rust, "The Origin and Development of the Cotton Picker," *West Tennessee Historical Society Papers* 7 (1953): 53.

13. John Rust, "The Origin and Development of the Cotton Picker," typescript, box 1, folder 1, 22.

14. Street, *New Revolution*, 131.

15. Quotes in *Collier's*, July 21, 1945, 24.

16. Ibid.

17. Charles Sayre, MOHP, XXVIII–XXXI.

18. E. D. White, *Proceedings of the Beltwide Cotton Mechanization Conference*, August 18–19, 1947, 39, NCC Technical Library, Memphis, Tennessee.

19. *Cotton Trade Journal*, August 22, 1947, 1; for transcript of panel discussion see White, *Proceedings*, August 18–19, 1947, NCC Technical Library.

20. *Cotton Trade Journal* August 15, 1947, 7.

21. Bill Pearson, interview by author, Sumner, Miss., January 5, 2006.

22. Congress, House, Special Committee on Cotton of the Committee on Agriculture, *Hearings, Study of Agricultural and Economic Problems of the Cotton Belt*, 80th Cong., 1st sess., pt. 2, October 10, 1947, 1019.

23. Ibid., 993–97; Pearson interview; Charles S. Aiken, *The Cotton Plantation South since the Civil War* (Baltimore: Johns Hopkins University Press, 1998), 102.

24. Congress, House, Special Subcommittee on Cotton, *Hearings, Study of Agricultural and Economic Problems of the South*, pt. 2, 1019.

25. James R. Supak, Carl G. Anderson, and William D. Mayfield, "Trends in Cotton Production: History, Culture, Mechanization, and Economics," in *Weeds of Cotton: Characterization and Control*, ed. Chester G. McWhorter and John R. Abernathy (Memphis, Tenn.: Cotton Foundation, 1992), 31.

26. R. Douglas Hurt, ed., "Cotton Pickers and Strippers," *Journal of the West* 30 (April 1991): 36.

27. Congress, House, Special Subcommittee on Cotton, *Hearings, Study of Agricultural and Economic Problems of the Cotton Belt*, pt. 2, 1004.

28. Ibid., 1005.

29. Karen Gerhardt Britton, *Bale o' Cotton: The Mechanical Art of Cotton Ginning* (College Station: Texas A&M University Press, 1992), 104–6; A. L. Vandergriff, *Cotton Ginning: An Entrepreneur's Story* (Lubbock: Texas Tech University Press, 1997).

30. Pearson interview.

31. Congress, House, Special Subcommittee on Cotton, *Hearings, The Study of Agricultural and Economic Problems of the Cotton Belt*, pt. 2, 990; Frank E, Smith, *The Yazoo River* (New York: Rinehart, 1954), 329–46.

32. For information on herbicides, I drew on Chester G. McWhorter and John R. Abernathy, eds., *Weeds of Cotton: Characterization and Control* (Memphis, Tenn.: Cotton Foundation, 1992).

33. Gale A. Buchanan, "Trends in Weed Control," in McWhorter and Abernathy, *Weeds of Cotton*, 61.

34. Chester G. McWorter and Charles T. Bryson, "Herbicide Use Trends in Cotton," in McWhorter and Abernathy, *Weeds of Cotton*, 241.

35. Buchanan, "Trends in Weed Control," 65.

36. Gilbert C. Fite, "Mechanization of Cotton Production since World War II," *Agricultural History* 54 (January 1980): 203.

37. *Cotton Trade Journal*, International Edition, 1947, 150.

38. Oscar Johnston to James B. Murphy, March 13, 1947, NCC-EVP files, box "Jan 1–Dec 31, 1947," folder M.

39. William D. Anderson to Stephen Pace, November 8, 1947, NCC-EVP files, box "Jan 1–Dec 31, 1947," folder "Circulars."

40. Quote in Thomas D. Clark, *The Emerging South* (New York: Oxford University Press, 1961), 90. In 1958 the *Cotton Trade Journal* reported that poultry exceeded cotton in importance in Georgia and tobacco in South Carolina, and that the small cotton farmer "is moving off the scene fast" (September 12, 1958, 1).

41. Clark, *The Emerging South*, 90.

42. Fite, *Cotton Fields No More*, 197.

43. Reprinted in Congress, Senate, Committee on Agriculture and Forestry, *Cotton Programs*, 85th Cong., 1st sess., March 12, June 24–25, 1957, 24.

44. Craig Heinicke, "African-American Migration and Mechanized Cotton Harvesting, 1950–1960," *Explorations in Economic History* 31 (1994): 502.

45. Pearson interview.

46. Oscar Johnston, "Will the Machine Ruin the South," *Saturday Evening Post*, May 31, 1947, 36.

47. Ibid., 98.

48. Street, *New Revolution*, ix.

49. Ibid., 249, 229.

50. Nicholas Lemann, *The Promised Land: The Great Black Migration and How It Changed America* (New York: Alfred A. Knopf, 1991), 50.

51. W. Petersen and Y. Kislev, "The Cotton Harvester in Retrospect: Labor Displacement or Replacement?" *Journal of Economic History* 46 (March 1986): 199.

52. Holley, *The Second Great Emancipation*, 183.

53. Pete Daniel, *Breaking the Land: The Transformation of Cotton, Tobacco, and Rice Culture since 1880* (Urbana: University of Illinois Press, 1985), 244.

54. Richard Wright, *Twelve Million Black Voices: A Folk History of the Negro in the United States* (New York: Viking, 1941), 43, 59, quote on 49.

55. *New York Times*, January 6, 1961.

56. Carroll Van Nest, ed., *The Tennessee Encyclopedia of History and Culture* (Nashville: Rutledge Hill Press, Tennessee Historical Society, 1998), 965–66.

57. B. B. King, *Blues All around Me: The Autobiography of B. B. King* (New York: Avon Books, 1996), 34.

58. Ibid., 72.

59. Janis F. Kearney, *Cotton Field of Dreams: A Memoir* (Chicago: Writing Our Press, 2004), 54.

60. Charles Sayre, MOHP, 43.

61. King, *Blues All around Me*, 36.

62. Frank Smith, *The Yazoo River*, 346.

63. Congress, House, Special Subcommittee on Cotton, *Hearings, Study of Agricultural and Economic Problems of the Cotton Belt*, pt. 2, 990.

64. Lemann, *The Promised Land*, 6, 7.

8. THE WHITE GOLD RUSH

1. Jack Stone, interview by author, Stone Land Company, Stratford, Calif., April 13, 2005.

2. Morgan Nelson, interview by author, Roswell, N.Mex., April 8, 2005.

3. Don Anderson, interview by author, Lubbock, Tex., April 4, 1998.

4. USDA Economic Research Service, *Statistics on Cotton and Related Data, 1930–67*, 4.

5. Janet M. Neugebauer, ed., *Plains Farmer: The Diary of William G. DeLoach, 1914–1964* (College Station: Texas A&M Press, 1991), 68.

6. Joseph C. McGowan, *History of Extra-Long Staple Cottons* (El Paso, Tex.: Hill Printing Company, 1961), 84.

7. Ibid., 98.

8. Thomas E. Sheridan, *Arizona: A History* (London: University of Arizona Press, 1995), 214.

9. For a review of the San Joaquin Valley, see Ellen Liebman, *California Farmland: A History of Large Agricultural Landholdings* (Totowa, N.J.: Rowman and Allanheld, 1983).

10. Ibid., 121.

11. Ibid., 96; Read Dunn, interview by author, Washington, D.C., October 14, 1999.

12. Lamar Fleming Jr. to Will Clayton, May 9, 1946, Alphabetical File, box 54, folder "Lamar Fleming," Clayton Papers, Harry S. Truman Library.

13. "Biggest Cotton Plantation," *Fortune*, March 1937, 156.

14. Anderson interview; George B. Tindall, *The Emergence of the New South, 1913–1945* (Baton Rouge: Littlefield Fund for Southern History of the University of Texas, 1967), 121–24; Rupert B. Vance, *Human Factors in Cotton Culture: A Study in the Social Geography of the American South* (Chapel Hill: University of North Carolina Press, 1929), 89.

15. Philip Haney, "The Cotton Boll Weevil in the United States: Impact on Cotton Production and the People of the Cotton Belt," In *Boll Weevil Eradication in the United States through 1999*, ed. Willard Dickerson et al. (Memphis, Tenn.: Cotton Foundation, 2001), 10.

16. USDA Economic Research Service, *Statistics on Cotton and Related Data, 1930–67*, 85–91.

17. Lewis T. Barringer to the President, August 15, 1946, White House official file, box 879, Harry S. Truman Papers.

18. *Cotton Trade Journal*, International Edition, 1948–49, 3.

19. John Turner, *White Gold Comes to California* (Bakersfield: California Planting Cotton Seed Distributors, 1981), 80.

20. *Cotton Trade Journal*, International Edition, 1948–49, 154–55.

21. Turner, *White Gold*, 89–91.

22. *Collier's*, June 23, 1951, 24.

23. James D. Black, "An Analysis of Cotton Production in the San Joaquin Valley, California" (Ph.D. diss., University of California, Los Angeles, 1956), 36.

24. Ibid., 37.

25. "Cotton in California," *Nation*, February 19, 1949, 211, 212.

26. Lamar Fleming Jr. to Will Clayton, November 11, 1944, box 20, folder 11-4-44, Clayton Papers, Rice University.

27. USDA Economic Research Service, *Statistics on Cotton and Related Data, 1930–67*, 84–91.

28. Jack Lichtenstein, *Field to Fabric: The Story of American Cotton Growers* (Lubbock: Texas Tech University Press, 1990), 20.

29. James H. Street, *The New Revolution in the Cotton Economy: Mechanization and Its Consequences* (Chapel Hill: University of North Carolina Press, 1957), 85.

30. William J. Briggs and Henry Cauthen, *The Cotton Man: Notes on the Life and Times of Wofford B. (Bill) Camp* (Columbia: University of South Carolina Press, 1983), 252–53.

31. Congress, Senate, Subcommittee of the Committee on Agriculture and Forestry, *Hearings, Cotton Marketing Quotas and Acreage Allotments* 81st Cong., 1st sess., June 14–15, 22, 27, 1949, 90–91.

32. Ibid., 51.

33. Street, *New Revolution*, 85.

34. Ibid., 86.

35. USDA Economic Research Service, *Statistics on Cotton and Related Data, 1930–67*, 1.

36. *Hearings, Cotton Marketing Quotas and Acreage Allotments*, 83rd Cong., 1st sess., June 30, July 1, 10–11 (July 1953), 57–67.

37. Ibid., 26.

38. Ibid., 10.

39. Ibid., 87.

40. Stone interview.

41. Reymond E. Blair, MOHP, CXIX (1979), 6.

42. Congress, Senate, Committee on Agriculture and Forestry, *Hearings, Cotton Marketing Quotas and Acreage Allotments*, 8.

43. John Samuel Ezell, *The South since 1865* (New York: Macmillan, 1963), 442.

44. Congress, Senate, Committee on Agriculture and Forestry, *Hearings, Cotton Marketing Quotas and Acreage Allotments*, 157–61.

45. Ibid., 18.

46. Ibid., 160; Neugebauer, *Plains Farmer*, 199.

47. Congress, Senate, Committee on Agriculture and Forestry, *Hearings, Cotton Marketing Quotas and Acreage Allotments*, 192.

48. Ibid., 166.

49. *Cotton Trade Journal*, January 9, 1953, 3.

50. *Cotton Trade Journal*, May 1, 1953, 1.

51. Congress, Senate, Committee on Agriculture and Forestry, *Hearings, Cotton Marketing Quotas and Acreage Allotments*, 165, 192.

52. Ibid., 166.

53. Lamar Fleming Jr., "Barriers to a Sound National Cotton Policy," in *Proceedings of the Seventh Annual Research Congress*, Dallas, Tex., July 8–9, 1946, 51.

54. Congress, Senate, Committee on Agriculture and Forestry, *Hearings, Cotton Marketing Quotas and Acreage Allotments*, 6.

55. *Cotton Trade Journal*, October 2, 1953, 2.

56. *Cotton Trade Journal*, November 6, 1953, 1.

57. Congress, Senate, Committee on Agriculture and Forestry, *Amending the Agricultural Adjustment Act of 1938, as Amended*, report 838, calendar 831, 83rd Cong., 2nd sess., 1.

58. Murray R. Benedict and Oscar Stein, *The Agricultural Commodity Programs: Two Decades of Experience* (New York: Twentieth Century Fund, 1956), 39.

59. James L. Whitten to Ezra Taft Benson, January 6, 1954, NCC-EVP files, box "Jan 1–Dec 31, 1954," folder "Acreage Allotments."

60. USDA Economic Research Service, *Statistics on Cotton and Related Data, 1930–67*, 1.

61. Benedict and Stine, *Agricultural Commodity Programs*, 41.

62. Thad Sitton and Dan K. Utley, *From Can See to Can't See* (Austin: University of Texas Press, 1997), 269.

63. Gilbert Fite, *Cotton Fields No More: Southern Agriculture, 1865–1980* (Lexington: University Press of Kentucky, 1984), 220.

64. *Nation*, May 5, 1951, 408.

65. Richard Street, *Photographing Farmworkers in California* (Stanford, Calif.: Stanford University Press, 2004), 176.

66. Stone interview; Nelson interview.

67. Ellis W. Hawley, "The Politics of the Mexican Labor Issue," *Agricultural History* 40 (July 1966): 157–76.

68. Pauline R. Kibbe, *Latin Americans in Texas* (Albuquerque: University of New Mexico Press, 1945), 178–79; http://www.justiceformypeople.org/drhector3.html (accessed June 15, 2009).

69. Richard B. Craig, *The Bracero Program: Interest Groups and Foreign Policy* (Austin: University of Texas Press, 1971), 182.

70. David Montejano, *Anglos and Mexicans in the Making of Texas, 1836–1986* (Austin: University of Texas Press, 1987), 273.

71. McGowan, *History of Extra-Long Staple Cotton*, 40.

72. Ibid., 42.

73. *Cotton Trade Journal*, June 13, 1958, 4.

74. Mitchell F. Landers to the President of the United States, Secretary of Agriculture Ezra Taft Benson, et al., August 3, 1954; J. Banks Young to Mitchell Landers, August 20, 1954, all in NCC-EVP files, box "Jan 1–Dec 31, 1954," folder "Extra Long Staple Cotton"; McGowan, *History of Extra-Long Staple Cotton*, 46.

75. Forest G. Hill, "Regional Planning and Development," in *Technology in Western Civilization*, ed. Melvin Kranzberg and Carroll W. Pursell Jr. (New York: Oxford University Press, 1967), 456.

76. Lamar Fleming Jr. to Will Clayton, August 31, 1950, Alphabetical File, box 96, folder 1, Clayton Papers, Harry S. Truman Library.

9. BOLL WEEVILS, WORMS, AND MOTHS

1. Philip Haney, "The Cotton Boll Weevil in the United States: Impact on Cotton Production and the People of the United States, *Boll Weevil Eradication in the United States through 1999*, ed. Willard Dickerson et al. (Memphis, Tenn.: Cotton Foundation, 2001), 11.

2. Ibid., 18.

3. William H. Skaggs, *The Southern Oligarchy* (New York: Devin-Adair, 1924; Negro Universities Press Reprint, 1969), 353.

4. Haney, "Cotton Boll Weevil," 14–15.

5. Knox Walker, interview by author, Texas A&M University, College Station, June 1, 1998.

6. Douglas Helms, "Technological Methods for Boll Weevil Control," *Agricultural History* 53 (January 1979): 294.

7. Douglas Helms, "Revision and Reversion: Changing Cultural Practices for the Cotton Boll Weevil," *Agricultural History* 54 (January 1980): 108–25.

8. Katie Dickie Stavinoha and Lorie A. Woodward, "Texas Boll Weevil History," in *Boll Weevil Eradication in the United States through 1999*, ed. Willard A. Dickerson et al. (Memphis, Tenn.: Cotton Foundation, 2001), 459.

9. Dorothy Scarborough, *In the Land of Cotton* (New York: Macmillan, 1936), 135–36.

10. http://songofthesouth.net/movie/lyrics/let-the-rain.html (accessed July 23, 2009). Thomas S. Hischak and Mark A. Robinson, *The Disney Song Encyclopedia* (Lanham, Md.: Rowman and Littlefield, 2009), 116.

11. Walker interview.

12. Perry L. Adkisson, Agricultural Oral History interview by Irvin M. May Jr., Office of Research Historian, Texas Agricultural Experiment Station, Texas A&M University, 15–16.

13. Committee on Cotton Insect Management, Board on Agricultural and Renewable Resources, Commission on Natural Resources, National Research Council, *Cotton Boll Weevil: An Evaluation of USDA Programs* (Washington, D.C.: National Academy Press, 1981), 20–21.

14. R. E. Frisbie, K. M. El-Zik, and L. T. Wilson, "Perspective on Cotton Production and Integrated Pest Management," in *Integrated Pest Management Systems and Cotton Production*, ed. R. E. Frisbie, K. M. El-Zik, and L. T. Wilson (New York: John Wiley and Sons, 1989), 5–7.

15. Committee on Cotton Insect Management, *Cotton Boll Weevil: An Evaluation of USDA Programs*, 22.

16. Richard L. Ridgway and Harry C. Mussman, "Integrating Science and Stakeholder Inputs: The Pivotal Years," in *Boll Weevil Eradication in the United States through 1999*, ed. Willard A. Dickerson et al. (Memphis, Tenn.: Cotton Foundation, 2001), 56.

17. Ibid., 56.

18. Ibid., 57.

19. Gerald H. McKibben, E. J. Villavaso, W. L. McGovern, and Bill Grefenstette, "United States Department of Agriculture—Research Support, Methods Development, and Program

Implementation," in *Boll Weevil Eradication in the United States through 1999*, ed. Willard A. Dickerson et al. (Memphis, Tenn.: Cotton Foundation, 2001), 125.

20. P. B. Haney and W. J. Lewis, USDA Agricultural Research Service, "Cotton Production and the Boll Weevil in Georgia—History, Cost of Control, and Benefits of Eradication," (Tifton, Georgia: Insect Biology and Population Management Research Laboratory), 16–17.

21. James W. Smith, "Boll Weevil Eradication: Area-Wide Pest Management," *Annals of Entomological Society of America* 91 (May 1998): 241; Stephen Yafa, *Big Cotton* (New York: Viking, 2006), 264–67.

22. McKibben et al., "United States Department of Agriculture," 121.

23. Adkisson, Agricultural Oral History interview, 7.

24. McKibben et al., "United States Department of Agriculture," 121.

25. Frank L. Carter, Cotton Nelson, Andrew G. Jordon, and J. Ritchie Smith, "U.S. Cotton Declares War on the Boll Weevil," in *Boll Weevil Eradication in the United States through 1999*, ed. Willard A. Dickerson et al. (Memphis, Tenn.: Cotton Foundation, 2001), 36–37.

26. *Dallas Morning News*, August 20, 2007, 12A.

27. Frank L. Carter, interview by author, Memphis, Tenn., June 3, 1998.

28. Truman McMahan, "The Fight against the Pink Bollworm in Texas," *Southwestern Historical Quarterly* 94 (July 1990): 63; L. F. Curl and R. W. White, "The Pink Bollworm," in *Insects: The Yearbook of Agriculture, 1952*, ed. USDA (Washington, D.C.: Government Printing Office, 1952), 505–11.

29. *Progress Bulletin*, January 1, 1950, 7.

30. Curl and White, "The Pink Bollworm," 511.

31. *Progress Bulletin*, January 1, 1950, 7.

32. *Progress Bulletin*, July 15, 1952, 5.

33. George S. Wells, *Garden in the West: A Dramatic Account of Science in Agriculture* (New York: Dodd, Mead, 1969), 142.

34. Ibid., 132, 144.

35. Ibid., 149.

36. http://www.uanews.org (accessed February 8, 2006).

37. http://www.cotton.org (accessed February 8, 2006); *Farm Press*, September 17, 2008, http://www.southwestfarmpress.com/cotton/pink-bollworm-0917.

38. Daniel Charles, *Lords of the Harvest: Biotech, Big Money, and the Future of Food* (Cambridge, Mass.: Perseus, 2001), 150.

39. *Cotton Farming*, April 1997, 28.

40. Jane Rissler, "Bt Cotton: Another Magic Bullet?" *Global Pesticide Campaigner* 7, no. 1 (March 1997), http://www.mindfully.org/GE/Another-Magic-Bullet.htm.

41. Charles, *Lords of the Harvest*, 174.

42. *Cotton Farming*, June 1998, 14–15; *Cotton Farming*, March 2000, 18–20.

43. *Cotton Farming*, January 1996, 42–43; Barry Aycock, interview by author, Parma, Mo., June 6, 1998.

44. *Progressive Farmer*, December 2000, 24–26.

45. *USA Today*, April 12, 2005, 4A.

46. Ibid; Rick Lavis, Arizona Cotton Growers Association, interview by author, Phoenix, Ariz., April 11, 2005.

47. Bill Lambrecht, *Dinner at the New Gene Café* (New York: Thomas Dunne, 2001), 124.

48. http://www.consumerfreedom.com (accessed March 1, 2006).

49. http://www.aces.edu (accessed February 21, 2006).

50. "PANUPS: No Carbofuran for Cotton," http://www.ibiblio.org/london/permaculture/mailarchives/sanet2/msg00479.html.

51. Statement of William Lovelady on behalf of the National Cotton Council before the Subcommittee on Department Operations, Nutrition, and Foreign Agriculture, June 15, 1998, http://agriculture.house.gov/hearings/105/h80624wb.htm (accessed February 21, 2006).
52. Ibid.
53. *Cotton Farming*, May 1998, 42.
54. http://www.consumersunion.org/food/pesticidedc698.htm.

10. MEMPHIS

1. Dearing's quote in *Commercial Appeal*, April 1, 1959, sec. 6, 21.
2. David Cohen, *God Shakes Creation* (New York: Harper and Brothers, 1935), 15.
3. Hortense Russom, interview by author, Campbell, Mo., May 16, 2005.
4. William Bearden, *Cotton: From Southern Fields to the Memphis Market* (Charleston, S.C.: Arcadia Publishing, 2005), 79.
5. William Johnston Britton, *Front Street: A Book of Poems* (Memphis, Tenn.: Dudley M. Weaver and Charles M. Ozier, 1948). Copy available in Memphis Public Library, Memphis Room.
6. Bearden, *Cotton*, 85.
7. William R. Ferris, *Mule Trader: Ray Lum's Tales of Horses, Mules, and Men* (Jackson: University Press of Mississippi, 1998); *Fortune*, March 1937, 126.
8. *Press Scimitar*, February 14, 1947, sec. 2, 13.
9. *Press Scimitar*, March 31, 1962, 2.
10. *Commercial Appeal*, January 26, 1936, sec. 2, 3.
11. Ibid.
12. *Press Scimitar*, July 28, 1951, 2.
13. *Commercial Appeal*, January 4, 1959, sec. 6, 21.
14. *Cotton Farming*, November–December 1976, 4.
15. George S. Bush, *An American Harvest: The Story of Weil Brothers Cotton* (Englewood Cliffs, N.J.: Prentice-Hall, 1982), 208.
16. Robert Gordon, *Can't Be Satisfied: The Life and Times of Muddy Waters* (Boston: Little, Brown, 2002). 32.
17. Francis Davis, *The History of the Blues* (Cambridge, Mass.: Da Capo Press, 2003), 42–43.
18. Ibid., 43.
19. B. B. King, *Blues All around Me: The Autobiography of B. B. King* (New York: Avon Books, 1996), 98–99.
20. Ibid., 57, 142.
21. Ibid., 33.
22. Frederic Ramsey Jr., *Been Here and Gone* (New Brunswick, N.J.: Rutgers University Press, 1960), 54.
23. Margaret McKee and Fred Chisenhall, *Beale Black and Blue: Life and Music on Black America's Main Street* (Baton Rouge: Louisiana State University Press, 1981), 93.
24. Scott Faragher and Katherine Harrington, *The Peabody Hotel* (Charleston, S.C.: Arcadia Publishing, 2002), 98.
25. Perre Magness, *The Party with a Purpose: Seventy-five Years of Carnival in Memphis, 1931–2006* (Memphis, Tenn.: Carnival Memphis Association, 2006) ,vi.
26. Ibid., 12.
27. Shields McIlluraine, *Memphis Down in Dixie* (New York: E. P. Dutton, 1948), 285.
28. Bearden, *Cotton*, 107.

29. For information on the Cotton Carnival, I drew on Magness, *Party with a Purpose*; and Bearden, *Cotton*.
30. Magness, *Party with a Purpose*, 111.
31. Ibid., 93.
32. Ibid., 206–10.
33. Ibid., 176; Beverly G. Bond and Janann Sherman, *Memphis in Black and White* (Charleston, S.C.: Arcadia Publishing, 2003), 109.
34. Ibid.
35. Quote in Magness, *Party with a Purpose*, 136.
36. Ibid., 149.
37. http://www.thecommercialappeal.com (accessed June 22, 2006).
38. ACCO Archives, box 33, BAH Study, folder "Management Organization and Survey," RGE-61, Anderson-Clayton Papers, Houston Public Library.
39. Forrest Orren Lax, "The Memphis Cotton Exchange: From Beginning to Decline" (M.A. thesis, Memphis State University, 1970), 129.
40. ACCO Archives, typescript, box 1, folder 15.
41. *Cotton Farming*, April 1990, 30.
42. ACCO Archives, box 1, folder 22.
43. Paul A. Ruh, "The Functions of a Cotton Merchant," paper presented at the ACSA International Cotton Institute at Rhodes College, Memphis, Tenn., July 12, 2005, 3, http://www.thecottonschool.com/Cotton%20Merchant.pdf (accessed September 20, 2006); *Cotton Farming*, April 1990, 38.
44. *Memphis Business Journal*, November 12, 2004, http://www.bizjournals.com/memphis/stories/2004/11/08/daily34.html (accessed September 20, 2006).
45. http://www.theseam.com (accessed September 20, 2006).
46. *Cotton Farming*, December 2001, 12.
47. http://www.cottonfarming.com/home/archive/2001_cfnews0531.html (accessed September 20, 2006).
48. Lax, "Memphis Cotton Exchange."

11. "THE FABRIC OF OUR LIVES"

1. Morgan Nelson, "Cotton Incorporated Problems: The First Decades," typescript, n.d., 1, Morgan Nelson Private Papers, Roswell, N.Mex.; Clifton Kirkpatrick, MOHP, CXLIV (1979), 26.
2. Nelson, "Cotton Incorporated Problems," 2.
3. Macon Edwards, MOHP, CXXXX (1979), 12.
4. Clifton Kirkpatrick, MOHP, 87–89.
5. Rhea Blake, MOHP, 106.
6. Clifton Kirkpatrick, MOHP, 91.
7. Ibid., 30.
8. Albert Russell, *U.S. Cotton and the National Cotton Council, 1938–1987* (Memphis, Tenn.: National Cotton Council, 1987), 56.
9. D. W. Brooks to Harold Young, July 12, 1955, NCC-EVP files, box "Jan 1–Dec 31, 1955," folder "Industry Wide Committee on Future of NCC."
10. Banks loaned to Baker's Producers Oil Company, and he made loans to growers. There was no secret about the flow of capital, but bankers wanted Baker's firm to carry the liability.
11. Timothy Curtis Jacobson and George David Smith, *Cotton's Renaissance: A Study in Market Innovation* (Cambridge, U.K.: Cambridge University Press, 2001), 124. I drew on this account for information about Cotton Incorporated.

12. Clifton Kirkpatrick, MOHP, 29.
13. Ibid., 39.
14. Quote appears in Russell, *U.S. Cotton*, 66; Jacobson and Smith, *Cotton's Renaissance*, 124.
15. Russell Griffin, MOHP, CXXIV (1979), 4.
16. Ibid., 3.
17. Ibid., 5.
18. Jacobson and Smith, *Cotton's Renaissance*, 129.
19. Ibid., 130.
20. Nelson, "Cotton Incorporated Problems," 3.
21. Rhea Blake, MOHP, 106.
22. Russell Griffin, MOHP, 5.
23. Ibid., 7.
24. *Commercial Appeal*, March 23, 1966, 31; Rhea Blake, MOHP, 108.
25. Russell, *U.S. Cotton*, 80–81.
26. Dabney Wellford, "Cotton Programs of the Federal Government, 1929–1967," July 1968, typescript, 25, Dabney Wellford Files, NCC Archives.
27. Arizona Cotton Growers Association, "Resolutions," February 20, 1968, box 14, folder 1, Arizona Cotton Growers Association, Arizona Historical Foundation, Arizona State University.
28. Jacobson and Smith, *Cotton's Renaissance*, 140–41.
29. Russell, *U.S. Cotton*, 87.
30. McDonald K. Horne, "The Economic Outlook for Cotton," A report before the Council's 31st annual meeting, January 27, 1969, typescript, 36, McDonald K. Horne Private Papers.
31. *Time*, June 30, 1967, 19; Paul Findley, "Let's Play the Billion-Dollar Farm Drain," *Reader's Digest*, March 1968, 72–76.
32. Macon Edwards, MOHP, 33.
33. Russell Griffin, MOHP, 10.
34. Ibid., 11–12.
35. Ibid., 13.
36. Edward Kirkpatrick, MOHP, 54.
37. Jacobson and Smith, *Cotton's Renaissance*, 155–56.
38. Ibid., 144.
39. Ibid.
40. Morgan Nelson, "Cotton the King," typescript, 6, Morgan Nelson Private Papers.
41. Jacobson and Smith, *Cotton's Renaissance*, 160.
42. Ibid., 163.
43. Stephen Yafa, *Big Cotton: How a Humble Fiber Created Fortunes, Wrecked Civilizations, and Put America on the Map* (New York: Viking, 2005), 230.
44. James Sullivan, *Jeans: A Cultural History of an American Icon* (New York: Gotham Books, 2006), 80–109.
45. Sullivan, *Jeans*, 7–8.
46. Jacobson and Smith, *Cotton's Renaissance*, 186.
47. Ibid., 190.
48. http://deltafarmpress.com/mag.farming_retiring_fabric_lives/index.html.
49. *Commercial Appeal*, April 1, 1979, B7.
50. *Press Scimitar*, May 11, 1979, 13.
51. Resolution by the Executive Committee of Cotton Incorporated, "Resolution," December 5, 1975, box 33, folder 3, Morgan Nelson Papers.

52. Written statement to Office Inspector General Special Agent, October 4, 1978, box 38, folder 3, Nelson Papers, New Mexico State University.
53. Morgan Nelson, interview by author, Roswell, N.Mex., December 16, 2006.
54. *Commercial Appeal,* June 13, 1978, 18.
55. *Commercial Appeal,* August 23, 1979, 1, 11.
56. Ibid., 11.
57. Jacobson and Smith, *Cotton's Renaissance,* 165.
58. Chauncey L. Denton Jr., MOHP, CXXIII (1979), 25.
59. Quote in Hadley C. Ford and Associates Inc., "Evaluation of Cotton Incorporated" (New York, 1976), vii; Hadley Ford, MOHP, CXLI (1979), 14.
60. Quotes in Hadley C. Ford, "Evaluation of Cotton Incorporated," vii, 4.
61. Ibid., 23.
62. Bill Pearson, telephone interview by author, Sumner, Miss., August 24, 2006.
63. Jacobson and Smith, *Cotton's Renaissance,* 291; *Delta Farm Press,* May 17, 2006, http://www.deltafarmpress.com.
64. Jacobson and Smith, *Cotton's Renaissance,* 303.
65. Mark D. Lange, "The Economic Outlook for Cotton," *Beltwide News,* January 5, 2000, 2–3, http://www.cottonfarming.com/home/archive/1999_bltwtide6 (accessed July 22, 2009).

12. THE TEXAS PLAINS

1. Lawrence L. Graves, *Lubbock: From Town to City* (Lubbock: West Texas Museum Association, 1986), 6.
2. Janet M. Neugebauer, *Plains Farmer: The Diary of William G. DeLoach* (College Station: Texas A&M University Press, 1991), 286.
3. Dick J. Reavis, "Way Out West in the Land of Cotton," *Texas Monthly,* February 1982, 109.
4. http://www.plainscotton.org/esw/stats/1928-Present.html (accessed August 30, 2006).
5. Guy Nickels, May 3, 1973, Oral History Collection, Southwestern Collection, Texas Tech University.
6. Jack Lichtenstein, *Field to Fabric: The Story of American Cotton Growers* (Lubbock: Texas Tech University Press, 1990), 21.
7. Ibid., 37; William N. Stokes Jr., *Oil Mill on the Texas Plains: A Study in Agricultural Cooperation* (College Station: Texas A&M Press, 1979), 18; Bently Baize, April 18, 1975, Oral History Collection, Southwestern Collection.
8. Rex Dunn, October 18, 1975, Oral History Collection, Southwestern Collection.
9. Roger Haldenby, interview by author, Lubbock, Tex., December 12, 2006.
10. Graves points out that installation of gas lines put farmers within reasonable reach of natural gas. See Graves, *Lubbock,* 11.
11. John Opie, *Ogallala: Water for a Dry Land,* 2nd ed. (Lincoln: University of Nebraska Press, 2000), 136.
12. Harry S. Walker, "The Economic Development of Lubbock," in *History of Lubbock,* ed. Lawrence L. Graves (Lubbock: West Texas Museum Association, 1962), 300–330.
13. Donald E. Green, *Land of Underground Rain: Irrigation on the Texas High Plains, 1910–1970* (Austin: University of Texas Press, 1973), 163.
14. E. L. Thaxton, June 2, 1972, Oral History Collection, Southwestern Collection.

15. Roy B. Davis, July 5, 1972, Oral History Collection, Southwestern Collection.
16. *ACCO Press*, August 1949, 9.
17. Ibid., 9–10.
18. Ibid., 11.
19. Earl Sears, interview by author, Memphis, Tenn., January 11, 2007.
20. J. Mark Welch, Conrad P. Lyford, and Kenneth Harling, "The Value of the Plains Cotton Cooperative Association," March 23, 2005, 9, http://www.aaec.ttu.edu/CERI.
21. Lichtenstein, *Field to Fabric*, 35.
22. Stokes, *Oil Mill on the Texas Plains*, 50–66.
23. Karen Gerhardt Britton, *Bale o' Cotton: The Mechanical Art of Cotton Ginning* (College Station: Texas A&M University Press, 1992), 119.
24. Welch, Lyford, and Harling, "Value of the Plains Cotton Cooperative Association," 8–9.
25. Emerson Tucker, interview by author, Lubbock, Tex., December 14, 2006.
26. Emerson Tucker, July 17, 1984, Oral History Collection, Southwestern Collection.
27. Tucker, interview by author.
28. Emerson Tucker, July 17, 1984, Oral History Collection, Southwestern Collection.
29. Katie Dickie Stavinola and Lorie A. Woodward, "Texas Boll Weevil History," in *Boll Weevil Eradication in the United States through 1999*, ed. Willard Dickerson et al. (Memphis, Tenn.: Cotton Foundation, 2001), 465.
30. Ibid., 466.
31. Don Anderson, interview by author, Lubbock, Tex., April 4, 1998.
32. Haldenby interview.
33. Stavinola and Woodward, "Texas Boll Weevil History," 466.
34. Haldenby interview.
35. Don Ethridge, interview by author, Lubbock, Tex., December 11, 2006.
36. Neuman Smith, September 29, 1982, Oral History Collection, Southwestern Collection.
37. Pauline R. Kibbe, *Latin Americans in Texas* (Albuquerque: University of New Mexico Press, 1946), 176.
38. *El Paso Times*, October 29, 1959, clipping in El Paso Valley Cotton Association Records, 1954–66, Official Files, box 1, Southwestern Collection.
39. El Paso Valley Cotton Association to Lyndon B. Johnson, April 25, 1958; *Washington Post*, October 25, 1959; *El Paso Times*, June 30, 1960; Robert C. Goodwin to Lyndon B. Johnson, December 15, 1960; all in El Paso Valley Cotton Records, box 1, folders "Alleged Tractor Violations" and "Foreign Labor."
40. Blake Gumprecht, "Lubbock on Everything: The Evocation of Place in the Music of West Texas," *Journal of Cultural Geography* 18 (1998): 255–75.
41. Ibid., 268.
42. Ibid., 258.
43. Ibid., 261–62.
44. Ibid., 267.
45. Ibid., 262.
46. http://www.songwritershalloffame.org (accessed August 25, 2009).
47. Joe Carr and Alan Munde, *Prairie Nights to Neon Lights: The Story of Country Music in West Texas* (Lubbock: Texas Tech University Press, 1995), 46.
48. Quotes in Nicholas Dawidoff, *In the Country of Country: People and Places in American Music* (New York: Pantheon Books, 1997), 300, 307.

13. THE QUESTION OF SUBSIDIES

1. A. Whitney Griswold, *Farming and Democracy* (New Haven, Conn.: Yale University Press, 1948), vi.

2. Thomas Inge, ed., *Agrarianism in American Literature* (New York: Odyssey Press, 1969), xiii.

3. Twelve Southerners, *I'll Take My Stand: The South and the Agrarian Tradition* (New York: Harper Torchbooks, 1962), xxix.

4. Don Paarlberg, *Food and Farm Policy: Issues of the 1980s* (Lincoln: University of Nebraska Press, 1980), 5.

5. Ibid., 17.

6. Ibid., 25.

7. Milton Brown to Brook Hayes, June 5, 1958, box 3093, folder "Cotton 1," record group 16, National Archives, College Park, Md.

8. J. W. Phillips, September 17, 1957, box 2907, folder "Cotton 1, Acreage Allotments," record group 16, National Archives, College Park, Md.

9. *ACCO Press*, August 1952, 2–4, box 36, ACCO Archives.

10. *New York Times*, July 29, 1956.

11. USDA Economic Research Service, *Statistics on Cotton and Related Data, 1930–67*, 5.

12. Ibid., 16.

13. *New York Times*, March 12, 1956.

14. *Life*, May 20, 1957, 46.

15. Correspondence and editorials in D&PL Records, "DPL-IX-History: 1956–1957," box 26, Special Collections, Mississippi State University.

16. *Nation*, October 27, 1969, 429; quote in *Nation*, June 16, 1969, 749.

17. *Farm Journal*, July 1969, 15.

18. Ibid.

19. Arizona Extension Service, *Arizona Farm News*, July 15, 1970; James R. Carter to Congressman Sam Rhodes, Sam Steiger, Morris Udall, December 7, 1971; for NCC opposition, see C. L. Denton to Producers Steering Committee, June 28, 1971; all in Arizona Cotton Growers Association Archives, box 84, folder 1, Arizona Historical Foundation.

20. W. C. Hamilton, "Payment Limitations: Pros and Cons as Seen by Farm Organizations," *American Journal of Agricultural Economics*, 51, no. 5 (1969): 1243–46.

21. National Economics Division, *The Cotton Industry in the United States: Farm to Consumer*, 14.

22. Ibid.; Harold G. Halcrow, *Food Policy for America* (New York: McGraw-Hill, 1977), 299.

23. "Borrowing and Credit, 1970–1979," http://www.access.gpo.gov/congress/senate/sen_agriculture/ch7.html (accessed May 13, 2010).

24. Quote appears in Paul Howe Harvey Jr., "The Farming Crisis in Texas from 1965 to 1996" (Ph.D. diss., Texas Christian University, 1997), 62.

25. Ibid.

26. Gaylon Booker, interview by author, Memphis, Tenn., June 6, 1998.

27. This calculation made from data by the USDA Foreign Agricultural Service, available at the Cotton Council Web site, http://www.cotton.org/econ/cropinfo/cropdata/country-statistics.cfm (accessed April 4, 2007).

28. *Cotton Farming*, January 1997, 55.

29. *New York Times*, May 14, 2001.

30. http://www.ers.usda.gov/publications/aib778 (accessed April 5, 2007).
31. Lamar Fleming Jr., "Federal Programs and Their Effects," *ACCO Press*, September 1957, 4, box 7, folder 8, ACCO Archives.
32. Ibid., 6.
33. Ibid., 8.
34. http://www.ewg.org/reports/doubledippers/execsumm.php (April 6, 2007).
35. *Atlanta Journal-Constitution*, October 1, 2006.
36. Ibid.
37. *New York Times*, August 5, 2003, A18.
38. Ibid.
39. "Foreign Crop Subsidies and Tariffs," April 2007, http://www.aaec.ttu.edu/CERI.
40. CRS, "Report for Congress," June 24, 2004, 2, http://www.nationalaglawcenter.org/assets/crs/RL32442.pdf (accessed April 9, 2007).
41. James R. Carter to Congressman Sam Steiger, June 11, 1971, box 84, folder 1, Arizona Cotton Growers Association, Arizona Historical Foundation.
42. CRS, "Report for Congress," 7.
43. Ibid., 17.
44. Chapter 1, "Overview of Payments and Payment Limitations," http://www.ers.usda.gov (accessed April 11, 2007).
45. Ibid., 19–20.
46. UNCTAD, http://ro.unctad.org/infocomm/anglais/cotton/market.htm (accessed April 11, 2007).
47. Chip Morgan, interview by author, Stoneville, Miss., January 4, 2006.
48. Jack Stone, interview by author, Stratford, Calif., April 13, 2005.
49. Earl Sears, interview by author, Memphis, Tenn., June 8, 1998.
50. *ACCO Press*, December 1948, 1.
51. A. A. Bell, "Diseases of Cotton," In *Cotton: Origin, History, Technology, and Production*, ed. C. Wayne Smith and J. Tom Cothren (New York: John Wiley and Sons, 1999), 566–67.
52. Rebecca Sharpless, "Technology behind a Mule: Breaking the Blacklands and the Cotton Empire," in *The Texas Blackland Prairie: Land, History, and Culture*, ed. Rebecca Sharpless and Joe C. Yelderman Jr. (Waco, Tex.: Baylor University Press, 1993), 155–56.
53. George C. Cortright, MOHP, 23.

14. CROP LIEN TO FUTURES

1. Rupert Vance, *Human Factors in Cotton Culture: A Study in the Social Geography of the American South* (Chapel Hill: University of North Carolina Press, 1929), 178.
2. Frank E. Smith, *The Yazoo River* (New York: Rinehart, 1954), 204.
3. The National Emergency Council, *Report on Economic Conditions of the South* (Washington, D.C.: Government Printing Office, 1938), 49–52.
4. For a discussion of credit in 1920s, see Murray R. Benedict, *Can We Solve the Farm Problem: An Analysis of Federal Aid to Agriculture* (New York: Twentieth Century Fund, 1955), 124–46.
5. Ibid., 161–62.
6. For a full history of the Farm Security Administration, see Sidney Baldwin, *Poverty and Politics: The Rise and Decline of the Farm Security Administration*; for Johnston's opposition, see Lawrence Nelson, *King Cotton's Advocate: Oscar G. Johnston and the New Deal* (Knoxville: University of Tennessee Press, 1999), 206–27.
7. Benedict, *Can We Solve*, 173.

8. *ACCO Press*, November 26, 1948, 26, ACCO Archives.

9. Ibid., 27.

10. "World Wide Operations of Anderson-Clayton," pamphlet, n.d., NCC Archives, folder "Anderson-Clayton I"; N. T. Ness, typescript, untitled, box 8, folder 18, ACCO Archives; *Abilene Reporter News*, October 30, 1955, 11D, clipping in box 19, folder 8, ACCO Archives.

11. "The Main Business of ACCO," pamphlet, folder "Anderson-Clayton I," NCC Archives.

12. Quotes in *ACCO Press*, August 1952, 3, ACCO Archives.

13. Ibid., 1–3.

14. "World Wide Operations of Anderson-Clayton," folder "Anderson-Clayton I," NCC Archives.

15. Typescript, December 27, 1978, 2, S. M. McAshan file, box 9, folder 3, ACCO Archives.

16. Ibid.

17. Booz, Allen and Hamilton Survey, May 1961, box 33, folder "Management Organization Survey," ACCO Archives.

18. Quotes in "Highlights of Marplan Survey, Marplan Study, Summary and Analysis of an Opinion Study," August 24, 1961, VI-3, V-2, V-P 11, box 33, ACCO Archives.

19. Ibid., V-P 11.

20. Quotes in "Closing of New Orleans Cotton Exchange," handwritten notes, n.d., box 1, folder 15, ACCO Archives.

21. S. M. McAshan file, typescript, box 9, folder 3, ACCO Archives.

22. Will Clayton to Hon. Sam Rayburn, March 12, 1959, Alphabetical File, 1945–60, folder "Agriculture—Cotton Cooperatives," Will Clayton Papers, Harry S. Truman Library.

23. S. M. McAshan file, December 27, 1978, box 9, folder 3, ACCO files.

24. Harry Baker, MOHP, CXI (1979), 15.

25. Jack Stone, interview by author, Stratford, Calif., April 13, 2005.

26. Quote in Ellen Liebman, *California Farmland: A History of Large Agricultural Landholdings* (Totowa, N.J.: Rowman and Allanheld, 1983), 90; Read Dunn, interview by author, Washington, D.C., October 14, 1999.

27. "Farm Credit Conditions," Research Division, Farm Credit District of Wichita, Kansas, 21, no. 6 (June 1956), box 29, folder 7, Morgan Nelson Papers.

28. Smith, *The Yazoo River*, 331.

29. Catherin M. Merlo, *Legacy of a Shared Vision: The History of Calcot, Ltd.* (Bakersfield, Calif.: Calcot, 1995), 47.

30. Ibid., 80; Gifford W. Hoag, *The Farm Credit System: A Short History of Financial Self Help* (Danville, Ill.: Interstate Printers and Publishers, 1976).

31. http://www.staplcotn.com (accessed February 26, 2007).

32. For discussion of commercial banks, see William G. Murray and Aaron G. Nelson, *Agricultural Finance* (Ames: Iowa State University Press, 1960), 276–94.

33. Bruce L. Gardner, *American Agriculture in the Twentieth Century: How It Flourished and What It Cost* (Cambridge, Mass.: Harvard University Press, 2002); *Cotton and Wool Situation and Outlook Yearbook* (November 2006), appendix table 11.

34. *New York Times*, July 23, 1986.

35. "History of the American Agriculture Movement," http://www.aaminc.org/history.htm (accessed March 1, 2007).

36. Gardner, *American Agriculture*, 197–99.

37. *Cotton Farming*, January 1997, 16.

38. *Hedging Cotton with Futures*, Cooperative Extension Service Publication 1140 (1978), Mississippi State University Publication.

39. *Southeast Farm Press*, June 2001, http://www.southeastfarmpress.com/mag/farming_new_marketing_program.
40. Phil Burnett, interview by author, Memphis, Tenn., June 5, 1998.
41. *Washington Post*, January 6, 1999.
42. *New York Times*, July 11, 2002, A16.
43. http://www.ewg.org/reports/blackfarmers/fulltable.php?rankby=STAT&numtodo=ALL (accessed May 4, 2007).
44. http://www.ratical.org/corporations/linkscopy/report.html (accessed May 4, 2007).

15. THE ROLE OF TEXTILES

1. Robert Blake, MOHP, 62.
2. Robert Jackson, MOHP, 84–85; Robert Jackson, interview by author, Clemson, S.C., August 15, 2005.
3. Jackson, interview by author; Jackson, MOHP, 179–82.
4. Quotes in Jackson, MOHP, 187, 188–90.
5. Ibid., 188.
6. Ibid., 191–92.
7. Ibid., 195.
8. Ibid., 193–94.
9. Ibid., 190–98.
10. USDA Economics Research Service, *Statistics on Cotton and Related Data, 1930–67*, 46–47.
11. Robert Jackson, MOHP, 47–52.
12. Ibid., 54–55.
13. Ibid., 60.
14. Ibid., 65–68.
15. Ibid., 105.
16. Ibid., 148.
17. Lewis T. Barringer, MOHP, 7.
18. Albert Russell, *U.S. Cotton and the National Cotton Council, 1938–1987* (Memphis, Tenn.: National Cotton Council, 1987), 70.
19. Ibid., 74; Barringer, MOHP, 13.
20. Russell, *U.S. Cotton*, 162.
21. Kitty G. Dickerson, *Textiles and Apparel in the Global Economy*, 3rd ed. (Upper Saddle River, N.J.: Merrill, 1999), 378.
22. Pietra Rivoli, *The Travels of a T-Shirt in the Global Economy* (Hoboken, N.J.: John Wiley and Sons, 2005), 136.
23. Robert Jackson interview; http://www.ncto.org.
24. Timothy Curtis Jacobson and George David Smith, *Cotton's Renaissance: A Study in Market Innovation* (Cambridge, U.K.: Cambridge University Press, 2001), 224–32.
25. Harvin Smith and Reijao Zhu, "The Spinning Process," in *Cotton: Origin, History, Technology, and Production*, ed. C. Wayne Smith and J. Tom Cothren (New York: John Wiley and Sons, 1999), 740–41.
26. Don Anderson, interview by author, Lubbock, Tex., April 4, 1998; Jack Lichtenstein, *Field to Fabric: The Story of American Cotton Growers* (Lubbock: Texas Tech University Press, 1990), 67.
27. Michael Stevens, interview by author, Lubbock, Tex., December 13, 2006.
28. Roy B. Davis, July 5, 1972, Oral History Collection, Southwestern Collection.

29. Stevens interview; Jack Towery, October 2, 1972, Oral History Collection, Southwestern Collection. From 1987 to 1993, the ITC was called the International Center for Textile Research and Development.

30. Lichtenstein, *Field to Fabric*, 67.

31. Ibid., 48–49.

32. Ibid., 73.

33. For a history of the American Cotton Growers, see Lichtenstein, *Field to Fabric*; Emerson Tucker, interview by author, Lubbock, Tex., December 14, 2006.

34. Don Ethridge, interview by author, Lubbock, Tex., December 11, 2006.

35. *Cotton Farming*, July 1996, 6.

36. John M. Rothgeb Jr., *U.S. Trade Policy: Balancing Economic Dreams and Political Realities* (Washington, D.C.: CQ Press, 2001), 223–24.

37. Rivoli, *Travels of a T-Shirt*, 122.

38. For an analysis of the dispute, see Gaylon Booker, *U.S. Cotton and the National Cotton Council, 1988–2007* (Memphis, Tenn.: National Cotton Council, 2007).

39. Quotes in http://www.cotton.org/news/releases/2003 (accessed February 14, 2007).

40. Rivoli, *Travels of a T-Shirt*, 162.

41. Ibid., 140–41.

42. Ibid., 214.

43. Ibid., 211–15.

44. http://www.ncto.org.

45. *New York Times*, April 23, 2005, B2.

46. *Nation*, October 5, 2005.

47. *New York Times*, December 27, 2001, A19.

48. *Dallas Morning News*, December 2, 2004, 8D.

49. PCCA Release, January 30, 2004, files of Emerson Tucker, Plains Cotton Cooperative Association, Lubbock, Tex.

50. *Dallas Morning News*, March 2, 2005, 8D.

51. Robert Jackson, interview by author, Clemson, S.C., September 27, 2002.

16. RESEARCH

1. For quote see *Proceedings of the Committee on Organization*, November 21, 1938, NCC-EVP files.

2. Early C. Ewing, "Early History of Delta & Pine Land Co., 1911–1967," typescript, 20–21, box 27, Special Collections, Mississippi State University.

3. C. Wayne Smith et al., "History of Cotton Cultivar Development in the United States," in *Cotton: Origin, History, Technology, and Production*, ed. C. Wayne Smith and J. Tom Cothren (New York: John Wiley and Sons, 1999), 146.

4. Ibid., 149; Charles Glover, interview by author, New Mexico State University, April 7, 2005.

5. Smith et al., "History of Cotton Cultivar," 153.

6. Ibid., 161.

7. David Brown, *Inventing Modern America: From the Microwave to the Mouse* (Boston: MIT Press, 2001), 28–32; *Cotton Farming*, March 1995, 6–7, 10.

8. http://www.ars.usda.gov (accessed June 19, 2009).

9. *Cotton Farming*, July 2000, http://www.cottonfarming.com (accessed September 26, 2006).

10. http://www.southwestfarmpress.com (accessed October 5, 2006).

11. W. C. McArthur, ed., Statistics and Cooperative Service, USDA Department of Agricultural Economics, *The Cotton Industry in the United States: Farm to Consumer* (April 1980), Texas Tech University, College of Agricultural Sciences Publication T-1-186, 14.

12. http://agnews.tamu.edu (accessed September 19, 2006).

13. Timothy Jacobson and George Smith, *Cotton's Renaissance: A Study in Market Innovation* (Cambridge: Cambridge University Press, 2001), 250.

14. Cameron Saffel, "From Wagon to Module: New Ways of Handling Harvested Cotton," *West Texas Historical Association* 73 (1997): 46–71.

15. H. H. Ramsey Jr., "Classing Cotton," In *Cotton: Origin, History, Technology, and Production*, ed. C. Wayne Smith and J. Tom Cothren (New York: John Wiley and Sons, 1999), 710–11.

16. C. F. Lewis and T. R. Richmond, "Cotton as a Crop," in *Advances in Production and Utilization of Quality Cotton: Principles and Practices* (Ames: Iowa State University Press, 1968), 15–16.

17. Lee S. Stith, "Dr. E. H. Pressley: His Contribution to Cotton Fiber Technology," *Agricultural History* 54 (January 1980): 185–89.

18. Ibid., 209; Jack Lichtenstein, *Field to Fabric: The Story of American Cotton Growers* (Lubbock: Texas Tech University Press, 1990), 41; *Cotton Farming*, February 1991, 41.

19. Calvin Turley, interview by author, Memphis, Tenn., March 14, 2003.

20. Lynette B. Wren, *Cinderella of the New South: A History of the Cottonseed Industry, 1855–1955* (Knoxville: University of Tennessee Press, 1995).

21. Box 8, folder 3, 18, ACCO Archives, Houston Public Library.

22. Rhea Blake, "Statement to Industry Wide Committee on Future of the National Cotton Council," July 18, 1955, typescript, 3, NCC-EVP files, box "Jan 1–Dec 31, 1955," folder "Industry Wide Committee, 'What Is the Cotton Foundation?'"

23. *Cotton Farming*, September 2006, 30.

24. Albert Russell, *U.S. Cotton and the National Cotton Council, 1938–1987*, 176.

25. http://www.planetark.com (accessed September 20, 2006).

17. CHALLENGES ANEW

1. Wm. Keith C. Morgan and Anthony Seaton, *Occupational Lung Diseases* (Philadelphia: W. B. Saunders, 1975), 277.

2. Charles Levenstein, George F. DeLaurier, and Mary Lee Dunn, *The Cotton Dust Papers: Science, Politics, and Power in the Discovery of Byssinosis in the U.S.* (Amityville, N.Y.: Baywood, 2002), 67.

3. Ibid.

4. Oscar Johnston to Ward Delaney, April 14, 18; May 3, 1945, NCC-EVP files, box "Jan 1–Dec 31, 1945," folder D.

5. Levenstein, DeLaurier, and Dunn, *Cotton Dust Papers*, 61.

6. *Nation*, March 15, 1971, 335.

7. Bennett M. Judkins, *We Offer Ourselves as Evidence: Toward Workers' Control of Occupational Health* (New York: Greenwood Press, 1986), 113.

8. J. C. Self to Duke Wooters, December 18, 1975, box 33, folder 3, Morgan Nelson Papers, Rio Grande Collection.

9. *Textile News*, Special Edition, August 11, 1980, 11.

10. *Commercial Appeal*, March 8, 1977, 8.

11. *Lubbock Avalanche Journal*, April 15, 1977, 12A.

12. *Lubbock Avalanche Journal*, May 8, 1977, 6A.

13. Ibid.
14. *Press Scimitar*, May 6, 1977, 13.
15. *New York Times*, May 15, 1977, 21.
16. *Press Scimitar*, June 12, 1978, 5.
17. Gaylon Booker to Ed Glade, March 31, 1981, NCC-EVP files, box "Jan 1–Dec 31, 1981," folder "Cotton Dust."
18. *Press Scimitar*, June 12, 1978, 4; and June 15, 1978, 5.
19. *Commercial Appeal*, June 15, 1978, 5.
20. Ibid.
21. *Business Week*, June 26, 1978, 135.
22. Ibid.
23. *Commercial Appeal*, June 20, 1978, 20.
24. *Press Scimitar*, August 26, 1978, 2.
25. *Press Scimitar*, September 11, 1978, 7.
26. *Commercial Appeal*, September 20, 1978, 42.
27. *Press Scimitar*, June 20, 1978, 6.
28. *A Conversation with Secretary Ray Marshall: Inflation, Unemployment, and the Minimum Wage* (Washington, D.C.: American Enterprise Institute for Public Policy Research, 1978), 16.
29. Ibid., 14.
30. Ibid., 17.
31. "Exposure to Cotton Dust and Respiratory Disease," *Journal of the American Medical Association* 244 (October 17, 1980): 1797–98.
32. Judkins, *We Offer Ourselves*.
33. Frank Mitchener, interview by author, Memphis, Tenn., October 20, 2006; "Memorandum: Cotton Dust," NCC-EVP files, box "Jan 1–Dec 31, 1981," folder "Cotton Dust."
34. Ibid.
35. Earl Sears to J. S. Francis, Arizona Cotton Ginners Association, March 18, 1981, NCC-EVP files, box "Jan 1–Dec 31, 1981," folder "Cotton Dust."
36. "Court Upholds OSHA," http://biotech.law.lsu.edu/cases/adlaw/cotton.htm (accessed October 25, 2006).
37. *New York Times*, October 8, 1981.
38. Committee on Byssinosis, Division of Medical Sciences, Assembly of Life Sciences, National Research Council, *Byssinosis: Clinical and Research Issues* (Washington, D.C.: National Academy Press, 1982), iv, 135.
39. *New York Times*, July 11, 1984.
40. Albert Russell, *U.S. Cotton and the National Cotton Council*, 150.
41. *Daily News Record*, August 30, 1984.
42. Phillip J. Wakelyn, James R. Supak, Frank Carter, and Bruce A. Roberts, "Public and Environmental Issues," in *Cotton Harvest Management: Use and Influence of Harvest Aids*, ed. James R. Supak and Charles E. Snipes (Memphis, Tenn.: Cotton Foundation, 2001), 278–81.
43. *Cotton Farming*, March 1997, 32.
44. http://agriculture.senate.gov/Hearings/Hearings_1997/wakelyn.htm (accessed November 10, 2006).
45. http://yosemite.epa.gov (accessed November 10, 2006).
46. http://www.sfgate.com (accessed November 8, 2006).
47. http://www.prweb.com (accessed November 8, 2006).
48. Stephen H. Crawford, J. Tom Cothren, Donna E. Sohan, and James R. Supak, "A History of Cotton Harvest Aids," in *Cotton Harvest Management: Use and Influence of Harvest Aids*, ed. James R. Supak and Charles E. Snipes (Memphis, Tenn.: Cotton Foundation, 2001), 15.

49. Albert R. Russell to Fred W. Clayton, July 16, 1976, box 31, folder 1, Morgan Nelson Papers.

50. Albert R. Russell to Robert W. Long, Assistant Secretary of Agriculture, July 15, 1976, box 31, folder 1, Morgan Nelson Papers.

51. *Cotton Farming*, June 2002, Archives.

52. http://deltafarmpress.com (accessed November 10, 2006).

53. *Cotton Farming*, July 2006, Archives; http://www.cottonfarming.com (accessed October 31, 2006).

54. John Opie, *Ogallala: Water for a Dry Land* (Lincoln: University of Nebraska Press, 1993), 188–89.

55. Donald Green, *Land of the Underground Rain: Irrigation on the Texas High Plains, 1910–1970* (Austin: University of Texas Press, 1973), 185.

56. Ibid., 189.

57. Ibid., 229.

58. Opie, *Ogallala*, 142, 195.

59. Robert Gottlieb, *A Life of Its Own: The Politics and Power of Water* (New York: Harcourt Brace and Jovanovich, 1988), 86.

60. *New York Times*, May 22, 1985.

61. *New York Times*, May 12, 1986.

62. Terry L. Anderson and Pamela S. Snyder, *Water Markets: Priming the Invisible Pump* (Washington, D.C.: Cato Institute, 1997); F. Andrew Schoolmaster, "Water Marketing and Water Rights Transfers in the Lower Rio Grande Valley, Texas," *Professional Geographer* 43 (August 1991): 292–304.

63. *New York Times*, May 12, 1986.

64. Gottlieb, *Life of Its Own*, 94.

65. *New York Times*, February 6, 1991.

66. Opie, *Ogallala*, 261.

67. http://www.water-ed.org/cabriefing (accessed November 3, 2006).

68. *Cotton Farming*, July 2006, Archives.

69. Opie, *Ogallala*, 326.

70. *Cotton Farming*, March 2005, Archives.

71. Ibid.

72. *Cotton Farming*, June 2002, Archives.

73. David Carle, *Drowning the Dream: California's Water Choices at the Millennium* (Westport, Conn.: Praeger, 2000), 173–83; "Accelerated Loss of Farmland to Development," http://clinton4.nara.gov (accessed November 16, 2006).

74. *Cotton Farming*, March 2005, Archives.

75. http://www.texasep.org/html/1nd/1nd_2agr_sprawl.thml (accessed November 10, 2006).

76. http://www.fb.org (accessed January 10, 2006).

18. THE GLOBALIZATION OF COTTON

1. Aven Beckert, "Emancipation and Empire: Reconstructing the Worldwide Web of Cotton Production in the Age of the American Civil War," *American Historical Review* 109 (December 2004): 1408, 1438.

2. Read Dunn, *Remembering* (Memphis, Tenn.: National Cotton Council, 1992), 54.

3. "Report: The Cotton Industry of China," September 1, 1947, NCC-EVP files, box "Jan 1–Dec 31, 1947," folder "Foreign Trade," 30.

4. Ibid., 29, 33.

5. Read Dunn, interview by author, Washington, D.C., October 14, 1999.

6. "Summary: The European Cotton Situation," June 1949, box "Jan 1–Dec 31, 1949," folder "Foreign Trade."

7. Dunn, *Remembering*, 54.

8. Murray R. Benedict, *Can We Solve the Farm Problem: An Analysis of Federal Aid to Agriculture* (New York: Twentieth Century Fund, 1955), 455.

9. Read Dunn Jr., "After the Marshall Plan—What?" (December 29, 1949), box "Jan 1–Dec 31, 1949," folder "Foreign Trade Committee," first quote 6, second quote 9.

10. Ibid., 17–18; Congress, Senate, Senator Kefauver of Tennessee speaking for the Atlantic Union, Concurrent Resolution 57, 81st Cong., 2nd sess., *Congressional Record* (March 13, 1950), vol. 96, pt. 3, 3204–15.

11. Read Dunn to W. D. Lawson, August 20, 1952, NCC-EVP files, box "Jan 1–Dec 31, 1952," folder "Foreign Trade Committee."

12. Robert C. Jackson, interview by author, Clemson, S.C., September 27, 2002.

13. Howard Stovall, "The Cotton Producers Interest in International Trade," August 18, 1952, 5, NCC-EVP files, box "Jan 1–Dec 31, 1952," folder "Foreign Trade Committee."

14. Ibid., 3.

15. Ibid.

16. Dunn, *Remembering*, 55.

17. Ibid., 57.

18. Ibid., 60.

19. Ibid., 94.

20. Ibid., 77.

21. Ibid., 81–83.

22. Ibid., 99.

23. D. Clayton Brown, "The International Institute for Cotton: The Globalization of Cotton since 1945," *Agricultural History* 74 (Spring 2000): 267.

24. Read Dunn Jr. to author, April 4, 1999.

25. Brown, "International Institute for Cotton," 268.

26. Ibid., 269.

27. Dunn, *Remembering*, 82.

28. H. Solomon and H. W. Kinnucan, "Effects of Non-price Export Promotion: Some Evidence for Cotton," *Australian Agricultural Economics Society* 37 (April 1993): 1–15.

29. Read Dunn to author, April 4, 1999.

30. "Report, Cotton Strategic Framework" (March 1999), 2, Cotton Council International Office, National Cotton Council.

31. Ibid., 12.

32. USDA Economic Research Service, *Statistics on Cotton and Related Data, 1930–67*, 63; http://www.cotton.org/econ/cropinfo/cropdata/summary.cfm (accessed September 18, 2007).

33. Read Dunn Jr. to author, November 17, 1999.

34. Dunn, *Remembering*, 83.

35. GATT Uruguay Round Highlights, USDA Office of Public Affairs, no. 20 (December 7, 1990), in NCC-EVP files, box 1990, "GATT."

36. NCC-EVP files, box 1990, folder "International Trade."

37. Wayne Labor to Phil Burnett, October 17, 1990, NCC-EVP files, box 1990, folder "International Trade."

38. William Gillon to Committee on Trade, April 28, 1994, NCC-EVP files, box 1994, folder "International Trade."

39. "Testimony of Bruce Brumfield, President, National Cotton Council of America, before the Agriculture Committee of the House of Representatives," April 20, 1994, typescript, NCC-EVP files, box 1994, folder "International Trade."

40. Bill Clinton to Patrick J. Leahy, September 30, 1994, NCC-EVP files, box 1994, folder "International Trade."

41. *Journal of Commerce*, December 2, 1994, clipping in NCC-EVP files, box 1994, folder "International Trade"; *Washington Post*, October 7, 1994, C1.

42. "Cultivating Poverty," Oxfam Briefing Paper, September 2002, http://www.oxfam.org.

43. Congressional Research Service, Report for Congress, July 11, 2005, 2, http://www.nationalaglawcenter.org/assets/crs/RL32571.pdf.

44. "U.S. Subsidies Create Cotton Glut That Hurts Foreign Cotton Farms," *Wall Street Journal*, June 26, 2002, http://online.wsj.com (accessed September 18, 2007).

45. *New York Times*, December 30, 2003.

46. *Cotton Farming*, http://www.cottonfarming.com (accessed April 3, 2006).

47. Oxfam Briefing Paper, "Cultivating Poverty: The Impact of U.S. Cotton Subsidies on Africa," http://www.oxfam.org (accessed February 17, 2006).

48. *Bridges Weekly Trade News Digest*, March 9, 2005, 1; http://www.ictsd.org (accessed April 6, 2006).

49. *Sydney Morning Herald*, http://www.smh.com (accessed April 3, 2006).

50. Ibid.

51. *USA Today*, December 13, 2005, 12A.

52. "U.S. Cotton Economic Situation and Issues Update," prepared by the National Cotton Council of America, March 2005, "Executive Summary," 12, http://www.cotton.org/issues/2005/upload/EconOutlookIssues.pdf (accessed April 3, 2006).

53. Ibid.

54. Congressional Research Service, CRS Web, "U.S. Agricultural Policy Response to WTO Cotton Decision" (January 18, 2006), 4; Gaylon Booker, *U.S. Cotton and the National Cotton Council, 1988–2000*, 127. In 2010 the United States agreed to pay Brazil $147 million to avoid trade sanctions that could be imposed as a result of the WTO ruling on cotton. See *Dallas Morning News*, April 10, 2010, 10.

55. *Cotton Farming*, September 2005, 25.

56. *Economist: Global Agenda*, December 19, 2005, http://www.economist.com (accessed December 19, 2006).

57. *Cotton Farming*, February 2006, 38.

58. Ibid.

59. Ibid., 28.

60. *Cotton Farming*, January 1994, 12.

61. *Cotton Farming*, September 2005, 25.

19. THE NEW COTTON CULTURE

1. FMC, "50 Years of Accomplishments, Part III, The Identity Emerges," pamphlet, n.d., files of EVP, box "Jan 1–Dec 31, 1988."

2. Gilbert C. Fite, *Cotton Fields No More: Southern Agriculture, 1865–1980* (Lexington: University of Kentucky Press, 1984), 231.

3. PCG E-Mail Services, April 6, 2007.

4. Harry Cline, "Short Stay Lasts Forty-five Years: Calcot's Tom Smith Closing Long Career," *Western Farm Press*, September 21, 2002, http://www.westernfarmpress.com/mag/farming_short_stay_lasts (accessed April 19, 2007).

5. National Cotton Council cotton crop database, http://www.cotton.org (accessed April 18, 2007).

6. Production statistics from the NCC at http://www.cotton.org (accessed April 24, 2007).

7. *Cotton Farming*, April 2007, 8–9, 12.

8. Ibid., 16.

9. Congressional Research Service, "Cotton Production and Support in the United States" (June 24, 2004), 14, http://www.nationalaglawcenter.org (accessed April 9, 2007).

10. Rick Lavis to author, telephone interview, April 20, 2007.

11. Keith Deputy to author, typescript, May 1, 2007.

12. Ibid.

13. *Fort Bend (Texas) Herald-Coaster*, September 23, 2001, 1.

14. *Dallas Morning News*, December 17, 2000, H1.

15. *Dallas Morning News*, June 2, 2001, F1.

16. *Fort Worth Star-Telegram*, July 25, 1999, F1.

17. *Progressive Farmer*, July 1999, 18–20.

18. *Fort Worth Star-Telegram*, July 25, 1999, F3.

19. Ibid.

20. Dorothy Scarborough, *In the Land of Cotton* (New York: Macmillan, 1936), 144.

GLOSSARY

AAM	American Agricultural Movement
ACCO	Anderson, Clayton and Company
ACG	American Cotton Growers
ACMI	American Cotton Manufacturers Institute
ARS	USDA Agriculture Research Service
BAH	Booz, Allen and Hamilton
BLA	Brown Lung Association
CALCOT	California Cotton Cooperative Association
CCC	Commodity Credit Corporation
CCI	Cotton Council International
CI	Cotton Incorporated
COAP	Cottonseed Oil Assistance Program
CPI	Cotton Producers Institute
DR-CAFTA	Dominican Republic Caribbean Area Free Trade Agreement
EFS	Engineered Fiber System
ELF	Earth Liberation Front
ERP	European Recovery Program
EU	European Union
EWG	Environmental Working Group
FAIR	Federal Agricultural Improvement Act
FAS	USDA Foreign Agricultural Service
FCA	Farm Credit Administration

FCS	Farm Credit Service
FFACT	Fiber, Fabric, and Apparel Coalition for Trade
FIFRA	Federal Insecticide, Fungicide, and Rodenticide Act
FQPA	Food Quality Protection Act
HVI	High-Volume Instrumentation
IIC	International Institute for Cotton
ITC	International Textile Institute
ITCB	International Textile and Clothing Bureau
ITMF	International Textile Manufacturers Institute
MFA	Multi Fibre Arrangement
NAFTA	North Atlantic Free Trade Agreement
NBFA	National Black Farmers Association
OSHA	Occupational Safety and Health Administration
PCA	Production Credit Association
PCCA	Plains Cotton Cooperative Association
PCG	Plains Cotton Growers
PCOM	Plains Cotton Oil Mill
PL 480	Agricultural Trade Development and Assistance Act
PMA	Production Management Administration
TIIC	Technical Industrial Intelligence Committees
UNACSTD	United Nations Advisory Committee on the Application of Science and Technology to Development
USPHS	U.S. Public Health Service
WTO	World Trade Organization

BIBLIOGRAPHIC ESSAY

There are manuscript and archival collections rich in cotton history. The records of the National Cotton Council are incomparable, but other archival gems are the Anderson-Clayton Papers, the W. R. Poage Papers, the records of the Arizona Cotton Growers Association, the Morgan Nelson Papers, the Delta and Pine Land archives, and the Southwestern Collection. The Harry S. Truman Library houses the papers of Harry S. Truman, Will Clayton, and Charles Brannan. Another collection of Clayton papers is available at Rice University. The largest archival collection is Record Group 16, the official files of the Secretary of Agriculture. Smaller collections include the John Rust Papers and the personal papers of MacDonald Horne, Read Dunn, and Morgan Nelson. The papers of Lewis T. Barringer are held by his son John Barringer.

Numerous studies focus on cotton, but the classics start with Rupert B. Vance, *Human Factors in Cotton Culture: A Study in the Social Geography of the American South*; Howard Odum, *Southern Regions of the United States*; and James Street, *The New Revolution in the Cotton Economy*.

Among the excellent studies dealing with cotton farming and culture are Charles Aiken, *The Cotton Plantation since the Civil War*; Karen G. Britton, *Bale o' Cotton: The Mechanical Art of Cotton Ginning*; Pete Daniel, *Breaking the Land: The Transformation of Cotton, Tobacco, and Rice Culture since 1880*; Gilbert Fite, *Cotton Fields No More: Southern Agriculture, 1865–1980*; Donald Holley, *The Second Great Emancipation: The Mechanical Cotton Picker, Black Migration, and How They Shaped the Modern World*; Jack T. Kirby, *Rural Worlds Lost: The American South, 1920–1960*; Lawrence Nelson, *King Cotton's Advocate: Oscar G. Johnston and the New Deal*; and George Tindall, *The Emergence of the New South, 1913–1945*.

Official histories have information unavailable elsewhere, particularly the two histories of the National Cotton Council, Albert Russell, *U.S. Cotton and* the National Cotton Council, 1938–1987; and Gaylon Booker, *U.S. Cotton and the National Cotton Council, 1988–2007*. Further information can be found in Timothy C. Jacobson and George D. Smith, *Cotton's Renaissance: A Study in Market Innovation*; Jack Lichtenstein, *Field to Fabric: The Story of American Cotton Growers*; and Catherine M. Merlo, *Legacy of a Shared Vision: The History of Calcot, Ltd.*

The two memoirs of Read Dunn Jr., *Mr. Oscar* and *Remembering*, provide an inside look at the origin of the National Cotton Council and the promotion of world trade in cotton.

Special studies on the technology of farming are essential for understanding the cultivation of cotton. They include Willard A. Dickerson, ed., *Boll Weevil Eradication in the United States through 1999*; Raymond Frisbee, El-Zik Kamal, and Ted Wilson, *Integrated Pest Management Systems and Cotton Production*; Chester G. McWhorter and John Abernathy, *Weeds of Cotton: Characterization and Control*; and C. Wayne Smith and Tom Cothren, eds., *Cotton: Origin, History, Technology, and Production*.

Recent and popular histories are Gerald Helferich, *High Cotton: Four Seasons in the Mississippi Delta*; and Stephen Yafa, *Big Cotton: How a Humble Fiber Created Fortunes, Wrecked Civilizations, and Put America on the Map*. A unique and valuable work on the Memphis Cotton Carnival is Perre Magness, *The Party with a Purpose: Seventy-five Years of Carnival in Memphis, 1931–2006*.

Government documents rank among the most valuable resources for studying the formation of policy and farm conditions. No document compares with the special study undertaken by the House Agriculture Committee and the National Cotton Council, U.S. House, Subcommittee of the Committee on Agriculture, *Hearings, Study of the Agricultural and Economic Problems of the Cotton Belt*, 80th Congress, 1st session, July 7–8, 1947; and U.S. House, Subcommittee of the Committee on Agriculture, *Hearings, Study of the Agricultural and Economic Problems of the Cotton Belt*, 80th Congress, 1st session, part 2, October 1947. For the complete testimonies of the special Cotton Conference in 1944, see U.S. House, Subcommittee of the Committee on Agriculture, *Hearings, Cotton*, 78th Congress, 2nd session, December 4–9, 1944.

The slippery topic of cotton culture can be studied from many perspectives. For the conditions of the cotton South before World War II, the classic studies include James Agee and Walker Evans, *Let Us Now Praise Famous Men*; David Cohen, *God Shakes Creation*; Charles S. Johnson, Edwin Embree, and W. W. Alexander, *The Collapse of Cotton Tenancy*; Herman C. Nixon, *Possum Trot: Rural Community, South*; and Arthur Raper, *Preface to Peasantry*. For exploring aspects of the cotton culture after 1945, see Joe Carr and Alan Munde, *Prairie Nights to Neon Lights: The Story of Country Music in West Texas*; James C. Cobb, *The Most Southern Place on Earth: The Mississippi Delta and the Roots of Regional Identity*; William R. Ferris, *Blues from the Delta*, B. B. King, *Blues All around Me: The Autobiography of B. B. King*; Billy Joe Shaver, *Honky Tonk Hero*; Thad Siton and Dan K. Utley, *From Can to Can't*; and James Sullivan, *Jeans: A Cultural History of an Icon*.

Questions of the environmental impact of cotton farming are explored in David Carle, *Drowning the Dream: California's Water Choices at the Millennium*; Rachel Carson, *Silent Spring*; Charles Daniel, *Lords of the Harvest: Biotech, Big Money, and the Future of Food*; Bill Lambrecht, *Dinner at the Gene Café: How Genetic Engineering Is Changing What We Eat, How We Live, and the Global Politics of Food*; and Henry I. Miller and Gregory Conko, *The Frankenfood Myth: How Protests and Politics Threaten the Biotech Revolution*.

To date the literature lacks an updated history of Latinos in the cotton culture, but the best description of their lives as migrants is found in Pauline R. Kibbe, *Latin Americans in Texas*. A valuable work for the pre-1945 era is Neil Foley, *The White Scourge: Mexicans, Blacks, and Poor Whites in the Texas Cotton Culture*. Other works include Leon Metz, *Border: The U.S.-Mexico Line*; and David Montejano, *Anglos and Mexicans in the Making of Texas, 1836–1986*.

Some of the most effective descriptions of the cotton culture come from fiction, and among the most outstanding examples are Erskine Caldwell, *Tobacco Road*; John Faulkner, *Dollar Cotton*; and Dorothy Scarborough, *In the Land of Cotton*.

Cited in the chapter endnotes are numerous books and articles of great importance; and Internet sources furnish information not available in print.

Oral history collections are essential, and the most notable is the Mississippi Oral History Project of the University of Southern Mississippi. An outstanding assortment of interviews is archived in the Southwestern Collection at Texas Tech University. The Columbia University Oral Collection and the Texas A&M University Agricultural Extension Service also have interviews pertaining to cotton.

I conducted personal interviews with the following individuals, who provided information and perspective unavailable in stored collections or printed material.

Adkisson, Perry
Anderson, Carl
Asley, Harrison
Aycock, Barry
Barringer, John
Booker, Gaylon
Brumfield, Bruce
Burnett, Phillip
Carter, Frank
Curlee, Jesse

Day, Kenny
Deputy, Keith
Dixon, Paul
Dulaney, H. G.
Eastland, Woods
Ethridge, Don
Frisbie, Ray
Gibson, Charles
Gillon, William
Glover, Charles
Haldenby, Roger
Hamilton, Allen
Harvey, Jr., Paul
Hickman, William T.
Houston, William (Bill)
Johnson, Fred
Jordon, Andrew
Lanclos, D. Kent
Lavis, Rick
Lyons, Daniel
Mayers, Drayton
Miller, Tom
Mitchener, Frank
Morgan, Chip
Nelson, Cotton
Nelson, Morgan
Pearson, William (Bill)
Pendergrass, J. Stan
Person, Janice
Russom, Dallas
Russom, Hortense
Sears, Earl
Smith, Tom
Stephen, Michael
Stone, Jack
Tucker, Emerson
Turley, Calvin
Walker, Knox
Welford, Dabney

Magazines and newspapers provided effective supplemental information, particularly the *Cotton Trade Journal*. Others include the Memphis *Commercial Appeal*, *Cotton Farming*, the Memphis *Press Scimitar*, and the *New York Times*.

INDEX

Abernathy, Thomas G., 161
Acala cotton, 55, 148, 152–54, 166, 240, 312–13, 315, 356, 375, 381
Acala 1517, 148, 152, 253, 275, 313–14
Acheson, Dean, 47, 60, 69, 98–99, 103–4
Adams, Harvey, 160
Adkisson, Perry, 173, 176–77, 244
Agricenter International, 196
Agricultural Adjustment Act 1938, 155, 157–60, 162; Act of 1956, 296; Section 22, 364–65
Agricultural Adjustment Administration (AAA), 21, 22, 25, 31–32, 35, 53, 56, 111, 122, 142, 272
Agricultural Council of Arkansas, 119, 160
Agricultural Export and Trade Expansion Act, 258
Agricultural Land Stewardship Program, 348
Ahlgren, Frank, 75, 197
Aiken, George, 162
Albin, Bob, 245
Aldrich, Ransom, 131
Alexander, William, 19, 23, 27
Allen, Frederick Lewis, 11–12
Allen, Terry, *Flatland Boogie*, 249
Allenburg, 210
Allis-Chalmers, 129
Amalgamated Clothing and Textile Wokers Union, 330, 331, 333
American Agricultural Movement, 283–84
American Cotton Growers (ACG), 303–4
American Cotton Manufacturers Institute, 291, 293–96, 298, 302, 331, 336, 354

American Cotton Manufacturing Association, 42
American Cotton Shippers Association, 32, 38, 79–80, 85, 97, 196, 227–28, 324, 352
"American Designer" series (NCC), 72
American Dilemma, 27
American Farm Bureau Federation, 32, 35, 38, 43, 76, 109, 121–22, 156, 162–63, 198, 217, 220, 231, 257, 297, 355
American Farmland Trust, 348
American Federation of Labor, 332
American Home Economics Association, 109
American Korean Foundation, 292
American Lung Association, 340
American Manufacturing Trade Action Committee (AMTAC), 306–7
American Peasant, 24, 28, 146, 201, 272, 372
American Society of Agricultural Engineers, 317
American Textile Manufacturers Institute (ATMI), 295
Anderson, Carl, 285
Anderson, Clayton and Company (ACCO), 38–39, 60, 70, 80–81, 90–91, 151, 208–9, 236, 260, 273–79, 289, 312, 321–22, 350, 354, 376; foods, 322
Anderson, Clinton P., 95, 111–12, 119, 121, 162
Anderson, Don, 148, 151, 244
Anderson, Frank, 70, 273
Anderson, Monroe, 273, 275
Anderson, W. D., 79, 139
Anthony, Mark, 79

429

INDEX

Arizona, 55, 149, 151–52, 155, 158, 162–63, 165–67, 178, 219, 256, 274–75, 278–79, 316, 322, 377, 379; University of, 314, 319, 341
Arizona Cotton Growers Associaton, 264, 377
Arizona Cotton Producers Association, 221
Arkansas Delta, 21–22, 40, 43, 173, 191
Asian flu, 260
Atlanta Journal Constitution, 262–63
Atlantic Charter, 73
Atlantic Constituion, 102
Atlantic Union, 353
Aycock, Barry, 184

Baker, Harry, 216, 279
Bank of America, 279
Bankhead, John H., 43, 113
Bankhead Act, 53
Bankhead-Brown Bill, 43
Banks for Cooperatives, 272
Barringer, Lewis T., 42, 68–70, 112, 123, 152, 296–97
Beach Boys, 6, 383
Beale Street, 190, 199–200, 202–3, 208, 211, 296
Beltwide Cotton Mechanization Conference, 131–32
Benedict, Murray, 76
Beneficials (insects), 171, 176
Benson, Ezra Taft, 161, 163, 167
Benzene Case, 338
Bergland, Bob, 288, 329
Bibb Manufacturing, 42, 79, 139
Bible Belt, 12–13
Bilbo, Theodore, 92, 98
Bingham, Eula, 331–33
Black lung, 326, 330
Blackburn, Norris, 36–37
Blacklands (Texas), 9, 108, 137, 149, 163, 241, 249, 266–67, 384
Blair, Raymond E., 159
Blake, Rhea, 65–66, 69, 75, 77, 156, 353; Marshall Plan, 94, 104; margarine, 107, 109, 113, 118–19; origin of CI, 215, 220, 223; origin of NCC, 33–34, 36–39, 43–44; textiles, 290, 296; textile payment plan, 85, 87
Bledsoe, Sam, 42–43, 69, 75, 80, 82, 88, 107–9
Block, John, 178

Blue jeans, 225–26
Blues, 5; cotton music, 18, 199–200, 211, 248
Boll weevil, 49, 52, 55, 62, 105, 169–83, 188, 214–15, 217, 313, 315, 373; "Boll Weevil Blues," 6; cotton damage, 10, 14–15, 19, 21, 29, 46, 378; poetry, 172; western states, 147–48, 151, 163
Boll Weevil Eradication Program (BWEP), 178, 187, 244–45
Boll worms, 63, 108, 163, 174, 176, 182, 275
Booker, Gaylon, 298, 342
Booze, Allen and Hamilton (BAH), 221, 225, 276, 303
Bork, Robert, 336
Boswell, J. G., 141, 222, 227
Bosworth, Joe, 328
Bouhuys, Arend, 328
Braceros, 161, 165–66, 247
Brando, Marlon, 225
Brannan, Charles F., 119–23
Brannan Plan, 119–22, 124
Brazil, 51, 86, 263, 264, 274, 360, 362, 366–67, 369
Brazzel, J. R., 175–76
Breman (Germany), 9, 51, 83–84, 273
Breman Cotton Exchange, 83
Bremerhaven, Germany, 84
Bretton Woods Conference, 73–75, 93, 95, 102, 258
British Loan Agreement, 83, 92–94, 97
Britton, William Johnston, 193
Brooks, D. W., 215–17
Brooks, Douglas, 36
Brown Lung Association (BLA), 330, 331, 337
Browner, Carol, 340
Brumfield, Bruce, 266, 365
Bryan, Walker, 314
Bt cotton, 181–86, 324, 341
Burlington Mills, 329, 337
Burnett, Chester (Howlin' Wolf), 200
Burnett, Phillip C., 187, 210, 260, 365, 370
Burnett, Sam, 382–83
Bush, George H. W., 336
Bush, George W., 306, 309
Business Week, 332–33, 368
Buxton Conference, 353–55, 357, 362
Byers, Jane, 320
Byrnes, Jimmy, 43
Byssinosis, 326–39, 375; "Monday Fever," 328

Calcot, 195, 210, 215, 277, 281, 285–86, 363, 376
Caldwell, Erskine, 16
California, 26, 55–56, 141, 147, 149, 156, 178, 185, 196, 274–75, 322, 340, 348, 374–75, 381; allotments, 162–65; gold rush, 150–52, 154–55; origin of CI, 219, 227; seed development, 312–14; water rights, 344–46, 377
California Agriculture Efficient Water Management Practices Act, 346
California Planting Cotton Seed Distributors, 154, 312–13
Campbell, Carl, 292
Campbell, Harvey, 315
Cancun (WTO), 367, 369
Cannon, C. A., 42, 292, 295
Cantrell, Roy, 342
Caprock Escarpment (Texas), 233–34, 237, 244, 250
Cargill, 208, 376
Carnegie Endowment for International Peace, 75–76, 78
Carson, Rachel, 177
Carter, Frank, 179
Carter, Hodding, II, 35, 255
Carter, Jimmy, 331–32
Cash, W. J., 11, 13, 24–25
Cellar, Emanuel, 109
Cheek, Tom, 58
China, 264, 266, 305–6, 308, 360, 362, 365, 367, 370, 384
Chinatown, 344
Church, Robert B., 206
Churchill, Winston, 113
Civil War, 4, 8–9, 13–14, 17, 60, 170, 190, 267, 270, 289, 290, 350
Clarksdale, Mississippi, 130
Classers (cotton), 190, 192–93, 243, 319–20; USDA, 244
Clayton, Will, 38–39, 49, 59–60, 70–71, 74, 110; origin of Marshall Plan, 80–85, 87, 90–94, 96, 100, 103
Clean Air Act, 339
Clemson University, 301–2
Cleveland, O. A., 286
Clinton, Bill, 143, 364
Cochise, 274–75
Cohn, David, 39, 191, 199
Coker, Robert, 82, 174, 188

Coker Pedigree Seed Company, 82, 293
Collapse of Cotton Tenancy, The, 19
Collier's magazine, 130–31, 153, 160
Combest, Larry, 245
Commercial Appeal, 41, 48, 58, 60, 75, 99, 140, 190, 194, 197–98, 203, 207, 255
Commercial banks, 281
Commission on the Application of Payment Limitations to Agriculture, 265
Commodity Credit Corporation (CCC), 31–32, 45, 67, 81, 111, 122, 155, 159, 162–63, 167, 210, 240, 243, 253, 257, 259–60, 262, 296, 304, 320; farming for loan, 167, 254; Western Europe, 83–84, 88, 91, 97
Common cause, 333
Congress of Industrial Organizations (CIO), 22, 113
Congressional Research Service (CRS), 264
Conran, J. V., 68
Consumers Union, 187
Conte, Silvio, 258
Cook, Everett, 73, 204
Cooke, Morris L., 118
Cooper, Hubert, 313
Cooperative Milk Producers Association, 110
Cooperatives, 240–45, 272–77, 280–81, 304, 376
Corcoran, Thomas, 69
Cortright, G. C., Jr., 160–61
Cotton Bloc (Congressional), 43, 110, 113, 116, 154, 156, 256
Cotton Board, 219–20, 224, 227–29, 231, 321
Cotton Carnival, 7, 72, 190, 203–8, 211, 331
Cotton Council International (CCI), 229, 356–63, 371
Cotton Economic Research Institute (CERI), 245–46, 263–64; alternative price analysis, 246; global fibers model, 246
Cotton Farming, 184, 196–97, 210, 212, 305, 342, 346–47, 377
Cotton flu, 185
Cotton Foundation, 322–23, 337
Cotton gins, 57, 135–37, 145, 161, 164, 169, 180, 188, 195, 220, 240–41, 247, 274–81, 284–85, 317–18, 333–34, 338–40, 347, 372, 378, 385
Cotton Incorporated (CI), 213, 224–28, 231–32, 284, 286, 299–300, 311, 316, 318, 320, 323, 329

Cotton Makers Jubilee, 206
Cotton Manufacturers Association, 290–91
Cotton Patch (Texas), 233, 235, 243, 247, 310, 383
Cotton Producers Institute (CPI), 218–24, 228, 242
Cotton Research and Promotion Act, 220, 230, 323
Cotton Research Conference, 71, 78, 97, 99, 104, 272
Cotton Row (Memphis), 68, 190–91, 198, 208, 210, 211, 243
Cotton Textile Institute, 291
Cotton Trade Journal, 24, 45, 58, 85–86, 91, 99–100, 161
Cotton USA, 361
Cottonseed oil, 106–7, 113, 169, 185, 188, 190, 196, 241, 250, 264, 302, 321
Cottonseed Oil Assistance Program (COAP), 365
Council of Economic Advisors (CEA), 331, 332, 336
Cox, A. B., 61
Crop lien, 10, 13, 19, 25, 28, 59
Crump, Ed, 202
Cultivation practices, 21, 23, 28–29, 34, 55, 63, 127–28, 173; chisel plow, 238
Cultural control, 171, 176, 178–79, 181, 316–17

Dairy Record, 106
Daniel, Margaret Truman, 69
Darker Phases of the South, 12
Davich, Ted, 175
Davidson, David, 56
Davis, Chester, 272–73
Davis, Dan, 241–303
Davis, John H., 159
Davis, Mac, 248
Davis, Manvel, 69
Davis, Roy, 217, 241, 300–2
Davis, Wolborn B., 79
DDT, 172–73, 177, 179–80
Dean, James, 225
Dearing, Gerald, 190, 197, 255
Debt crisis, 282–84
Deer Creek (Mississippi), 65, 190
Deloach, William, 234
Delta and Pine Land Company (D&PL), 32, 45, 65, 93, 127, 132, 145, 186, 194, 216, 255, 293, 301; Bt cotton, 182–83; Johnston and, 140, 151; seed breeding, 312, 321
Delta Boll Weevil Laboratory, 171
Delta Chamber of Commerce, 33–37
Delta Cooperative Farm (Arkansas), 129
Delta Council, 52, 67, 87, 98, 104, 132, 156, 160, 198, 243, 255, 330, 333
Delta Teachers College (Mississippi), 98
Democratic National Committee Platform, 109–10
Dent, Fred, 295
Deputy, Alison and Keith, 380–82
Dewey, Thomas F., 119, 216
Diamond, Neil, 225
DiCaprio, Leonardo, 226
Dickerson, Kitty, 299
Dickerson, R. C., 79
DiSalle, Michael V., 123
Dixon, Roger, 91
Dollar Cotton, 17
Dominican Republic–Caribbean Area Free Trade Agreement (DR-CAFTA), 369–70
Dorman, Clarence, 57
Draper, William, 89
Drinker, Philip, 327
Du Bois, W. E. B., 27
Duke University, 329, 335, 337
Dunavant, Billy, 209, 227–28, 285, 324; Dunavant Enterprises, 210, 376
Dunn, Read, 33, 37–38, 42, 67, 87–90, 93, 101, 103, 121, 198, 292, 296, 351–61
Dunn, Rex, 236–37

Earth Liberation Front, 185
Eastland, James, 69, 162, 216, 292, 331
Eastland, Woods, 369, 376
Economic Conditions of the South, 23, 106, 270
Economic Cooperation Administration (ECA), 122
Ed Sullivan Show, 214
Edwards, Macon, 214, 223
Eisenhower, Dwight D., 82–83, 85, 167, 255, 294
Eizenstat, Stuart E., 331–32
El Paso Cotton Association, 247
Ellender, Allan J., 160
Ely, Joe, 248–49
Engineered Fiber System, 230–31
Enterprise, Alabama, 171

Environmental Protection Agency (EPA), 177, 186–88, 339–42, 373, 376
Environmental Working Group (EWG), 287–88, 368
Ethridge, Dean, 298
Ethridge, Don, 245
European Community (EC), 364
European Cooperative Administration (ECA), 351–52
European Recovery Program (ERP), 102–3
European Union (EU), 298, 353, 363, 367, 369
Everett, C. K., 50
Ewing, Early, 312
Export-Import Bank, 71, 79, 88, 97, 351
Extra-long staple (ELS), 149, 156, 166–67, 313–14, 360, 375

"Fabric of Our Lives," 3, 226
Factors (credit), 270–71, 286
Fair Labor Standards Act, 113
Farm Aid (Willie Nelson), 383–84
Farm Credit Act: 1933, 272, 280, 282; 1987, 284
Farm Credit Administration, 53, 282
Farm Credit Board, 304
Farm Credit System, 282, 284
Farm Security Administration, 19, 22–23, 53, 119, 129, 201, 280
Farm Security and Rural Investment Act, 265
Farmers Home Administration (FmHA), 258, 284
Farmers Union (National), 32, 47, 54, 116, 119, 121, 156
Farrington, C. C., 111
Faulkner, John, 17
Faulkner, William, 25–26, 191, 201, 252
Feaster, Carl, 314
Feber, Edna, 18
Federal Agriculture Improvement and Reform Act (FAIR), 259
Federal Farm Board, 53
Federal Farm Loan Act, 271
Federal Farm Loan Board, 271
Federal Insecticide, Fungicide, and Rodenticide Act (FIFRA), 341
Federal Intermediary Credit Banks, 280, 282
Federal Land Banks, 274, 282
Feed sack clothes, 73, 375

Fiber, Fabric, and Apparel Coalition for Trade (FFACT), 298–99
Fields, Sally, 6
Findley, Paul, 222, 228, 255
Fine Cotton Spinners and Doublers Association, Ltd., 93, 327
Finer, Herman, 76
Finley, Murray H., 331
Fisher, A. E., 29
Flannagan, John, 48
Flatlanders, 248–49
Fleming, Lamar, 60, 80–81, 85, 90–91, 99, 104, 151, 161, 168, 260–61, 274, 278, 280, 354
Food Quality Protection Act (FQPA), 186–87, 341
Fordney-McCumber Tariff, 91
Fortune, 32, 153
Forward-contracting, 209, 285
Fox, Sally, 315
Franklin, John Hope, 27
Freedom Village, 143
Freeman, Orville, 69, 222
Frito-Lay, 322
Front Street (Memphis), 192–93, 204, 209–11
Fulbright, William, 69, 110
Fulmer, Hampton, 107

Gable, Clark, 226
Gaines, J. C., 176, 244
Garnett, Gwynn, 355
Garrard, W. M., 354
General Accounting Office (GAO), 358–59
General Agreement on Tariffs and Trade (GATT), 95–96, 100, 295, 298, 363–65
Gerdes, Frances L., 131
German Textile Industry Association, 83
Germany, 52, 60, 80–95, 97, 99, 111, 352, 355, 357–58
Gershwin, George, 18
Gillon, William, 369
Gilmore, Jimmy Dale, 250
Glamour, 375
Glover, Danny, 6
Go Down, Moses, 25
Goldwater, Barry, 158–60
Goodyear Tire and Rubber, 149–50
"Governor's Lady Series" (NCC), 72
Grady, Henry, 9
Grapes of Wrath, 26, 140, 154

Graves, Bibb, 38, 106
Gray, Borden, 336
Greenspan, Alan, *Age of Turbulence*, 118
Greenville Delta Star, 35
Grew, Joseph, 100
Griffin, Russell, 217–21, 223
Grisham, John, 6, 191
Guenther, George C., 331
Guthrie, Woody, 6

Haldenby, Roger, 376
Hammond, James, 327
Hancock, Butch, 248–49
Hand, James, 33–34, 37, 52, 58, 140
Handy, W. C., 199–200, 202, 206
Hannagan, Bob, 58
Harding, Warren G., 169
Hardscramble, 127, 140
Harper, Roy, 68
Harper's Bazaar, 72, 167
Harrison, George J., 152, 312
Hawley-Smoot Tariff, 35, 52, 91, 96, 354
Hayes, Brook, 253
Helm, Neal, 68
Henderson, Leon, 43
Henry, C. G., 110
Herbicides, 128, 134, 141, 144, 196–97, 247, 261, 286, 315–16, 339, 347, 376, 378–79; Dalapon, 138; Dinoseb, 137; Diuron, 138; Chlorpropham, 137
Hickman, Francis, 85–86, 99, 196–97
Hohenberg Bros., 86, 208, 210, 286, 376
Hold Autumn in Your Hand, 24
Hollings, Fritz , 294
Holly, Buddy, 248
Hoover, Herbert, 95
Hope, Clifford, 356
Hopkins, Harry, 24
Hopson Plantation, 130–31, 140, 145, 196
Horne, MacDonald (Mac), 42, 66–67, 68, 88, 116–18, 220, 222
Horton, Tommy, 342
Hull, Cordell, 38, 71, 95, 363
Huskey, G. A., 318
Hutson, J. B., 82, 111–12
HVI classing, 242–43, 246, 304, 319–20, 376

I Magnin, 72
Illinois Central Railroad, 140

Imperial Valley (California), 153, 344–45
In the Land of Cotton, 15
Industry Wide Committee on Future of Cotton Council, 215–16
Industrywide Cotton Dust Committee (NCC), 329–30
Integrated Pest Management (IPM), 176–77, 181, 183, 341, 386
International Bank for Reconstruction and Development (World Bank), 74
International Cotton Advisory Committee (ICAC), 359
International Harvester, 57, 130–31, 196
International Institute for Cotton (IIC), 358–60, 371
International Monetary Fund (IMF), 74–75, 309
International Teamsters Union, 113–15
International Textile and Clothing Bureau (ITTCB), 299
International Textile Center, 301–2
International Trade Organization (ITO), 95–96
Irrigation, 173, 254, 273–74, 311, 317, 380–83, 385; Central Valley Project, 150, 344; Texas High Plains, 236–39; water wars, 343–46; West, 153–58
Ives, Burl, 6

J. C. Penney, 294
Jackson, Burris, 67, 98, 115
Jackson, Robert (Bob), 43, 67, 69, 82–88, 103, 291–96, 310, 351, 354
Japan, 52, 60, 95, 97, 99, 112, 152, 291–95, 297, 300, 353–55, 357, 359, 368
Jazz Age, 16
Jefferson, A. T., 253
Jefferson, Blind Lemon, 18, 201
Jefferson, Thomas, 252
Jeffersonian Agrarianism, 63, 251
Jennings, Waylon, 249
Jesness, Oscar B., 76
John Deere Company, 130
Johnson, Lonnie, 201
Johnson, Luther A., 87
Johnson, Lyndon, 220, 245, 247, 297
Johnson, Robert, 130
Johnston, Oscar, 32, 44–45, 51, 55, 58, 65–67, 71, 82, 101, 103, 106, 151, 211, 213, 215, 217,

290, 311, 327, 358; Bretton Woods, 73–75, 78; British Loan Agreement, 92–95; CIO, 114–15; mechanization, 131, 138–39; migration, 140, 144; origin of NCC, 33, 35, 36–42, 198; textiles, 295–96
Jones, J. K. "Farmer," 318
Jones, Jesse, 71
Jones, Joseph, 98
Jordon, Hamilton, 332
Journal of American Medical Association (JAMA), 335
Journal of Cotton Science, 323
Judkins, Bennett, 335

Kearney, Janice, 143–44
Kearney, Thomas, 314
Kefauver, Estes, 353
Kennan, George, 100
Kennedy, John F., 216, 294
Kennedy, Russell, 217
Kennedy Seven Point Textile Program, 295
Keynes, John Maynard, 74, 92
King, Allen, 316
King, B. B., 5, 143, 145, 200–1, 383
King Cotton Hotel (Memphis), 190, 211
King Cotton March, 18
Kirklin, W. G., 159
Kirkpatrick, Clifton, 213, 215, 218
Kline, Allan B., 121
Knipling, E. F., 174–75
Knowland, W. F., 98
Kohler, Wily, 83
Korean War, 103, 121–23, 139, 157, 159, 213, 291, 293, 296–97, 305, 352–55
Kroese, W. T., 358
Ku Klux Klan (KKK), 13, 26–27, 143, 237
Kuchel, Thomas H., 158–59

Lambright, Joe, 236
Land Conservation Act, 348
Landor, Walter, 225
Lange, Dorothea, 201
Lange, Mark, 369–70
Latino culture, 246–48
Lavis, Rick, 380
Le Havre, France, 51, 80, 273
Leache, Richard, 36, 38
Leahy, Patrick J., 366
Ledbetter, Huddie "Lead Belly," 5–6

Lever Brothers, 73
Levi Strauss and Company, 303–4, 310
Liberty League, 70–71
Life Magazine, 153, 255
Lilienthal, David, 116
Linder, Thomas, 133
Lipscomb, Edward, 67, 72–73, 354–55
Lipscomb, J. N., 24
Littlefield, Texas, 302, 304, 309
Liverpool, England, 9, 51, 79, 85, 273; Cotton Exchange, 363
Livi, Camillo, 80–81, 84
Llano Estacado (Texas), 149, 233–35, 240, 246–49
Lockett, Aubrey, 293
Long, Huey, 21
Los Angeles Chamber of Commerce, 161
Lubbock, Texas, 195, 223, 233, 236–38, 241, 245, 250, 272, 319, 343–44
Lubbock Avalanche Journal, The, 331

MacArthur, Douglas, 291–92
Macha, H. A., 235–36
Madden, Ray J., 258
Magness, Perre, 204
Mahon, George H., 87, 245, 256, 330
Maids of Cotton, 72–73, 204–5, 207, 214, 298, 357, 359
Manchester Cotton Association, 79
Mann, Lon, 332
March of Time, 22, 28
Margarine (oleo), 7, 105–11, 116, 185, 321–22
Margarine Manufacturers Association, 107
Market Promotion Program, 259, 361
Marks and Spencer, 357, 359
Marplan, 277–78
Marshall, F. Ray, 331–34
Marshall, George, 98–102
Marshall Plan, 100, 102–3, 155, 350, 352–53
Martin, V. G., 24
Mayflower Hotel (Washington, D.C.), 43, 69
McAshan, A. M., 276, 279
McCaffrey, J. T., 131
McCarthy, Joseph, 122
McMurtry, Larry, *Lonesome Dove*, 234
McNeil, Zinn, 318
McWilliams, Carey, 154, 165
Meals, M. R., 195
Meany, George, 333

Mechanical cotton picker, 7, 19, 57, 124, 128–31, 135–36, 140, 195–96, 217, 238, 312, 316–17, 327, 339, 378, 380; West, 154, 158, 238, 247

Mechanization, 34, 46, 54, 62, 65, 90, 117, 141, 191, 195, 197, 211, 234, 267, 272–73, 323, 372; Cotton Conference, 56–59; impact, 144–46, 203, 286; modules, 317–19; push-pull, 126–27, 142; western, 151, 153, 161, 164, 166

Memphis Cotton Exchange, 36, 40, 190, 193, 197, 199, 204, 208–10

Memphis Cotton Exchange Museum, 211

Memphis in May International Festival, 208

Mencken, H. L., 13

Mesilla Valley (New Mexico), 147, 150, 166, 313, 380

Mexican Americans, 19, 164

Migration, 14, 45, 54–55, 60, 62, 76, 126, 142, 146, 286, 374

Milan, Italy, 51, 80, 273, 359

Milliken, Roger, 295, 306–7, 366

Mississippi Agricultural Council, 35

Mississippi Agricultural Experiment Station, 57, 128, 255

Mississippi Mafia, 168

Mississippi State College (University), 117, 132, 175

Mississippi State Commissioners of Agriculture, 35

Mitchell, H. L., 23, 26, 47, 133

Mitchell, Joni, 347

Mitchener, Frank, 183, 335–36

Mobley, Leonard, 351

Modules, 320, 380; rickers, 317–18; University of Arkansas, 318

Molyneaux, Peter, 78

Mondale, Walter, 332

Monroe, Marilyn, 225–26

Monsanto, 182–83, 292, 301

Morales, Santana, 20

Morgan, Chip, 266

Morganthau Plan, 87

Motion control, 242, 319

Muddy Waters, 200

Mules, 57, 105, 126, 128, 132–34, 136–38, 140, 145, 151, 170–71, 194, 196, 200, 237, 239, 256, 264, 271–73, 286–87, 372, 374

Multi-Fiber Agreement, 295, 298, 365

Murchison, Claudius T., 48, 290, 354

Murphy, Charles, 69–70

Murray, Philip, 118

Musharraf, Pervez, 309

Myers, Charles, 295

Myrdal, Gunnar, 27

Nader, Ralph, 328, 332, 370

Nasser, Gamal, 167

National Academy Sciences, 186

National Association for the Advancement of Colored People (NAACP), 27

National Association of Manufacturers (NAM), 27

National Association of Wheat Growers, 257

National Black Farmers Association (NBFA), 287

National Consumers League, 47

National Cotton Council (NCC), 43–44, 46, 60, 71–73, 77, 91, 97, 104, 119, 120, 136, 152, 156, 167–68, 181, 186–87, 196–98, 207, 212, 244, 246, 279, 284, 316, 320, 321–23, 333, 335–36, 338, 342, 375–76; Beltwide Conference, 174, 184, 323–24, 334; boll weevils, 174–78; British Loan Agreement, 93–94; Byssinosis, 328–31; CIO, 113–16; field service, 44, 107, 218, 264, 284; globalization, 350–51, 356–58; Korean War, 122–23; margarine, 106–11; Marshall Plan, 101–3; mechanization, 131–32, 144; origin of, 31–42, 311; origin of CI, 213–32; staffing (1945), 64–68; subsidies, 256, 259, 263; textiles, 290, 294–98; WTO, 364–69

National Cotton Women's Committee, 324

National Cottonseed Products Association, 38, 196, 324

National Council of Jewish Women, 109

National Council of Trade Organizations, 299

National Emergency Council, 23

National Grange, 58, 76, 109, 156, 257

National Institute for Occupational Safety and Health, 329

National Research Council of National Academy of Sciences, 327

National Sugarbeet Growers Federation, 257

National Textile Center, 301

National Wool Growers Association, 257

Neiman Marcus, 72
Nelson, Donald, 44
Nelson, Morgan, 148, 213–14, 227
Nelson, Willie, 249, 383–84
Nematodes, 230, 320–21
Neugebauer, Randy, 245
New England Journal of Medicine, 328
New Mexico, 55, 149, 153, 155, 162, 165, 181, 213, 219, 233, 247, 275, 279, 313–14, 380–81
New York Times, 260, 263, 331, 368; *Style Magazine*, 375
New Yorker, 167
Nixon, Herman, 24, 62
Nixon, Richard, 223, 258, 295, 303, 305, 330, 331
Norris, P. K., 50
North Atlantic Free Trade Agreement (NAFTA), 349, 369–70
North Carolina Industrial Commission, 337
No-till farming, 316–17, 379
Nourse, Edwin G., 116

Occupational Safety and Health Administration (OSHA), 303, 329–30, 332, 334, 336–38
O'Daniel, W. Lee, 92
Oden, Syndor, 80
Odum, Howard, 20, 23, 117
Office of Management and Budget (OMB), 337
Office of Price Administration (OPA), 43, 112–14
Office of Price Stablization (OPS), 123–24
Ogallala Aquifer, 150, 237, 342, 344, 377
Old Red (mechanical picker), 196
O'Neal, Edward, 38
O'Neal, Jack, 217
Operation Wetback, 165
Orbison, Roy, 248, 383
Organic Cotton, 315–16
Osaka, Japan, 51, 300, 359
Owens Bros., 194–95
Owens River Valley (California), 344–45
Oxfam, 366, 368–69

Paarlberg, Don, 252
Pace, Stephen, 47–48, 58, 60, 66, 111, 118, 123, 139
Pace Committee, 48, 53, 55–56, 58, 62–63, 66–67, 116, 133

Painted House, The, 6
Pakistan, 264, 309, 351, 360
Parker, Jim, 301
Parnell, Calvin, 317–18
Patterson, Robert, 95
Patton, James, 116, 119
Payne, George, 38
Peabody Hotel (Memphis), 39–41, 63, 72, 106, 115–16, 194, 198–99, 203, 354
Pearson, Bill, 137, 230, 321
Pecos Valley, 313
Peebles, Robert, 314
Pellagra Belt, 12, 15
Percy, William, 25, 34, 191
Pereira, Peter, 360
Perry, George Sessions, 24
Pfiffenberger, George, 243–44
Pigford, Timothy, 287
Pima cotton, 149–50, 176, 179–82, 188, 215, 315, 341
Pink bollworm, 55, 170, 176, 179–82, 188, 215, 315, 431
Places in the Heart, 6
Plains Cooperative Oil Mill (PCOM), 241, 277–78, 300–1
Plains Cotton Cooperative Association (PCCA), 195, 215, 241–42, 245, 286, 303–4, 309–10, 319, 363, 376
Plains Cotton Growers (PCG), 243–45, 301, 330, 377
Pledger, D. A., 127
Poage, Robert (Bob), 69, 87, 108–10, 256, 356
Poage-Fulbright Bill, 110
Pool Sales Contracts, 285
Porgy and Bess, 18
Porter, Paul, 43
Possum Trot: Rural Community, South, 24
Poteet, Bob, 245
Potsdam Conference, 71
Poultry farming, 139, 283, 357, 378
Preface to Peasantry: A Tale of Two Blackbelt Counties, 19
President's Committee on Farm Tenancy, 22
Presley, Elvis, 200, 205
Press Scimitar, 66, 197, 198
Pressley, E. H., 319
Producer Steering Committee (PSC), 221
Producers Cotton Oil Company, 217–18, 279

INDEX

Production Credit Association (PCA), 277, 280–82
Public Law 480, 355–56, 358, 361
Pyle, Howard, 159

Ranchers Cotton Oil, 215
Randolf, Walter, 123, 297
Raper, Arthur, 19, 23, 48
Rayburn, Sam, 63, 103, 115, 201, 279
Rayner, Ron, 316
Reagan, Ronald, 259, 298, 333, 335–37
Reciprocal Trade Act (1934), 95
Reichstelle für Textilwerke, 83
Requiem for a Nun, 26
Resettlement Administration (RA), 119
Retail Industry Trade Action Coalition (RITAC), 299
Reynolds, Barry, 318
Rhodes, Marion, 159
Richard, Wynn, 72
Rio Grande Valley (Texas), 20, 148–49, 165, 169, 178, 246–47, 303, 345, 348–49, 383
Risk-debt complex, 382–83
Rivers, Mendel, 256
Riverside County (California), 55, 153
Rivoli, Pietra, 308
Roberts, Jay, 182
Rockabilly, 248–50
Rockefeller, Nelson, 100
Roosevelt, Franklin D., 22, 36, 74, 204
Root rot, 137, 267
Rotor spinning, 300, 303
Rural Credits (Development) Act: 1923, 281; 1972, 258
Rural Electrification Administration (REA), 108, 139, 239
Rusk, Howard, 292
Russell, Albert, 67, 284, 297, 372
Russell, Richard, 115
Rust, John, 130, 136, 146

Salt River (Arizona), 147, 149–50, 179
San Franciso Chronicle, 340
San Joaquin Valley (California), 55, 63, 147–48, 150–53, 158, 216, 218, 222, 227, 313, 340, 344, 348, 374
Sandburg, Carl, 6, 172
Sayre, C. R. (Jerry), 69, 132, 144, 156, 216–17, 229

Sayre, Francis B., 38, 255
Scarborough, Dorothy, 15–16, 172; *In the Land of Cotton*, 386
Schnittker, John, 359
Schrage, Peter, 328
Schultz, Charles, 331–32
Seam, 210
Sears, Earl, 240, 266, 333, 336
Seed development, 235–37, 240, 274–75, 312–16
Shafter, California, 152, 154, 312–13
Sharecropper, 18–22, 24–29, 31–32, 36, 40, 43, 52, 54–55, 57, 68, 142, 206, 270–71; Beale Street, 199–200; blues, 201–2, 248; contract farming, 4; mechanization, 125–27, 129, 132, 138, 140–41, 144, 146, 211, 372; migration, 45, 58, 62, 135, 137, 195, 286, 374; poverty, 9–11, 13–16, 54, 63; West, 149, 151, 154, 163, 171, 190–92
Shaver, Billie Joe, 6
Show Boat, 18
Sieker, Herbert O., 337
Silent Spring, 177
Silverman, Leslie, 327
Simplicity Patterns (feed sack clothes), 73
Skaggs, William H., 14–16, 169
Skip-row farming, 317
Smith, B. F., 255, 333
Smith, C. C., 47, 81, 85, 111–12
Smith, Ellison (Cotton Ed), 31–32, 70
Smith, Frank, 270
Smith, Neuman, 247
Smith, Russell, 54
Smith, Tom, 376
Snydor, John W., 92
Soil Bank, 139, 253–54
Soil Conservation Service, 54
Song of the South, 172
Sousa, John Philip, 17
Southern Oligarchy, The, 14, 169
Southern Pacific Railroad, 151
Southern Regions of the United States, 20
Southern Renaissance, 15
Southern Tenant Farmers Union (STFU), 21–23, 33, 133, 191
Southern way of life, 5, 8, 15–16, 24, 30, 52, 63, 126, 133, 146, 162
Southwest Irrigated Cotton Growers Association, 38, 195

Soviet Union, 80–81, 93, 111, 167, 351, 360, 385
Spinlab, 319
Staplcotn, 132, 195, 281, 285, 354, 363, 369, 376
Stark, Lloyd C., 68
Starlink, 184–85
State Support Program (CI), 323
Steinbeck, Charles, 26, 140
Stevens, Boswell, 217, 220
Stewart, Ernest B., 42, 67, 72
Stone, Jack, 147–48, 151, 159, 266, 279, 284
Stoneville (Seed), 182
Stoneville, Mississippi, 33, 129, 131
Storm-proof cotton, 235–36
Stovall, Howard, 354–55
Stowe, Harriet Beecher, 17
Strauss, Robert S., 332
Street, James, 141
Stroman, G. N., 313–14
Subsidies, 7, 27–28, 32, 53–54, 59, 76, 117, 120–22, 232, 251–69, 297, 308, 349, 379, 384, 386; nonrecourse loan, 253–57, 259–60, 262, 296; opposition to, 222–23, 366–69; Step 1-2-3, 259, 366, 368–69; WTO, 365–69
Sulphur Springs Valley (Arizona), 274
Supima, 167–68, 375, 377

Taft-Hartley Act, 115–16, 120
Tannenbaum, Frank, 12–13, 15–16, 24
Targeted Export Assistance (TEA), 361
Teague, Olin, 256
Technical Industrial Intelligence Committee (TIIC), 82, 84
Texas High Plains Underground Water Conservation District, 343, 346
Texas Plains, 20, 39, 57, 66, 147–48, 150, 152–53, 155, 175, 233–50, 274, 278–79, 330, 343–44, 346, 374, 376–77, 382; AAM, 283–84; mechanization, 132, 135, 141, 317; migrants, 165; textiles, 300, 302–3, 308, 310
Texas Tech University, 235, 242, 244–45, 320
Textile Apparel Enforcement Act, 298
Textile Labor, 327
Textile Payment Plan, 85, 87, 90, 95, 97, 111
Textile Research Center, Texas Tech University, 300–1, 303–4
Textile Workers Union, 327
Thaiamerican, 303

Thaxton, E. L., 238
Thomas, Norman, 21
Thomas, Tol, 34
Thorp, Willard, 87
Tiger Economies, 260, 366
Tobacco Road, 16, 18, 26
Tobin, Dan, 114
Toker, Philip, 67, 106
Tolley, H. R., 67, 106
Town and Country, 167
Treadmill (insecticide), 174, 180
True, S. M., 245
Truman, Harry S., 48, 68–71, 74, 95–98, 103, 108–9, 112, 115, 119, 121, 203, 216, 353
Truman War Investigation Committee, 44, 48, 69
Tucker, Emerson, 242–43, 303, 319
Tumlinson, Jim, 176
Turcotte, Edgar, 314
Turner, John, 312–13
Two-price cotton (subsidies), 117, 296–97

Uncle Tom's Cabin, 17
United Nations, 64, 95
United Nations Advisory Committee on the Application of Science and Technology, 360
U.S. House Committee on Agriculture, 33, 47, 108, 116–18, 356, 365
U.S. Immigration Service, 165
U.S. Interbureau Committee on Post War Programs, 53
U.S. Publilc Health Service (USPHS), 327–28
U.S. Senate Committee on Agriculture and Forestry, 158, 366
USDA (Department of Agriculture), 122–23, 132, 157, 162–63, 167, 170, 174–75, 212–13, 220–21, 224, 228–29, 242–44, 246, 255, 262, 282, 284, 287, 311–16, 319, 324, 328, 342, 348, 350, 352, 356, 364; ARS, 133, 138, 174, 352; BAE, 58–59, 63, 116–17, 159; Bureau of Entomology, 179–80; Commission on Small Farms, 288; Delta Station, 34, 131, 135–36; ERS, 245; FAS, 228, 356–58; Fiber Testing Laboratory, 302; Marketing Service, 196; NRCS, 348; PMA, 161; Shafter, 152

Vaccaro, Tony, 70
Van Fleet, James, 292
Van Sickel, John, 75–76
Vance, Rupert, 5, 21, 23, 33, 117, 270
Vandenberg, Arthur, 31
Vandenberg Bill, 102
Vanderbilt Agrarians, 24, 58, 252
Veneman, Ann M., 287
Venson, R. Q., 206
Verett, Steve, 376, 383
Verticillium Wilt, 148, 163, 312–15, 320; Fusarium, 312
Vogue, 72, 167, 375

Wakelyn, Philip J., 340
Wallace, Henry A., 22, 24, 32
War Mobilization Board (WMB), 166
War Production Board (WPB), 44, 112
War Surplus Property Administration, 49
Washday (insecticides), 173, 176, 188
Washington Metropolitan Club, 87, 96
Water Conservation District No. 1 (Texas), 343
Webb, Walter P., 106
Weedsizzer (flame-thrower), 134, 148
Welch, Claude, 67
Welch, Frank, 117, 132–33
Welden, Keith, 217
Well, O. V., 47, 49
Wellford, Dabney, 222
Wells, Sumner, 100
Welty, Eudora, 191
West, Sid, 75
West Africa, 367
Western Cotton Growers Association, 156
White, E. D., 80–83, 85, 87, 97, 111–12, 132, 351–52
White, Harry Dexter, 74
White, Hugh, 38
White Gold, 4, 127, 147, 152, 194, 204, 211, 312
Whitney, Eli, 3, 57, 135, 211
Whitten, James, 163, 174
Wickard, Claude, 47, 54, 57, 130
Widenbaum, Murray, 336
Wieczerek-Zeul, Heidemarie, 367
Wilkes, Lambert H., 318
Williams, Tennessee, 207
Williamson, H. H., 71
Wills, Bob, 249
Witts, Glen, 242
Woofter, T. J., 29
Wooters, Duke, 223–26, 228–30
World Cotton Day, 367
World Trade Organization (WTO), 7, 246, 261, 305–6, 309, 353, 364–69
Wren, Grover C., 330
Wright, Richard, 142–43
Wynn, W. T. (Billy), 33–34, 37–38, 98

Yazoo Delta (Mississippi), 9, 33–34, 40, 43, 58, 87, 99, 131, 140–41, 150–51, 190, 194, 199–200, 211, 233, 248, 367, 373–74
Yazoo River (Mississippi), 40, 190
Young, Harold, 38

www.ingramcontent.com/pod-product-compliance
Lightning Source LLC
Chambersburg PA
CBHW030600230426
43661CB00053B/1786